Surveys in
Applied Mathematics
Volume 2

SURVEYS IN APPLIED MATHEMATICS

Series Editors

JOSEPH B. KELLER, *Stanford University, Stanford, California*
DAVID W. McLAUGHLIN, *New York University, New York, New York*
GEORGE C. PAPANICOLAOU, *Stanford University, Stanford, California*

VOLUME 1   Joseph B. Keller, Robert M. Lewis, David W. McLaughlin,
             Edward A. Overman II, and George C. Papanicolaou

VOLUME 2   Mark Freidlin, Sergey Gredeskul, Andrew Marchenko, Leonid Pastur,
             and John K. Hunter

A Continuation Order Plan is available for this series. A continuation order will bring delivery of each new volume immediately upon publication. Volumes are billed only upon actual shipment. For further information please contact the publisher.

Surveys in
Applied Mathematics

Volume 2

Mark Freidlin

University of Maryland
College Park, Maryland

Sergey Gredeskul

Ben Gurion University of the Negev
Beer-Sheva, Israel

Andrew Marchenko

Moscow State University of Transportation
Moscow, Russia

Leonid Pastur

B. Verkin Institute for Low Temperature Physics
Kharkov, Ukraine

and

John K. Hunter

University of California
Davis, California

Springer Science+Business Media, LLC

ISSN 1082-622X

ISBN 978-1-4613-5821-3   ISBN 978-1-4615-1991-1 (eBook)
DOI 10.1007/978-1-4615-1991-1

Contributors

Mark Freidlin • University of Maryland, College Park, Maryland 20742

Sergey A. Gredeskul • Department of Physics, Ben-Gurion University of the Negev, Beer-Sheva, 84105 Israel, Internet: sergeyg@bgumail.bgu.ac.il

John K. Hunter • Department of Mathematics and Institute of Theoretical Dynamics, University of California, Davis, California 95616

Andrew V. Marchenko • Moscow State University of Transportation, 15 Obraztsova Str., Moscow, Russia

Leonid A. Pastur • B. Verkin Institute for Low Temperature Physics and Engineering, Ukrainian Academy of Sciences, 47 Lenin Ave., Kharkov, 310164, Ukraine

Preface

Surveys in Applied Mathematics is a series of volumes, each of which contains expositions of several topics in mathematics and their applications. They are written at a level accessible to advanced graduate students and interested nonspecialists, but they also contain the results of recent research.

Volume I consists of three articles. The first is the classic paper of J. B. Keller and R. M. Lewis, "Asymptotic Methods for Partial Differential Equations: The Reduced Wave Equation and Maxwell's Equations." The second is by D. W. McLaughlin and E. A. Overman on "Whiskered Tori for Integrable Pde's: Chaotic Behavior in Near Integrable Pde's." This is a systematic analytical and numerical study of near integrable wave equations, including the sine-Gordon equations and the perturbed nonlinear Schrödinger equation. The third article is by G. Papanicolaou on "Diffusion in Random Media." It is an introductory survey of homogenization methods for the diffusion equation with random diffusivity.

This second survey volume also consists of three articles. The first, by M. Freidlin, on "Wave Front Propagation for KPP-Type Equations," analyzes nonlinear front propagation for a large class of semilinear partial differential equations using probabilistic methods. The second, by S. Gredeskul, A. Marchenko, and L. Pastur on "Particle and Wave Transmission in One-Dimensional Disordered Systems," treats in detail wave localization phenomena in one- dimensional random media. The third is by J. Hunter on "Asymptotic Equations for Nonlinear Hyperbolic Waves." It is an extensive introduction to certain model equations for nonlinear wave phenomena.

Other survey volumes are in preparation.

Joseph B. Keller
Stanford

David W. McLaughlin
New York

George C. Papanicolaou
Stanford

Contents

1

Wave Front Propagation for KPP-Type Equations

Mark Freidlin

1.1. INTRODUCTION

The following equation was considered in [15]:

$$\frac{\partial u(t,x)}{\partial t} = \frac{D}{2}\frac{\partial^2 u}{\partial x^2} + f(u), \qquad t > 0,\ x \in R^1,$$

$$u(0,x) = \chi^-(x) = \begin{cases} 1, & x \le 0 \\ 0, & x > 0. \end{cases} \tag{1.1.1}$$

Here $D > 0$ and $f(u) = c(u)u$, where the function $c(u)$ is supposed to be Lipschitz continuous, positive for $u < 1$ and negative for $u > 1$, and such that $c = c(0) = \max_{0 \le u \le 1} c(u)$. Let us denote the class of such functions $f(u)$ by F_1.

It is easy to see from (1.1.1) that $u(t,x)$ for each $t \ge 0$ is a strictly monotone function decreasing from 1 as $x \to -\infty$ to 0 as $x \to \infty$. Thus there exists a unique $m(t)$, $t > 0$, such that $u(t,m(t)) = 1/2$. It was proved in [15] that $\lim_{t\to\infty} t^{-1}m(t) = \sqrt{2cD}$, and that $u(t,m(t) + z) \to v(z)$ as $t \to \infty$, where the

MARK FREIDLIN • University of Maryland, College Park, Maryland 20742.

Surveys in Applied Mathematics, Volume 2. Edited by Mark Freidlin, Sergey A. Gredeskul, John K. Hunter, Andrew V. Marchenko, and Leonid A. Pastur, Plenum Press, New York, 1995.

function $v(z)$, $-\infty < z < \infty$, is the solution of the problem

$$\frac{D}{2}v''(z) + \alpha v'(z) + f(v(z)) = 0, \qquad -\infty < z < \infty,$$

$$\lim_{z \to \infty} v(z) = 0, \qquad \lim_{z \to \infty} v(z) = 1, \qquad v(0) = \frac{1}{2} \qquad (1.1.2)$$

for $\alpha = \sqrt{2cD}$. Problem (1.1.2) is solvable for $\alpha \geq \sqrt{2cD}$, and the solution is unique. Roughly speaking it means that the solution of problem (1.1.1) behaves for large t as a running wave $v(x - \alpha t)$. It can be characterized by the shape $v(z)$ and by the speed $\alpha = \sqrt{2cD}$.

One can introduce the asymptotic speed independently of the shape:

The number α^* is called the asymptotic speed as $t \to \infty$ for the problem (1.1.1) if for any $h > 0$

$$\lim_{t \to \infty} \sup_{x > (a^*+h)t} u(t,x) = 0, \qquad \lim_{t \to \infty} \inf_{x < (\alpha^*-h)t} u(t,x) = 1.$$

It follows from [15] that such α^* exists and is equal to $\sqrt{2cD}$. The notion of asymptotic speed can be introduced in a similar way in a more general situation.

Consider a tube $R^1 \times G$, where G is a bounded domain in R^r with a smooth boundary ∂G, and the following problem:

$$\frac{\partial u(t,x,y)}{\partial t} = \frac{D}{2}\Delta_{x,y}u - b\frac{\partial u}{\partial x} + f(u), \qquad t > 0, \ x \in R^1, \ y \in G$$

$$\left.\frac{\partial u(t,x,y)}{\partial n}\right|_{t>0,x\in R^1,y\in\partial G} = 0, \qquad u(0,x,y) = \chi^-(x). \qquad (1.1.3)$$

Here $n = n(y)$ is the outward normal to ∂G at point $y \in \partial G$; $\Delta_{x,y}$ is the Laplacian in x and y; $f \in F_1$ as before; D is positive.

As in the one-dimensional case α^* is called the asymptotic speed as $t \to \infty$ if for any $h > 0$

$$\lim_{t \to \infty} \sup_{x > (a^*+h)t,y\in G\cup\partial G} u(t,x,y) = 0, \qquad \lim_{t \to \infty} \inf_{x < (\alpha^*-h)t,y\in G\cup\partial G} u(t,x,y) = 1.$$

Equation (1.1.3) describes the evolution of particles which diffuse with diffusivity D in the flow having velocity b and take part in the "chemical reaction" governed by the nonlinear term $f(u)$. Of course, one cannot expect that some asymptotic speed will be established if D or b depends on x arbitrarily. Let $D = \mathrm{const}$ and b be independent of x. If $b = \mathrm{const}$ then it follows from the results of [15] that the asymptotic speed for problem (1.1.3) is equal to $\alpha^* = b + \sqrt{2cD}$. Now let the velocity of the flow b depend on the point of the cross section of the tube: $b = b(y)$. In the linear case $f(u) \equiv 0$ one can check that

$$\alpha^* = \bar{b} = \frac{1}{|G|}\int_G b(y)\,dy$$

where $|G|$ is the volume of the domain G. The last statement is the result of averaging in the y-variables: the uniform distribution is the invariant measure for the diffusion governed by $(D/2)\Delta_y$ in G with the normal reflection on the boundary.

If $b(y) \neq$ const and $f \in F_1$ one could expect that because of the same averaging the asymptotic speed will be equal to $\bar{b} + \sqrt{2cD}$. But it turns out that it is not the case. The real asymptotic speed is bigger than $\bar{b} + \sqrt{2cD}$ if $b(y) \not\equiv \bar{b}$. Besides the term $\sqrt{2cD}$, which is the result of the interaction between the diffusion in the x-direction and the nonlinear term, we will have in the asymptotic speed another positive summand which is caused by the interplay between the nonlinear term, deviations $b(y)$ from $\bar{b}$, and by the diffusion in the y-variables. The computation of the asymptotic speed in the case of a nontrivial velocity profile $b(y)$ of the flow in the tube is one of the goals of this article. We consider this question in Section 1.2. Then we consider also some generalizations of the problem (1.1.3), such as the case when not only the drift but also the diffusivity and the nonlinear term depend on $y \in G \cup \partial G$. For example, let $D =$ const, $b \equiv 0$, $f = f(y, u)$, and $f(y, u) = c(y, u)u \in F_1$ for each $y \in G \cup \partial G$. If $c(y) = c(y, 0) =$ const then the asymptotic speed is equal to $\sqrt{2cD}$. If $c = c(y)$ depends on y some effective coefficient $\tilde{c}$ should replace c in this formula. It turns out that $\tilde{c}$ is the eigenvalue, corresponding to the positive eigenfunction for the problem

$$\frac{D}{2}\Delta_y \varphi(y) + c(y)\varphi(y) = \lambda \varphi(y), \qquad y \in G, \ \frac{\partial \varphi}{\partial n}\Big|_{\partial G} = 0.$$

Such an eigenvalue exists and is unique. If $c(y) = c =$ const then $\tilde{c} = c$. The asymptotic speed in this case is equal to $\sqrt{2\tilde{c}D}$.

We consider also the following problem:

$$\frac{\partial u(t, x, y)}{\partial t} = \frac{D}{2}\Delta_{x,y}u, \qquad t > 0, \ x \in R^1, \ y \in G,$$

$$\frac{\partial u}{\partial n} + f(y, u)|_{t>0, x \in R^1, y \in \partial G} = 0, \qquad u(0, x, y) = \chi^-(x, y).$$

Here $f \in F_1$, $D =$ const > 0. It turns out that the nonlinear term in the boundary conditions also provides a positive asymptotic speed. These generalizations are considered in Section 1.3. Initial functions of a more general form are considered there as well.

The establishment of the asymptotic speed occurs not only in the case when the equation is invariant with respect to all shifts along the variable x. It is enough, for example, if the equation is periodic in x. Such problems are briefly considered in Section 1.3 (see also [13] and [15]). An asymptotic speed will be established if the coefficients of the equation and the nonlinear term form a random field with characteristics invariant with respect to shifts as well [13], [17].

The asymptotic speed describes the motion of the transition zone, between areas where the solution is close to 0 or to 1. The motion of such a zone can be described in a more general situations for problems with a small parameter. Usually the "thickness" of the transition zone tends to zero as the parameter vanishes, and we can speak of the motion of a surface which is the interface between these two zones or the wave front (compare [4], [6], [7]). To clarify this statement, let us consider an equation of the type of (1.1.1) with small diffusivity:

$$\frac{\partial u(t, x)}{\partial t} = \frac{\varepsilon^2}{2}D(x)\Delta u + f(x, u), \qquad x \in G \subset R^r$$

$$\frac{\partial u}{\partial n}\Big|_{\partial G} = 0, \qquad u(0, x) = g(x) \geq 0.$$

We assume that G has a smooth boundary ∂G. In particular G can be the tube considered above. Let G_0 be the support of the initial function. We assume that the closure of G_0 coincides with the closure of its interior. The nonlinear term is such that $f(x, \cdot) \in F_1$ for each $x \in R^r$ so that $f(x, u) = c(x, u) \cdot u$, $c(x) = c(x, 0) = \max_{0 \leq u \leq 1} c(x, u)$.

It follows from [15] that the area of large values of $u(t, x)$ propagates with the speed of order ε as $\varepsilon \downarrow 0$. Therefore it is useful to introduce new time $t \to t/\varepsilon$. Let $u^\varepsilon(t, x) = u(t/\varepsilon, x)$. We have the following problem for $u^\varepsilon(t, x)$:

$$\frac{\partial u^\varepsilon(t, x)}{\partial t} = \frac{\varepsilon D(x)}{2} \Delta u^\varepsilon + \frac{1}{\varepsilon} f(x, u^\varepsilon), \qquad x \in G$$

$$\left. \frac{\partial u^\varepsilon}{\partial n} \right|_{\partial G} = 0, \qquad u^\varepsilon(0, x) = g(x) \geq 0.* \tag{1.1.4}$$

We are interested now in the behavior of the solution of problem (1.1.4) as $\varepsilon \downarrow 0$. Denote:

$$R_{0t}(\varphi) = \int_0^t \left[c(\varphi_s) - \frac{1}{2D(\varphi_s)} |\dot{\varphi}_s|^2 \right] ds, \qquad \varphi \in C_{0t}, \ \varphi \text{ is absolutely continuous,}$$

$$V(t, x) = \sup_{\varphi_0 = x, \varphi_t \in G_0} \min_{0 \leq a \leq t} R_{0a}(\varphi).$$

It is easy to see that $V(t, x)$ is a continuous nonpositive function. One can prove that

$$\lim_{\varepsilon \downarrow 0} u^\varepsilon(t, x) = 0, \qquad \text{if } V(t, x) < 0;$$

$$\lim_{\varepsilon \downarrow 0} u^\varepsilon(t, x) = 1, \qquad \text{if the point } (t, x) \text{ belongs to the interior of the set}$$

$$\{(s, y) : V(s, y) = 0\}.$$

Thus the boundary of the set $\{x \in R^r : V(t, x) < 0\}$ is the position of the wave front at time t. This boundary separates areas in R^r where $u^\varepsilon(t, x)$ tends to 0 and to 1 as $\varepsilon \downarrow 0$. One can conclude from this result that if $c(x) = c = \text{const}$ then the wave front moves according to the Huygens principle, and the corresponding velocity field is isotropic, homogeneous, and equal to $\sqrt{2c}$ if calculated in the Riemannian metric corresponding to the form $ds^2 = D^{-1}(x)|dx|^2$. If $D(x)\Delta$ in equation (1.1.4) is replaced by an elliptic operator $\sum_{i,j=1}^n a^{ij}(x)(\partial^2/\partial x^i \partial x^j)$ then the same statement is true if we consider the Riemannian metric corresponding to this operator (see [4], [6]). As it is shown there, if $c(x) \neq \text{const}$ the wave front may have jumps and some other peculiarities. The general case is considered in [4], [6], [7], [3].

The problem of wave front propagation for the equation with slowly changing coefficients and nonlinear term

$$\frac{\partial u(t, x)}{\partial t} = \frac{D(\varepsilon x)}{2} \Delta u + f(\varepsilon x, u), \qquad x \in R^r, \ t > 0,$$

also can be reduced to (1.1.4) after rescaling space and time: the function $u^\varepsilon(t, x) = u(t/\varepsilon, x/\varepsilon)$ satisfies equation (1.1.4).

In (1.1.3), (1.1.4) the diffusion matrix is isotropic. In Section 1.4 we consider the anisotropic case. Let us consider the problem in the tube $R^1 \times G$:

$$\frac{\partial u(t,x,y)}{\partial t} = \frac{\varepsilon}{2}\Delta_y u + \frac{D(y)}{2}\frac{\partial^2 u}{\partial x^2} + f(y,u), \qquad t > 0,\; y \in G,\; x \in R^1,$$

$$\left.\frac{\partial u}{\partial n}\right|_{t>0,x\in R^1,y\in\partial G} = 0, \qquad u(0,x,y) = \chi^-(x). \tag{1.1.5}$$

Here $f(y,\cdot) \in F_1$ for each $y \in G \cup \partial G$, $D(y) > 0$, $0 < \varepsilon \ll 1$, and for brevity we put $b(y) \equiv 0$. The diffusivity in (1.1.5) along the tube (along the x-variable) is of order 1 when the diffusivity across the tube (in the y-variables) is of order $\varepsilon \ll 1$. We now have two large parameters ε^{-1} and t. The behavior of the solution depends on the relation between these parameters as $\varepsilon^{-1}, t \to \infty$. If $t \to \infty$ but $\varepsilon t \to 0$ the system does not have enough time for mixing up in the y-variables, and, roughly speaking, each $y \in G \cup \partial G$ will have its own asymptotic speed proportional to $\sqrt{2c(y)D(y)}$, where $c(y) = c(y,0) = \partial f(y,u)/\partial u|_{u=0}$.

In the case $t \approx \varepsilon^{-1} \to \infty$ one can expect that the mixing in y-variables will result in the establishing of a common asymptotic speed for all $y \in G \cup \partial G$. It turns out that that really is the case. But what is unexpected is that the common speed of the front in the tube can be bigger than the front speed for each fixed $y \in G \cup \partial G$. This effect is due to the fact that the particles governed by (1.1.5) can use one part of the cross-section $G \cup \partial G$ for motion and another part for multiplication. We calculate the common asymptotic speed in Section 1.4.

In (1.1.5) the diffusion across the tube is small in comparison with the diffusion coefficient along the tube. In Section 1.4 we briefly consider the case when the diffusion coefficient in x is small compared with the coefficient of the diffusion in y. Propagation of wave fronts in a tube with absorbing walls is studied in Section 1.5. It turns out that in this case the front cannot propagate if the tube is too narrow. There exists a critical "size" of the cross-section such that the front propagates if this "size" is bigger than the critical one.

The wave front propagation in a layer with slowly changing characteristics is studied in Section 1.6. After proper rescaling of space and time the equation has the form

$$\frac{\partial u^\varepsilon(t,x,y)}{\partial t} = \frac{\varepsilon}{2}\sum_{i,j=1}^r a^{ij}(x,y)\frac{\partial^2 u^\varepsilon}{\partial x^i \partial x^j} + \frac{D(x,y)}{2}\frac{\partial^2 u}{\partial y^2} + \frac{1}{\varepsilon}f(x,y,u^\varepsilon)$$

$$t > 0, \qquad x \in R^r, \qquad |y| < 1, \qquad \left.\frac{\partial u^\varepsilon}{\partial y}(t,x,y)\right|_{y=\pm 1} = 0$$

$$u^\varepsilon(0,x,y) = g(x,y) \geq 0, \qquad f(x,y,\cdot) \in F_1. \tag{1.1.6}$$

Denote $G_0 = \operatorname{supp} g$ and assume that the closure of G_0 coincides with the closure of its interior. Let $\bar{G}_0 \subset R^r$ be the projection of G_0 on R^r. Assume that $\partial f(x,y,u)/\partial u|_{u=0} = c = \text{const}$, and let the coefficients in (1.1.6) be independent of y for a while. Then the wave front propagates according to the Huygens principle, such that in the Riemannian metric $\rho(\cdot,\cdot)$ corresponding to the form $ds^2 = \sum_{i,j=1}^r a_{ij}(x)\,dx^i\,dx^j$, $(a_{ij}) = (a^{ij})^{-1}$, the velocity field is homogeneous and isotropic and equal to $\sqrt{2c}$:

$$\lim_{\varepsilon \downarrow 0} u^\varepsilon(t,x) = \begin{cases} 1, & \text{if } \rho(x, \bar{G}_0) < t\sqrt{2c} \\ 0, & \text{if } \rho(x, \bar{G}_0) > t\sqrt{2c} \end{cases} \qquad (1.1.7)$$

If the diffusion coefficients depend on $y \in [-1, 1]$, the equality (1.1.7) is also true, but the metric ρ should be replaced by a Finsler metric $\hat{\rho}$, corresponding to the family of forms $ds_y^2 = \sum_{i,j=1}^r a_{ij}(x,y)\, dx^i\, dx^j$, $|y| \leq 1$. The unit sphere in the tangent space for this metric at the point x is the convex hull of the ellipsoids $\left\{ z : \sum_{i,j=1}^r a_{ij}(x,y) z_i z_j \leq 1 \right\}$, $y \in [-1, 1]$.

If $\partial f(x, y, u)/\partial u|_{u=0} = c(x, y)$ in (1.1.6) depends on $x \in R^r$ the limiting behavior can be described as well. In this case the wave front can have jumps and, in general, loses the Markov property of the motion: the knowledge of the position of the front at time s does not determine, in general, the position at time $t > s$.

A number of effects caused by a slowly changing drift are considered in Section 1.6 as well.

A class of systems of PDEs is considered in Section 1.7. This class is a natural generalization of one KPP equation. We show some results concerning wave front propagation for the solutions of those systems.

Reaction-diffusion equations in narrow branching channels are considered briefly in Section 1.8. Semilinear equations on graphs with corresponding gluing conditions in the vertices arise for the limits of solutions of RDEs in narrow channels. Variation of the cross-section of the narrow channel leads to an additional drift term in the equation.

In the conclusion of this section I would like to say several words about the methods which we use for studying the wave front propagation. As is well known with any second order elliptic, maybe degenerated, operator

$$\mathrm{L}u = \frac{1}{2} \sum_{i,j=1}^r a^{ij}(x) \frac{\partial^2 u}{\partial x^i \partial x^j} + \sum_{i=1}^r b^i(x) \frac{\partial u}{\partial x^i}, \qquad x \in R^r$$

one can connect a Markov diffusion process in R^r (see, for example, [7]). To define a diffusion process in R^r means to define a family of probability measures P_x in the space C_{0T} of continuous functions on $[0, T]$, $0 < T < \infty$, with values in R^r, depending on the parameter $x \in R^r$. The measure P_x is concentrated on $\{\varphi \in C_{0T} : \varphi_0 = x\}$. The integral of a functional ξ on C_{0T} with respect to the measure P_x is denoted $E_x \xi$. The process corresponding to $\frac{1}{2}\Delta$ is called the Wiener process and the corresponding measure is called the Wiener measure in C_{0T}. One should actually speak of the Wiener family of measures μ_x depending on $x \in R^r$, but since the measures μ_x are shifts of the measure μ_0 one can consider only $\mu_0 = \mu$. The Wiener measure is a Gaussian measure in C_{0T} with mean zero and correlation function $E W_s^i W_t^j = \delta^{ij}(s \wedge t)$, $W_s = (W_s^1, \ldots, W_s^r) \in C_{0T}$; $\delta^{ij} = 0$ for $i \neq j$ and $\delta^{ii} = 1$.

The family P_x corresponding to the general operator L can be constructed with the help of stochastic differential equations: Let $\sigma(x)$ be a matrix such that $\sigma(x)\sigma^*(x) = (a^{ij}(x))$, $b(x) = (b^1(x), \ldots, b^r(x))$. Consider the following ordinary differential equation

$$dX_t = \sigma(X_t)\, dW_t + b(X_t)\, dt, \qquad X_0 = x \in R^r. \qquad (1.1.8)$$

Under some natural assumptions on the coefficients this equation has a unique continuous solution for each $x \in R^r$ and for almost all (with respect to the Wiener measure in C_{0T}) functions W_t, $0 \le t \le T$. That means that (1.1.8) for any $x \in R^r$ defines a map $A_x : C_{0T} \to C_{0T}$, $A_x W. = X.$. The measure P_x can be defined as follows: $P_x(\mathcal{E}) = \mu(A_x^{-1}(\mathcal{E}))$.

The solution of the Cauchy problem connected with L as well as the solution of the Dirichlet problem can be written as functional integrals with respect to the measures P_x. Consider, for example, the Cauchy problem

$$\frac{\partial u(t,x)}{\partial t} = \mathrm{L}u + c(t,x)u, \qquad x \in R^r, \ t > 0, \ u(0,x) = g(x). \qquad (1.1.9)$$

Then the following Feynman–Kac formula holds:

$$u(t,x) = E_x g(X_t) \exp \left\{ \int_0^t c(t-s, X_s) \, ds \right\}. \qquad (1.1.10)$$

We exploit this formula and its generalizations in this article. If we consider equation (1.1.9) not in the whole space R^r but in a domain $D \subset R^r$ with Neumann conditions on the boundary, the process X_t in (1.1.10) should be replaced by the process with reflection in the boundary which is governed by the operator L inside D. The results on representation of the solutions of initial-boundary problems as functional integrals can be found in [7].

Formula (1.1.10) together with (1.1.8) shows in a very explicit way the dependence of the solution on the coefficients of the equation and on initial conditions. That is why such representations turn out to be very convenient for studying different initial-boundary problems connected with the operator L.

We deal in this article with semilinear equations connected with L. Let the coefficient c in (1.1.9) depend on the solution: $c = c(x, u(t,x))$. Then from (1.1.10) we have the following equality:

$$u(t,x) = E_x g(X_t) \exp \left\{ \int_0^t c(X_s, u(t-s, X_s)) \, ds \right\}. \qquad (1.1.11)$$

We can consider (1.1.11) as an equation for $u(t,x)$, $t \ge 0$, $x \in R^r$. For example, if $c(x,u)$ is Lipschitz continuous in u one can derive from (1.1.11) the existence and uniqueness of the solution even in the case of the degenerate operator L. We will use equations of the type of (1.1.11) for studying asymptotic properties of the solution as $t \to \infty$ or as a parameter included in L tends to 0. Different kinds of limit theorems for stochastic processes allow us to study the asymptotic behavior of the solution of (1.1.11).

Limit theorems for large deviations will be especially useful for us. Let us consider a metric space (X, ρ), and let μ^h, $h > 0$, be a family of probability measures on the Borel σ-field of X. Let $\lambda(h)$ be a positive function, $\lim_{h \downarrow 0} \lambda(h) = \infty$. We say that $\lambda(h)S(x)$ is the action function (rate function) for the family μ^h as $h \downarrow 0$ if the following properties hold:

1. The set $\phi(s) = \{x \in X : S(x) \le s\}$ for each $s \ge 0$ is compact.
2. For any $\delta > 0$, $\gamma > 0$ and any $x \in X$ there exists $h_0 > 0$ such that

$$\mu^h \{y : \rho(x,y) < \delta\} \ge \exp\{-\lambda(h)[S(x) + \gamma]\}$$

for all $h \leq h_0$.

3. For any $\delta > 0$, any $\gamma > 0$, and any $s > 0$ there exists h_0 such that

$$\mu^h\{y : \rho(y, \phi(s)) \geq \delta\} \leq \exp\{-\lambda(h)(s - \gamma)\}$$

for $h \leq h_0$.

For example, if P_x^ε are the measures in C_{0T} corresponding to the operator $\frac{\varepsilon}{2}\Delta$, then $\lambda(\varepsilon) = \varepsilon^{-1}$ and

$$S(\varphi) = \begin{cases} \dfrac{1}{2}\displaystyle\int_0^T |\dot{\varphi}_s|^2 ds, & \text{if } \varphi_0 = x, \ \varphi_s \text{ is absolutely continuous,} \\ +\infty, & \text{for the rest of } C_{0T}. \end{cases}$$

The logarithmic asymptotics of many interesting probabilities and expectations can be expressed in terms of action functionals. For example for a "good enough" set $A \subset X$

$$\lim_{h \downarrow 0} \lambda^{-1}(h) \ln \mu^h(A) = -\inf\{S(x) : x \in A\}.$$

For any continuous function $F(x)$

$$\lim_{h \downarrow 0} \lambda^{-1}(h) \ln \left\{ \int_X e^{\lambda(h)F(x)} \mu^h(dx) \right\} = \sup\{F(x) - S(x) : x \in X\}.$$

The results concerning the action functions and limit theorems for the probabilities of large deviations can be found in [11], [17].

Finally, I would like to note that we are not going for the most general statements. We restrict ourselves to the simplest situation where the effect in which we are interested can be observed.

1.2. WAVE FRONTS IN A FLOW WITH NONTRIVIAL VELOCITY PROFILE

Let G be a bounded domain in R^r with smooth boundary ∂G, $n(y)$ be the outward normal to ∂G at $y \in \partial G$. Consider the following problem in the tube $R^1 \times (G \cup \partial G)$ (Figure 1):

$$\frac{\partial u(t, x, y)}{\partial t} = \frac{D}{2}\Delta_{x,y}u - b(y)\frac{\partial u}{\partial x} + f(u), \qquad t > 0, \ x \in R^1, \ y \in G,$$

$$\frac{\partial u(t, x, y)}{\partial n(y)}\bigg|_{t>0, x \in R^1, y \in \partial G} = 0,$$

$$u(0, x, y) = \chi^-(x) = \begin{cases} 1, & x \leq 0 \\ 0, & x > 0. \end{cases} \tag{1.2.1}$$

We assume that $b(y)$ is continuously differentiable in G up to the boundary, $D > 0$, $f(\cdot) \in F_1$. Our goal is to calculate the asymptotic speed as $t \to \infty$.

To formulate the results we need some auxiliary construction. Suppose for a while that

$$\bar{b} = \frac{1}{|G|}\int_G b(y)\, dy = 0, \qquad b(y) \not\equiv \bar{b},$$

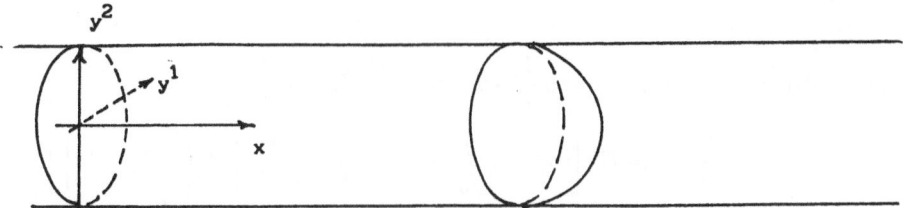

FIGURE 1.1.

where $|G|$ is the volume of the domain G, and consider the eigenvalue problem

$$\frac{1}{2}\Delta_y\varphi_\alpha(y) + \alpha b(y)\varphi_\alpha(y) = \lambda(\alpha)\varphi_\alpha(y),$$

$$y \in G, \quad \left.\frac{\partial\varphi_\alpha}{\partial n}\right|_{\partial G} = 0. \tag{1.2.2}$$

Here $\alpha \in R^1$ is a parameter. The problem (1.2.2) has a discrete spectrum. Let $\lambda = \lambda(\alpha)$ be the eigenvalue corresponding to the positive eigenfunction $\varphi_\alpha(y)$. As is well known, such eigenfunction exists and $\lambda(\alpha)$ is simple real and bigger than the real parts of all other eigenvalues. The eigenvalue $\lambda(\alpha)$ is continuously differentiable in $\alpha \in R^1$.

One can write down the two representations for $\lambda(\alpha)$. The first representation is a simple corollary of (1.2.2): Dividing (1.2.2) by $\varphi_\alpha(y)$ (note that $\varphi_\alpha(y) > 0$, $y \in G \cup \partial G$) and taking into account that $\bar{b} = 0$ we have

$$\lambda(\alpha) = \frac{1}{2|G|}\int_G \frac{\Delta\varphi_\alpha(y)\,dy}{\varphi_\alpha(y)}.$$

Since $\partial\varphi_\alpha/\partial n|_{\partial G} = 0$, using the Green's formula, we derive that

$$\lambda(\alpha) = \frac{1}{2|G|}\int_G \frac{|\nabla\varphi_\alpha(y)|^2}{\varphi_\alpha^2(y)}\,dy. \tag{1.2.3}$$

To introduce the second representation denote by y_t the Wiener process in G with normal reflection on the boundary. Then (see, for example, [11], Chap. 7) we can write

$$\lambda(\alpha) = \lim_{t\to\infty} t^{-1}\ln E_y \exp\left\{\alpha\int_0^t b(y_s)\,ds\right\}, \tag{1.2.4}$$

where E_y is the sign of expectation with respect to the probability measure in the space of trajectories y_s, $0 \le s \le t$, starting at the point $y \in G \cup \delta G$.

Using the Hölder inequality and (1.2.4) one can check that $\lambda(\alpha)$ is a convex function:

$$\lambda(p\alpha + q\beta) = \lim_{t\to\infty} t^{-1} \ln E_y \exp\left\{(\alpha p + \beta q) \int_0^t b(y_s)\, ds\right\}$$

$$= \lim_{t\to\infty} t^{-1} \ln E_y \left[\exp\left\{\alpha \int_0^t b(y_s)\, ds\right\}\right]^p \left[\exp\left\{\beta \int_0^t b(y_s)\, ds\right\}\right]^q$$

$$\leq p \lim_{t\to\infty} t^{-1} \ln E_y \exp\left\{\alpha \int_0^t b(y)\, ds\right\}$$

$$+ q \lim_{t\to\infty} t^{-1} \ln E_y \exp\left\{\beta \int_0^t b(y)\, ds\right\}$$

$$= p\lambda(\alpha) + q\lambda(\beta), \qquad p + q = 1, \qquad p,\, q > 0, \qquad \alpha,\, \beta \in R^1.$$

It follows from (1.2.3) that $\lambda(\alpha) \geq 0$ and $\lambda(\alpha) = 0$ if and only if $\alpha = 0$. Denote $B_+ = \max_{y\in G\cup\partial G} b(y)$, $B_- = \min_{y\in G\cup\partial G} b(y)$. Then $\lambda(\alpha) \leq B_+\alpha$ for $\alpha \geq 0$, and $\lambda(\alpha) \leq B_-\alpha$ for $\alpha \leq 0$.

Let us consider the Legendre transformation $L(\beta)$ of $\lambda(\alpha)$:

$$L(\beta) = \sup_\alpha [\alpha\beta - \lambda(\alpha)].$$

It is easy to check that $L(\beta)$ is finite in a neighborhood of the origin and equal to $+\infty$ for $\beta \notin [B_-, B_+]$. Taking into account the properties of $\lambda(\alpha)$ we conclude that $L(0) = \max_\alpha(-\lambda(\alpha)) = 0$ and $L(\beta) > -\lambda(0) = 0$ for $\beta \neq 0$.

Denote by $H(\alpha)$ the Legendre transformation of the convex function $L(\beta) + \beta^2/2$:

$$H(\alpha) = \sup_\beta \left[\alpha\beta - \frac{\beta^2}{2} - L(\beta)\right] = \frac{\alpha^2}{2} - \inf_\beta \left[\frac{(\alpha-\beta)^2}{2} + L(\beta)\right].$$

From the last equality we conclude that $H(\alpha) < \alpha^2/2$ for $\alpha \neq 0$ and $H(0) = 0$. One can check that since $L(\beta)$ is equal to $+\infty$ outside $[B_-, B_+]$ the function $H(\alpha)$ increases not faster than linearly as $|\alpha| \to \infty$.

The function $\alpha^2/2 - H(\alpha)$ is the Legendre transformation of the function $\beta^2/2 + \lambda(\beta)$:

$$\sup_\alpha \left(\alpha\beta - \frac{\alpha^2}{2} + H(\alpha)\right) = \sup_\alpha \left(\alpha\beta - \inf_x \left[\frac{(\alpha-x)^2}{2} + L(x)\right]\right)$$

$$= \sup_{\alpha,x} \left[\alpha\beta - \frac{(\alpha-x)^2}{2} - L(x)\right]$$

$$= \sup_x \left[\sup_\alpha \left(\alpha\beta - \frac{(\alpha-x)^2}{2}\right) - L(x)\right]$$

$$= \sup_x \left(x\beta + \frac{\beta^2}{2} - L(x)\right)$$

$$= \frac{\beta^2}{2} + \sup_x [x\beta - L(x)]$$

$$= \frac{\beta^2}{2} + \lambda(\beta).$$

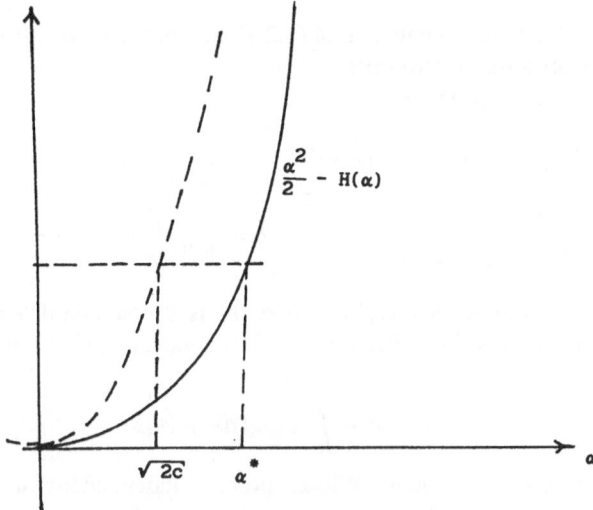

$$\frac{\alpha^2}{2} - H(\alpha)$$

$$\sqrt{2c} \qquad \alpha^*$$

$$\alpha$$

FIGURE 1.2.

The function $\beta^2/2 + \lambda(\beta)$ is strictly convex and increases faster than any linear function as $\beta \to +\infty$.

For any $c \geq 0$ consider the equation

$$\frac{\alpha^2}{2} - H(\alpha) = c. \tag{1.2.5}$$

Since $H(0) = 0$ and $H(\alpha)$ increases not faster than linearly as $\alpha \to \infty$ the left side of (1.2.5) continuously changes from 0 to $+\infty$ as α increases from 0 to $+\infty$ (Figure 2). Since $\alpha^2/2 - H(\alpha)$ is the Legendre transformation of a smooth strictly convex function $\beta^2/2 + \lambda(\beta)$, it is also strictly convex. Thus for each positive c the equation (1.2.5) has a unique positive root α, which is bigger than $\sqrt{2c}$.

So, for each continuous function $b(y)$, $y \in G \cup \partial G$, such that $\int_G b(y)\,dy = 0$, and any $c > 0$ we defined a number $\alpha^* = \alpha^*[b(\cdot), c]$ which is the positive root of the equation (1.2.5).

Now we are in a position to formulate the main result of this section.

Theorem 1.1. *Let $u(t, x, y)$ be the solution of problem (1.2.1), $f \in F_1$, and $c = df(u)/du|_{u=0}$. Then the asymptotic speed for problem (1.2.1) is equal to $\hat{\alpha} = \bar{b} + D\alpha^*[D^{-1}\tilde{b}(\cdot), c/D]$, where*

$$\bar{b} = \frac{1}{|G|} \int_G b(y)\,dy, \qquad \tilde{b}(y) = b(y) - \bar{b}.$$

That means that for any $h > 0$

$$\lim_{t \to \infty} \sup_{x > (\hat{\alpha}+h)t, y \in G \cup \partial G} u(t, x, y) = 0, \qquad \lim_{t \to \infty} \inf_{x < (\hat{\alpha}-h)t, y \in G \cup \partial G} u(t, x, y) = 1.$$

Proof. Without loss of generality we can assume that $D = 1$ and $\bar{b} = 0$: If it is not the case one should consider a new function $\hat{u}(t, x, y) = u(tD^{-1}, x - \bar{b}t, y)$. The function $\hat{u}(t, x, y)$ satisfies an equation of (1.2.1) type but with the diffusion coefficient equal to 1 and with a mean zero drift.

So we consider the problem

$$\frac{\partial u(t, x, y)}{\partial t} = \frac{1}{2}\Delta_{x,y}u + b(y)\frac{\partial u}{\partial x} + f(u), \qquad u(0, x, y) = \chi^-(x)$$

$$\left.\frac{\partial u}{\partial n}\right|_{t>0, x\in R^1, y\in\partial G} = 0, \qquad \bar{b} = \frac{1}{|G|}\int_G b(y)\, du = 0. \tag{1.2.6}$$

Consider the Markov process (X_t, Y_t) where Y_t is the standard Wiener process in G with instantaneous normal reflection on the boundary, and X_t is defined by the equation

$$X_t = x + \int_0^t b(y_s)\, ds + W_t,$$

where W_t is the one-dimensional Wiener process independent of Y_s. Using the Feynman–Kac formula one can write down the following integral equation in the space of trajectories for the solution of problem (1.2.6):

$$u(t, x, y) = E_{x,y}\chi^-(X_t)\exp\left\{\int_0^t c(u(t - s, X_s, Y_s)\, ds)\right\},$$

$$c(u) = u^{-1}f(u), \qquad t > 0, \qquad x \in R^1, \qquad y \in G \cup \partial G. \tag{1.2.7}$$

Here $E_{x,y}$ is the sign of expectation with respect to the distribution of the trajectories (X_s, Y_s), $0 \le s \le t$, starting at the point (x, y).

Since $f \in F_1$, $\max_{0\le u} c(u) = c(0) = c$, and thus we conclude from (1.2.7):

$$0 \le u(t, x, y) \le E_{x,y}\chi^-(X_t)e^{ct} = e^{ct}P_{x,y}\{X_t \le 0\}, \tag{1.2.8}$$

where $c = f'(0)$. To estimate $P_{x,y}\{X_t \le 0\}$ for large t and $x \approx zt$, $z > 0$, let us introduce the action function for the deviations of order 1 of $\xi_t = t^{-1}\int_0^t b(Y_s)\, ds$ and $(1/t)W_t$ from their mean values as $t \to \infty$.

As it follows from ([11], Chap. 7) for any $z_1 < z_2$

$$P_y\{\xi_t \in (z_1, z_2)\} \approx \exp\{-t \cdot \min_{z_1 \le z \le z_2} L(z)\}, \qquad t \to \infty,$$

where "$\approx$" is the sign of logarithmic equivalence and $L(z)$ is the Legendre transformation of the eigenvalue $\lambda(\alpha)$ of problem (1.2.2) considered above.

For the Gaussian random variable $(1/t)W_t$ we have

$$P\left\{\frac{1}{t}W_t \in (z_1, z_2)\right\} \approx \exp\left\{-\frac{t}{2}[\min_{z_1 < z < z_2} z^2]\right\}, \qquad t \to \infty.$$

Since ξ_t and W_t are independent we conclude that the action function for the pair $((1/t)W_t, \xi_t)$ is equal to the sum $t(z_1^2/2 + L(z_2))$. According to Theorem 3.3.1 from [11] the action function for $\xi_t + (1/t)W_t$ is equal to

$$t \cdot \min_{z_1}\left\{\frac{(z - z_1)^2}{2} + L(z_1)\right\} = \left(\frac{z^2}{2} - H(z)\right)t,$$

where $H(z)$ was introduced above. Using this action function we can calculate the logarithmic asymptotic of $P_{x,y}\{X_t \leq 0\}$ for $x = zt$ and $t \to \infty$:

$$\lim_{t \to \infty} t \ln P_{zt,y}\{X_t < 0\} = - \left[\frac{z^2}{2} - H(z) \right].$$

From (1.2.8) and the last equality we have

$$0 \leq u(t, zt, y) \leq \exp \left\{ t \left[c - \left(\frac{z^2}{2} - H(z) \right) \right] + o_t(1) \right\}. \qquad (1.2.9)$$

Inequality (1.2.9) shows that

$$\lim_{t \to \infty} \sup_{x > (\alpha^* + h)t, y \in G \cup \partial G} u(t, x, y) = 0. \qquad (1.2.10)$$

if $h > 0$ and α^* is the positive root of the equation

$$\frac{z^2}{2} - H(z) = c.$$

To prove the second equality in the definition of asymptotic speed denote $t^{-1} = \varepsilon$ and consider the function

$$u^\varepsilon(s, x, y) = u(s/\varepsilon, x/\varepsilon, y).$$

The function $u^\varepsilon(s, x, y)$ satisfies the following equation:

$$\frac{\partial u^\varepsilon(t, x, y)}{\partial t} = \frac{\varepsilon}{2} \frac{\partial^2 u^\varepsilon}{\partial x^2} + \frac{1}{2\varepsilon} \Delta_y u + b(y) \frac{\partial u^\varepsilon}{\partial x} + \frac{1}{\varepsilon} f(u^\varepsilon),$$
$$u^\varepsilon(0, x, y) = \chi^-(x). \qquad (1.2.11)$$

Let Y_t^ε be the process in $G \cup \partial G$ governed by $(1/2\varepsilon)\Delta_y$ inside G with normal reflection on ∂G, and let X_t^ε be the solution of the equation

$$X_t^\varepsilon - x = \sqrt{\varepsilon} W_t + \int_0^t b(Y_s^\varepsilon) \, ds,$$

where W_t is one-dimensional Wiener process.

The solution of (1.2.11) satisfies the following equation:

$$u^\varepsilon(t, x, y) = E_{x,y} \chi^-(X_t^\varepsilon) \exp \left\{ \frac{1}{\varepsilon} \int_0^t c(u^\varepsilon(t - s, X_s^\varepsilon, Y_s^\varepsilon)) \, ds \right\}. \qquad (1.2.12)$$

We can conclude from (1.2.10) that

$$\lim_{\varepsilon \downarrow 0} u^\varepsilon(t, x, y) = 0 \qquad (1.2.13)$$

for $x > \alpha^* t$, $y \in G \cup \partial G$ uniformly in any compact subset of $\{(t, x, y) : x > \alpha^* t, y \in G \cup \partial G, t > 0\}$.

Let us show that $\lim_{\varepsilon \downarrow 0} \varepsilon \ln u^\varepsilon(t, \alpha^*t, y) = 0$ uniformly in $0 \le t \le T < \infty$, $y \in G \cup \partial G$. It follows from [11] (Chaps. 3 and 7) that the action functionals for the families of processes $\sqrt{\varepsilon}\, W_t$ and $\int_0^t b(Y_s^\varepsilon)\, ds$, $0 \le t \le T$, as $\varepsilon \downarrow 0$ in the space of continuous functions have correspondingly the form:

$$\frac{1}{\varepsilon} S_1(\varphi) = \begin{cases} \dfrac{1}{\varepsilon} \displaystyle\int_0^T |\dot\varphi_s|^2\, ds, & \varphi \text{ is absolutely continuous, } \varphi_0 = 0, \\ +\infty, & \text{for the rest of } C_{0T}; \end{cases}$$

$$\frac{1}{\varepsilon} S_2(\varphi) = \begin{cases} \dfrac{1}{\varepsilon} \displaystyle\int_0^T L(-\dot\varphi_s)\, ds, & \varphi \text{ is absolutely continuous, } \varphi_0 = 0, \\ +\infty, & \text{for the rest of } C_{0T}. \end{cases}$$

Since $\sqrt{\varepsilon}\, W_t$ and Y_t^ε are independent we can conclude that the action functional $\varepsilon^{-1} S_{0T}(\varphi)$ for the family $X_t^\varepsilon = \sqrt{\varepsilon}\, W_t + b(Y_s^\varepsilon)\, ds$ as $\varepsilon \downarrow 0$ in C_{0T} is given as follows:

$$\varepsilon^{-1} S_{0T}(\varphi) = \begin{cases} \dfrac{1}{\varepsilon} \displaystyle\int_0^T \left[\dfrac{\dot\varphi_s^2}{2} - H(-\dot\varphi_s) \right] ds, & \varphi_s \text{ is absolutely continuous,} \\ +\infty, & \varphi_0 = 0, \text{ for the rest of } C_{0T}. \end{cases}$$

This statement follows from Theorem 3.3.1 of [11].

Choose $\delta \in (0, 1]$, and denote by φ_s^*, $0 \le s \le t$, the piecewise linear function connecting the points $(0, \alpha^*t)$, (δ, α^*t), $(t - \delta, 2\alpha^*\delta)$, $(t, -\alpha^*\delta)$ in the plane (s, x).

Taking into account (1.2.13) we conclude from (1.2.12) that

$$u^\varepsilon(t, \alpha^*t, y) = E_{\alpha^*t, y} \chi^-(X_t^\varepsilon) \exp \left\{ \frac{1}{\varepsilon} \int_0^t c(u^\varepsilon(t - s, X_s^\varepsilon, Y_s^\varepsilon))\, ds \right\}$$

$$\ge E_{\alpha^*t, y} \exp \left\{ \frac{1}{\varepsilon} \int_0^t c(u^\varepsilon(t - s, X_s^\varepsilon, Y_s^\varepsilon))\, ds \right\}$$

$$\cdot I_{\mathcal{E}_\delta}[X_\cdot^\varepsilon] \tag{1.2.14}$$

where $I_{\mathcal{E}_\delta}[X_\cdot^\varepsilon]$ is the indicator function of the set

$$\mathcal{E}_\delta = \{ \varphi \in C_{0t},\ \varphi_0 = \alpha^*t,\ \sup |\varphi_s - \varphi_s^*| < \delta^2 \}.$$

If $X^\varepsilon \in \mathcal{E}_\delta$ then X_s^ε for $\delta \le s \le t - \delta$ belongs to the subset of $\{(s, x) : s > 0,\ x > \alpha^*s\}$ where $u^\varepsilon(t, x, y)$ tends to 0 as $\varepsilon \downarrow 0$. Therefore $c(u(t - s, X_s^\varepsilon, Y_s^\varepsilon)) > c - h$ for such X_s^ε and s and any Y_s^ε, if ε is small enough, $c = c(0)$. Thus we can conclude from (1.2.14), that

$$u^\varepsilon(t, \alpha^*t, y) \ge P_{\alpha^*t, y} \{ X_\cdot^\varepsilon \in \mathcal{E}_\delta \} \exp \left\{ \frac{1}{\varepsilon} [ct - (2\delta c + h)t] \right\}. \tag{1.2.15}$$

From the properties of the action functional we have: for any $h > 0$ and $y \in G \cup \partial G$

$$P_{\alpha^*t, y} \{ X_\cdot^\varepsilon \in \mathcal{E}_\delta \} \ge \exp \left\{ \frac{1}{\varepsilon} [S_{0t}(\varphi^*) + h] \right\}$$

$$\times \exp \left\{ -\frac{1}{\varepsilon} \left[\frac{\alpha^{*2}t}{2} - H(\alpha^*)t + 2h \right] \right\} \tag{1.2.16}$$

if ε is small enough. From (1.2.15) and (1.2.16) we derive that

$$u^\varepsilon(t, \alpha^* t, y) \geq \exp\left\{ -\frac{1}{\varepsilon} \left[ct - S_{0t}(\varphi^*) + 4(c\delta + h)t \right] \right\}.$$

Since $\dot{\varphi}_s^* = -\alpha^*$ for $\delta < s < t - \delta$, taking into account the definition of α^*, we can conclude that for ε small enough

$$u^\varepsilon(t, \alpha^* t, y) \geq \exp\left\{ -\frac{A \cdot (\delta + h)t}{\varepsilon} \right\},$$
$$y \in G \cup \partial G, \qquad\qquad (1.2.17)$$

where A is a constant.

One can see from the maximum principle that $0 \leq u^\varepsilon(t, x, y) \leq 1$ for $t \geq 0$, $x \in R^1$, $y \in G \cup \partial G$. Since δ and h in (1.2.17) are arbitrary positive constants, we can conclude from the last inequality and (1.2.17) that

$$\lim_{\varepsilon \downarrow 0} \varepsilon \ln u^\varepsilon(t, \alpha^* t, y) = 0 \qquad\qquad (1.2.18)$$

uniformly in $y \in G \cup \partial G$ and $0 \leq t \leq T$. Using equation (1.2.7), the strong Markov property, and properties of the function $c(u)$ one can derive from (1.2.18) that

$$\lim_{\varepsilon \downarrow 0} u^\varepsilon(t, x, y) = 1 \qquad\qquad (1.2.19)$$

uniformly in (t, x, y) belonging to any compact subset of

$$\{(t, x, y) : t \geq 0, \ x > \alpha^* t, \ y \in G \cup \partial G\}.$$

The proof of this statement can be carried out in a standard way (compare with [7], Theorem 6.2.1), and we omit it.

Coming back to the function $u(t, x, y)$ we can conclude from (1.2.19), that for any $h > 0$

$$\lim_{t \to \infty} \inf_{x < (\alpha^* - h)t, y \in G \cup \partial G} u(t, x, y) = 1. \qquad\qquad (1.2.20)$$

The statement of Theorem 1 follows from (1.2.10) and (1.2.20).

Remark. As is well known, one can write down a variational representation for the eigenvalue $\lambda = \lambda(\alpha)$ of problem (1.2.2):

$$\lambda(\alpha) = \inf_{\varphi: \frac{\partial \varphi}{\partial n}|_{\partial G} = 0, \|\varphi\|_{L_G^2} = 1} \int_G \left[\frac{1}{2} \varphi(y) \Delta \varphi(y) + \alpha \beta(y) \varphi^2(y) \right] dy.$$

Using this representation of $\lambda(\alpha)$ we can derive the following representation for

$(\alpha^2/2) - H(\alpha)$:

$$\frac{\alpha^2}{2} - H(\alpha) = \sup_{\beta}\left[\beta\alpha - \frac{\beta^2}{2} - \lambda(\beta)\right]$$

$$= \sup_{\beta}\left[\beta\alpha - \frac{\beta^2}{2} - \inf_{\|\varphi\|_{L^2_G}=1}\int_G\left[\frac{1}{2}\varphi\Delta\varphi + \beta b\varphi^2\right]dy\right]$$

$$= \sup_{\beta,\|\varphi\|_{L^2_G}=1}\left[\beta\alpha - \frac{\beta^2}{2} - \int_G\left[\frac{1}{2}\varphi\Delta\varphi + \beta b\varphi^2\right]dy\right]$$

$$= \sup_{\varphi:\|\varphi\|_{L^2_G}=1,\,\frac{\partial\varphi}{\partial n}|_{\partial G}=0}\left\{\sup_{\beta}\left[\beta\alpha - \frac{\beta^2}{2} - \beta\int_G b\varphi^2\,dy\right]\right.$$

$$\left.-\frac{1}{2}\int_G\varphi\Delta\varphi\,dy\right\}$$

$$= \sup_{\varphi:\|\varphi\|_{L^2_G}=1,\,\frac{\partial\varphi}{\partial n}|_{\partial G}=0}\left\{\frac{1}{2}\left(\alpha - \int_G b\varphi^2\,dy\right)^2 - \frac{1}{2}\int_G\varphi\Delta\varphi\,dy\right\}.$$

Thus the equation for α^* has the form:

$$\sup_{\varphi:\|\varphi\|_{L^2_G}=1,\,\frac{\partial\varphi}{\partial n}|_{\partial G}=0}\left[\left(\alpha^* - \int_G b(y)\varphi^2(y)\,dy\right)^2 - \int_G\varphi(y)\Delta\varphi(y)\,dy\right] = 2c,$$

So, if $b(y) \equiv \bar{b}$ the asymptotic speed $\hat{\alpha}$, is given by the formula from [15]: $\hat{\alpha} = \bar{b} + \sqrt{2cd}$. If $b(y) \not\equiv \bar{b}$, as we can see from Figure 2 the asymptotic speed $\bar{b} + \alpha^*$ is bigger than $\bar{b} + \sqrt{2cd}$. That means that besides the increase of the speed due to the interaction of the diffusion in the x-direction and the nonlinear term, the interplay between the diffusion in the y-direction and $b(y) \neq \bar{b}$ together with the nonlinear term leads to an additional increase of the speed. We consider now the case when $\tilde{b}(y) = b(y) - \bar{b}$ is small. In this case we can give more explicit formulas for the additional increase of the speed.

Suppose that $b(y) = \bar{b} + \delta\tilde{b}(y)$, $\delta \ll 1$, and let us calculate the main term as $\delta \downarrow 0$ of $\alpha^* = \hat{\alpha} - \bar{b}$. We restrict ourselves to the case $D = 1$ and assume that $b(y)$ is smooth enough. As was shown before, $\alpha^2/2 - H(\alpha)$ is the Legendre transformation of $\beta^2/2 + \lambda(\beta)$, where $\lambda(\beta) = \lambda^\delta(\beta)$ is the first eigenvalue of the problem

$$\frac{1}{2}\Delta\varphi^\delta(y) + \beta\delta\tilde{b}(y)\varphi^\delta(y) = \lambda^\delta(\beta)\varphi^\delta(y), \qquad y \in G, \quad \left.\frac{\partial\varphi^\delta}{\partial n}\right|_{\partial G} = 0. \quad (1.2.21)$$

Since problem (1.2.21) has a discrete spectrum and $\tilde{b}(y)$ is smooth, the main term of $\lambda^\delta(\beta)$ as $\delta \downarrow 0$ can be calculated using the standard perturbation theory (see [14]).

We can look for $\lambda^\delta(\beta)$ and φ^δ in the following form:

$$\varphi^\delta = \varphi^{(o)}(y) + \beta\delta\varphi_1 + \delta^2\varphi_2 + o(\delta^2),$$
$$\lambda^\delta(\beta) = \delta^2\lambda_1(\beta) + o(\delta^2), \qquad \delta \downarrow 0.$$

The expansion for $\lambda^\delta(\beta)$ starts from the term of order δ^2 since $\lambda^o(\beta) \equiv 0$ and $\lambda^\delta(\beta) \geq 0$. Substituting these expansions in (1.2.21) and equating the terms with equal degrees of the small parameter we derive

$$\varphi^{(o)} = \text{const}$$

$$\frac{1}{2}\Delta\varphi_1(y) + \tilde{b}(y)\varphi_0 = 0, \qquad y \in G, \qquad \left.\frac{\partial\varphi_1}{\varphi n}\right|_{\partial G} = 0; \qquad (1.2.22a)$$

$$\frac{1}{2}\Delta\varphi_2(y) + \beta^2\tilde{b}(y)\varphi_1 = \lambda_1(\beta), \qquad \left.\frac{\partial\varphi_2}{\varphi n}\right|_{\partial G} = 0. \qquad (1.2.22b)$$

Since $\int_G \tilde{b}(y)\,dy = 0$ the equation (1.2.22a) is solvable. We find the main term $\delta^2\lambda_1(\beta)$ of $\lambda^\delta(\beta)$ as $\delta \downarrow 0$, from the solvability condition for the equation (1.2.22b). This equation is solvable if $\lambda_1(\beta) - \beta^2\tilde{b}(y)\varphi_1$ is orthogonal to any constant. This condition together with (1.2.22a) gives that

$$|G| \cdot \lambda_1(\beta) = \beta^2 \int_G \tilde{b}(y)\varphi_1(y)\,dy = -\frac{\beta^2}{2}\int_G \Delta\varphi_1(y) \cdot \varphi_1(y)\,dy = \frac{\beta^2}{2}\|\varphi_1\|_D^2,$$

where $\|\varphi_1\|_D$ is the Dirichlet norm:

$$\|\varphi_1\|_D^2 = \int_G \sum_{k=1}^r \left(\frac{\partial\varphi_1}{\partial y_k}\right)^2 dy.$$

We have now

$$\frac{\beta^2}{2} + \lambda(\beta) = \frac{\beta^2}{2}\left(1 + \frac{\delta^2\|\varphi_1\|_D^2}{|G|}\right) + o(\delta^2), \qquad \delta \downarrow 0.$$

Thus we have

$$\frac{\alpha^2}{2} - H(\alpha) = \frac{\alpha^2}{2\left(1 + \frac{\delta^2\|\varphi_1\|_D^2}{|G|}\right)} + o(\delta^2),$$

and we get the following value for the solution $\alpha^* = \alpha^*(\delta)$ of the equation $(\alpha^2/2) - H(\alpha) = c$:

$$\alpha^* = \sqrt{2c\left(1 + \frac{\delta^2\|\varphi_1\|_D^2}{|G|}\right)} + o(\delta^2), \qquad \delta \downarrow 0.$$

The asymptotic speed is then given as follows:

$$\hat{\alpha} = \bar{b} + \sqrt{2c} + \delta^2\sqrt{2c}\frac{\|\varphi_1\|_D^2}{2|G|} + o(\delta^2), \qquad \delta \downarrow 0.$$

It consists of three terms: The first, $\bar{b}$, is due to the average drift; the second, $\sqrt{2c}$, is KPP term; and the third, $(\delta^2\sqrt{2c}/2|G|)\|\varphi_1\|_D^2 + o(\delta^2)$, is due to the nontrivial profile of the speed of the current.

Consider as an example the case $G = (-1, 1) \in R^1$. Then the equation for $\varphi_1(y)$ has the form

$$\frac{1}{2}\varphi_1''(y) + \tilde{b}(y) = 0, \qquad -1 < y < 1, \ \varphi'(\pm 1) = 0.$$

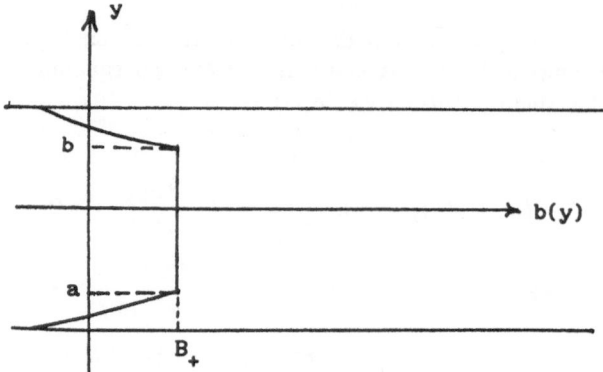

FIGURE 1.3.

Solving this equation we find that

$$\varphi_1(y) = -2\left(\int_0^y \int_0^x \tilde{b}(z)\, dx + y \int_{-1}^0 \tilde{b}(z)\, dz\right),$$

$$\|\varphi_1\|_D^2 = 4 \int_{-1}^1 \left(\int_{-1}^x \tilde{b}(z)\, dz\right)^2 dx.$$

Thus if the drift in the equation (1.2.11) is equal to $b(y) = \bar{b} + \delta\tilde{b}(y)$, the increase of the asymptotic speed, due to the interplay between the diffusion in the y-direction, the drift, and the nonlinear term, is equal to

$$\delta^2\sqrt{2c} \int_{-1}^1 \left(\int_{-1}^x \tilde{b}(z)\, dz\right)^2 dx + o(\delta^2), \qquad \delta \downarrow 0.$$

Let us mention, finally, one more effect. Assume for brevity that we have no diffusion in the x-direction:

$$\frac{\partial u(t,x,y)}{\partial t} = \frac{1}{2}\Delta_y u - b(y)\frac{\partial u}{\partial x} + f(u), \qquad x \in R^1,\ y \in G \subset R^r$$

$$\left.\frac{\partial u}{\partial n}\right|_{\partial G} = 0, \qquad u(0,x,y) = \chi^-(x).$$

Let $\bar{b} = (1/|G|)\int_G b(y)\, dy = 0$, $f \in F_1$, and $b(y)$ has the form as in Figure 3: $b(y)$ achieves its maximum B_+ on a closed domain $\hat{G} \subset G$, $\hat{G} = [a,b]$ in Figure 3. It is easy to see that the asymptotic speed cannot be bigger than B_+. The bigger $c = f'(0)$ the closer the asymptotic speed to B_+. If $\hat{G}$ has the Lebesgue measure zero in R^r, the speed in Figure 3 will always be less than B^+. But if $\hat{G}$ is the closure of an open set $(\hat{G})$ in R^r there exists a critical constant c_{crit} such that if $c = f'(0) > c_{\text{crit}}$ the front moves with the speed B_+. If $c < c_{\text{crit}}$ then the speed is less than B_+. Denote by τ the exit time from the domain $\hat{G}$: $\tau = \inf\{t : y_t \notin \hat{G}\}$, and let λ be the first eigenvalue of the operator $(1/2)\Delta$ in $\hat{G}$ with zero boundary conditions. It is well known that $\ln P_y\{\tau > t\} \approx -\lambda t$ as $t \to \infty$. Using this fact it is not difficult to check that $c_{\text{crit}} = \lambda$.

For example, if $G = [a,b] \in R^1$, then $\lambda = \pi^2/(b-a)^2$. If $c = f'(u) > \pi^2/(b-a)^2$ the asymptotic speed for the problem (1.2.23) is equal to B_+. In particular, we can see that even when the average speed of the flow $\bar{b} = (1/|G|) \int_G b(y)\, dy$ is negative but $b(y)$ is positive in some points, the front can propagate in the positive direction if c is large enough.

1.3.   GENERALIZATIONS AND REMARKS

1. Suppose that not only the drift depends on $y \in G$ but the diffusion coefficient and the nonlinear term also:

$$\frac{\partial u(t,x,y)}{\partial t} = \frac{D(y)}{2}\Delta_{x,y}u - b(y)\frac{\partial u}{\partial x} + f(y,u), \qquad t > 0,\ x \in R^1,\ y \in G,$$

$$\left.\frac{\partial u}{\partial n}\right|_{t>0,x\in R^1,y\in\partial G} = 0, \qquad u(0,x,y) = \chi^-(y). \tag{1.3.1}$$

We assume as before that the coefficients are smooth enough, and $f(y,\cdot) \in F_1$ for each $y \in G \cup \partial G$; $f(y,u) = c(y,u)u$, $c(y,0) = c(y)$, $D(y) > 0$.

To calculate the asymptotic speed consider the following eigenvalue problem:

$$\frac{D(y)}{2}\Delta_y(y) + [\alpha_1 D(y) + \alpha_2 b(y) + \alpha_3 c(y)]\varphi(y) = \lambda \cdot \varphi(y),$$

$$y \in G, \qquad \left.\frac{\partial\varphi}{\partial n}\right|_{\partial G} = 0; \qquad \alpha_1,\ \alpha_2,\ \alpha_3 \in R^1.$$

Let $\lambda = \lambda(\alpha_1,\alpha_2,\alpha_3)$ be the eigenvalue corresponding to the positive eigenfunction and let $L(z_1,z_2,z_3)$ be the Legendre transformation of $\lambda(\alpha_1,\alpha_2,\alpha_3)$:

$$L(z) = \sup_{\alpha=(\alpha_1,\alpha_2,\alpha_3)} [(\alpha,z) - \lambda(\alpha)].$$

The function $L(z)$, $z \in R^3$, as well as $\lambda(\alpha)$ is a convex function.

Consider the equation

$$\sup_{z_1,z_2,z_3}\left[z_3 - \frac{(v-z_2)^2}{2z_1} - L(z_1,z_2,z_3) \right] = 0, \qquad v \in R^1. \tag{1.3.2}$$

One can prove that there exists a root v^* of this equation, and

$$\tilde{b} = \int_G \frac{b(y)}{D(y)}dy \left(\int_G D^{-1}(y)\,dy \right)^{-1} \le v^* < \infty.$$

Theorem 1.2. *The asymptotic speed for problem (1.3.1) is equal to v^*, i.e., for any $h > 0$, we have*

$$\lim_{t\to\infty}\ \sup_{x>(v^*+h)t, y\in G\cup\partial G} u(t,x,y) = 0, \qquad \lim_{t\to\infty}\ \sup_{x>(v^*-h)t, y\in G\cup\partial G} u(t,x,y) = 1.$$

Proof. The proof is similar to the proof of Theorem 1. Using the Feynman–Kac formula we can write down the following equation:

$$u(t,x,y) = E_{x,y}\chi^-(x_t)\exp\left\{ \int_0^t c(y_s, u(t-s,x_s,y_s))\,ds \right\}. \tag{1.3.3}$$

Here y_t is the Markov process in $G \cup \partial G$ with normal reflection on the boundary, corresponding to the operator $(D(y)/2)\Delta_y$. The process x_t is defined by the equation

$$x_t - x = \int_0^t \sqrt{D(y_s)}\, dW_s - \int_0^t b(y_s)\, ds. \qquad (1.3.4)$$

Taking into account that $c(y, u) \le c(y)$ one can conclude from (1.3.3) that

$$0 \le u(t, x, y) \le E_{x,y}\chi^-(x_t) \exp\left\{ \int_0^t c(y_s)\, ds \right\}.$$

Because of the self-similarity of the Wiener process the stochastic integral in (1.3.4) has the same distribution as the process $\tilde{W}\left(\int_0^t D(y_s)\, ds \right)$, where $\tilde{W}_t$ is some Wiener process.

The function $L(z_1, z_2, z_3)$ defined above is the normalized action function for the family of random variables $\left(\int_0^t D(y_s)\, ds, \int_0^t b(y_s)\, ds, \int_0^t c(y_s)\, ds \right)$ as $t \to \infty$. The normalizing coefficient is equal to t (see [11], Chap. 7). Therefore we have

$$\lim_{t\to\infty} t^{-1} \ln E_{vt,y}\chi^-(x_t) \exp\left\{ \int_0^t c(y_s)\, ds \right\}$$
$$= \sup_{z_1,z_2,z_3} \left[z_3 - \frac{(v - z_2)^2}{2z_1} - L(z_1, z_2, z_3) \right] = \mathcal{A}(v).$$

One can check that the equation $\mathcal{A}(v) = 0$ has a root $v^* \ge \hat{b}$ and such a root is unique. The function $\mathcal{A}(v)$, as v increases, changes sign at v^* from plus to minus. Taking that into account we can conclude that for any $h > 0$

$$\lim_{t\to\infty} \sup_{x>(v^*+h)t,y\in G\cup\partial G} u(t, x, y) = 0.$$

The second statement of Theorem 2 can be proved similarly to the proof of the corresponding statement in Theorem 1.

Example. Let only $f = f(y, u)$ depend on y and assume for brevity that $b \equiv 0$. Then problem (1.3.1) has the form

$$\frac{\partial u(t, x, y)}{\partial t} = \frac{D}{2}\Delta_{x,y}u + f(y, u) \qquad y \in G,\ x \in R^1,\ t > 0,$$
$$\left.\frac{\partial u}{\partial n}\right|_{t>0,x\in R^1,y\in\partial G} = 0, \qquad u(0, x, y) = \chi^-(x).$$

One can expect that the behavior of the solution of this problem for large t would be, in a sense, the same as in an effective space-homogeneous equation:

$$\frac{\partial \tilde{u}(t, x)}{\partial t} = \frac{D}{2}\frac{\partial \tilde{u}}{\partial x^2} + \tilde{f}(\tilde{u}), \qquad \tilde{u}(0, x) = \chi^-(x),\ \tilde{f} = \tilde{c}(\tilde{u}) \cdot \tilde{u}.$$

As one can conclude from Theorem 2 the asymptotic speed is equal to $\sqrt{2D\lambda}$, where λ is the eigenvalue of the problem

$$\frac{D}{2}\Delta\varphi(y) + c(y)\varphi(y) = \lambda \cdot \varphi(y), \qquad y \in G,\ \left.\frac{\partial\varphi}{\partial n}\right|_{\partial G} = 0,$$

corresponding to the positive eigenfunction. It is easy to see that

$$\lambda \geq \bar{c} = \frac{1}{|G|} \int_G c(y)\,dy, \qquad c(y) = \lim_{u \to 0} u^{-1} f(y, u),$$

and the inequality is strict if $c(y) \not\equiv \bar{c}$. So λ is actually the "effective fitness coefficient" for this problem.

2. We consider now linear parabolic equations with nonlinear boundary conditions in the tube. We restrict ourselves for brevity with the classic heat equation

$$\frac{\partial u(t, x, y)}{\partial t} = \frac{1}{2}\Delta_{x,y} u, \qquad x \in R^1,\ y \in G \subset R^r,\ t > 0$$

$$\frac{\partial u}{\partial n} + f(y, u)\bigg|_{t>0, x \in R^1, y \in \partial G} = 0, \qquad u(0, x, y) = \chi^-(x). \tag{1.3.5}$$

We assume that $f(y, \cdot) \in F_1$ for each $y \in \partial G$. Then $f(y, u) = c(y, u)u$ and $c(y) = c(y, 0) = \max_{0 \leq u \leq 1} c(y, u)$.

Denote by $Z_t = (x_t, y_t)$ the Wiener process in the tube $(-\infty, \infty) \times (G \cup \partial G)$ with normal reflection on the boundary. Let ξ_t be the local time for the process Z_t on the boundary. The process (Z_t, ξ_t) satisfies the following equation:

$$dZ_t = dW_t + \chi_{\partial G}(Z_t) n(Z_t)\,d\xi_t, \qquad Z_0 = (x, y),\ \xi_0 = 0,$$

Here W_t is the Wiener process in R^{r+1} and $\chi_{\partial G}(\cdot)$ is the indicator function of the set $R^1 \times \partial G$. As is well known (see, for example, [7], Section 1.1.6), this stochastic equation defines (Z_t, ξ_t) in a unique way for each $x \in R^1$, $y \in G \cup \partial G$.

Using the generalized Feynman–Kac formula we can write the following equation for the solution $u(t, x, y)$ of (1.3.5):

$$u(t, x, y) = E_{x,y} \chi^-(x_t) \exp\left\{ \int_0^t c(y_s, u(t - s, x_s, y_s))\,d\xi_s \right\}. \tag{1.3.6}$$

Taking into account that x_t and y_t are independent processes and that $c(y, u) \leq c(y)$ we derive from (1.3.6)

$$0 \leq u(t, x, y) \leq P_x\{x_t \leq 0\} E_y \exp\left\{ \int_0^t c(y_s)\,d\xi_s \right\}. \tag{1.3.7}$$

We have for the one-dimensional Wiener process x_t

$$-\lim_{t \to \infty} t^{-1} \ln P_{\alpha t}\{x_t \leq 0\} = \frac{\alpha t^2}{2}. \tag{1.3.8}$$

The logarithmic asymptotic of the expectation in the right side of (1.3.7) is given as follows:

$$\lim_{t \to \infty} t^{-1} \ln E_y \exp\left\{ \int_0^t c(y_s)\,d\xi_s \right\} = \lambda, \tag{1.3.9}$$

where λ is the eigenvalue of the problem

$$\frac{1}{2}\Delta\varphi(y) = \lambda\varphi(y), \qquad y \in G,\ \frac{\partial\varphi(y)}{\partial n(y)} + c(y)\varphi(y)\bigg|_{y \in \partial G} = 0,$$

corresponding to the positive eigenfunction. Such an eigenvalue exists and is unique. Combining (1.3.7), (1.3.8), and (1.3.9) we can conclude that for any $h > 0$

$$\lim_{t \to \infty} \sup_{x > (\sqrt{2\lambda}+h)t, y \in G \cup \partial G} u(t, x, y) = 0. \tag{1.3.10}$$

Using the large deviation estimates for the processes x_t and $\int_0^t c(y_s) \, d\xi_s$ one can prove that

$$\lim_{t \to \infty} t^{-1} \ln u(t, t\sqrt{2\lambda}, y) = 0 \tag{1.3.11}$$

uniformly in $y \in G \cup \partial G$. Taking into account that $0 \leq u(t, x, y) \leq 1$ for $t \geq 0$, $x \in R^1$, $y \in G \cup \partial G$, and using the standard arguments (compare [7], Chap. 6) we conclude from (1.3.11) that

$$\lim_{t \to \infty} \inf_{x < (\sqrt{2\lambda}-h)t, y \in G \cup \partial G} u(t, x, y) = 1 \tag{1.3.12}$$

for any $h > 0$.

Bounds (1.3.10) and (1.3.12) give us the following result.

Theorem 1.3. *The asymptotic speed for problem (1.3.5) exists and is equal to $\sqrt{2\lambda}$.*

Example. Let $G = (-1, 1) \subset R^1$. Then we have the eigenvalue problem:

$$\frac{1}{2}\varphi''(y) = \lambda\varphi(y), \qquad -1 < y < 1,$$
$$\varphi'(-1) + c_-\varphi(-1) = 0, \qquad -\varphi'(1) + c_+\varphi(1) = 0,$$

where $c_\pm = \partial f / \partial u(\pm 1, 0)$. The solution of this problem has the form

$$\varphi(y) = Ae^{y\sqrt{2\lambda}} + Be^{-y\sqrt{2\lambda}},$$

where the constants A, B, λ should be found from the boundary conditions; the coefficients A and B are defined up to a factor. Denote $v = \sqrt{2\lambda}$. We have the following equations:

$$(Av + c_-A)e^{-v} = (Bv - c_-B)e^v, \qquad (Bv + c_+B)e^{-v} = (Av - c_+A)e^v.$$

One can derive from this system the equation for v:

$$\frac{(v - c_+)(v - c_-)}{(v + c_+)(v + c_+)} = e^{-4v};$$

the maximal root v^* of this equation should be taken.

In particular, if $c_- = c_+ = c$ the equation has a simpler form (see Figure 4)

$$\frac{v - c}{v + c} = e^{-2v}$$

The root v^* is, according to Theorem 3, the asymptotic speed in our problem.

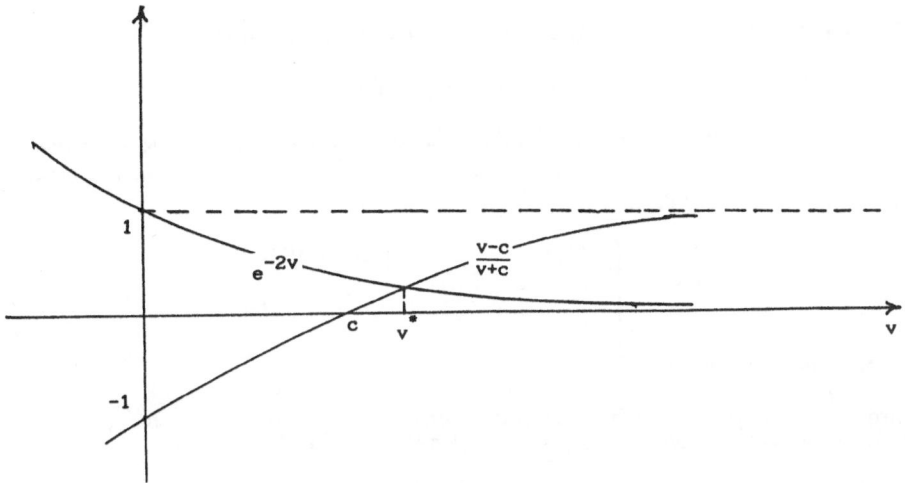

FIGURE 1.4.

3. We considered until now the indicator $\chi(x)$ of the set $\{(x, y) : x \leq 0,\ y \in G \cup \partial G\}$ as the initial function for equation (1.2.1). Now we discuss how the initial function of more general form can influence the asymptotic speed. Denote by $\mathcal{E}_0$ the support of the initial function $g(x, y) \geq 0$, $\mathcal{E}_0 \subset \{(x, y) : x \in R^1,\ y \in G \cup \partial G\}$, and assume that the closure of $\mathcal{E}_0$ coincides with the closure of its interior.

If $\mathcal{E}_0$ is a compact set one can in a natural way introduce two asymptotic speeds for equation (1.2.1); asymptotic speed $\hat{\alpha}_+$ for the propagation of the front to $+\infty$, and asymptotic speed $\hat{\alpha}_-$ for the propagation of the front to $-\infty$. The asymptotic speed $\hat{\alpha}_+$ will be the same as in Theorem 1. The speeds $\hat{\alpha}_+$ and $\hat{\alpha}_-$ can be different. The last one can be calculated in a similar way as $\hat{\alpha}_+$. Note that even if $\bar{b} > 0$ ($\bar{b} < 0$) the fronts can move in both directions, to $+\infty$ and to $-\infty$.

If $\mathcal{E}_0$ is not compact the asymptotic speed can increase. Let us supplement the equation (1.2.1) with an initial condition $u(0, x, y) = g(x, y) \geq 0$ and assume that g is bounded,

$$\inf_{y \in G \cup \partial G} g(x, y) > g_0 > 0 \tag{1.3.13}$$

in a neighborhood of the point $x = 0$, and

$$\lim_{x \to \infty} x^{-1} \ln g(x, y) = -\mu < 0 \tag{1.3.14}$$

uniformly in $y \in G \cup \partial G$. Let, for brevity, $\bar{b} = (1/|G|) \int_G b(y)\, dy = 0$.

Denote by $A(\mu)$ the Legendre transformation of the function $(v^2/2) - H(v)$, and assume that $v^* = v^*(\mu)$ is defined by the condition

$$A(\mu) = \sup_v \left(\mu v - \frac{v^2}{2} + H(v) \right) = \mu v^* - \frac{v^{*2}}{2} + H(v^*).$$

Recall that actually $A(\mu) = \mu^2/2 + \lambda(\mu)$, where $\lambda(\mu)$ is the first eigenvalue of problem (1.2.2), and the equation for $v^* = v^*(\mu)$ can be rewritten as follows:

$$\frac{1}{2}(\mu - v^*)^2 = H(v^*) - \lambda(\mu).$$

Theorem 1.4. *Consider equation (1.2.1) with the initial function*

$$u(0, x, y) = g(x, y) \geq 0$$

such that (1.3.13) and (1.3.14) are fulfilled. Let α^* be the positive root of the equation $\alpha^2/2 - H(\alpha) = c$ (see Section 1.2). Then the asymptotic speed $\hat{\alpha}_+$ in the positive direction is given as follows:

$$\hat{\alpha}_+ = \begin{cases} \alpha^*, & \text{if } \alpha^* < v^*(\mu); \\ \dfrac{c}{\mu} + \dfrac{\mu}{2} + \dfrac{\lambda(\mu)}{\mu}, & \text{if } \alpha^* \geq v^*(\mu). \end{cases}$$

It means that for any $h > 0$

$$\lim_{t \to \infty} \sup_{x > (\hat{\alpha}_+ + h)t, y \in G \cup \partial G} u(t, x, y) = 0, \quad \lim_{t \to \infty} \inf_{0 < x < (\hat{\alpha}_+ - h)t, y \in G \cup \partial G} u(t, x, y) = 1.$$

We will prove only the first statement. Given the first statement, the second one can be proved in the standard way.

Let, as before, Y_t be the Wiener process in G with normal reflection on ∂G, and

$$X_t = \alpha t + W_t - \int_0^t b(Y_s)\, ds.$$

Denote by $[x]^+ = x \vee 0$. Using the facts that $f(\cdot) \in F_1$ and that $g(x, y)$ is nonnegative and bounded, we can conclude from the Feynman–Kac formula:

$$0 \leq u(t, \alpha t, y) = E_{\alpha t, y} g(X_t, Y_t) \exp\left\{ \int_0^t c(u(t - s, X_s, Y_s)\, ds \right\}$$

$$\leq K E_{\alpha t, y} \exp\left\{ ct - \mu \cdot [X_t]^+ \right\} = K e^{ct} E_{\alpha t, y} \exp\{-\mu \cdot [X_t]^+\}. \quad (1.3.15)$$

We have found the action functional for the family of random variables X_t as $t \to \infty$ in Section 1.2. Using the large deviation principle we derive that

$$\lim_{t \to \infty} t^{-1} \ln E_{\alpha t, y} \exp\{-\mu[X_t]^+\} = \max_{z_1, z_2} \left\{ -[\mu\alpha - \mu z_1 - \mu z_2]^+ - \frac{z_1^2}{2} - L(z_2) \right\}$$

$$= \max_{z, v} \left\{ -[\mu\alpha - \mu V]^+ - L(z) - \frac{(V - z)^2}{2} \right\}$$

$$= \max_{v \leq \alpha} \max_z \left[-\mu\alpha + \mu V - \frac{(V - z)^2}{2} - L(z)] \right]$$

$$\vee \max_{v \geq \alpha} \max_z \left[-\left(\frac{(v - z)^2}{2} + L(z) \right) \right]$$

$$= \max_{V \leq \alpha} \left[-\mu\alpha - \left(\frac{v^2}{2} - H(v) \right) \right]$$

$$\vee \max_{V \geq \alpha} \left[-\left(\frac{v^2}{2} - H(v) \right) \right]$$

$$= \begin{cases} -\mu\alpha + A(\mu), & \text{if } v^*(\mu) \leq \alpha, \\ -\left[\dfrac{\alpha^2}{2} - H(\alpha) \right], & \text{if } v^*(\mu) > \alpha. \end{cases} \quad (1.3.16)$$

Denote by $\varepsilon(\alpha)$ the function in the right side of (1.3.16). From (1.3.15) one can derive that $\lim_{t\to\infty} u(t, \alpha t, y) = 0$ if $c + \varepsilon(\alpha) < 0$. The number $\hat{\alpha}_+$, introduced in Theorem 4, is the positive root of the equation $\varepsilon(\alpha) = -c$. One can derive from here the first statement of the theorem. The second statement can be proved similar to the proof of the second statement in Theorem 1.

It is easy to check that $\hat{\alpha}_+ \geq \alpha^*$ and $\hat{\alpha}_+$ strictly increases as $\mu \downarrow 0$, $\lim_{\mu\downarrow 0} \hat{\alpha}_+ = +\infty$.

We have a similar effect of increasing of the asymptotic speed if the initial function $g(x, y)$ has the values 1 and 0, and the support of $g(x, y)$ goes to $+\infty$ becoming narrower and narrower: Let $m(x)$, $x > 0$, be the Lebesgue measure of the set $\{y \in G : g(x, y) = 1]$, and assume that

$$\lim_{x\to\infty} x^{-1} \ln m(x) = -\mu < 0.$$

Then the closer μ is to zero the bigger will be the asymptotic speed in the positive direction.

4. We mention shortly here some generalization of the problem considered in Section 1.2. Let G be a compact metric space and Y_t be a Markov process on G with the generator A_y. Let X_t be a Markov process on R^1 with the generator A_x, independent of the process Y_t. We assume that both these processes are right continuous and have the Feller property.

Consider the Cauchy problem

$$\frac{\partial u(t, x, y)}{\partial t} = A_x u + A_y u + b(y)\frac{\partial u}{\partial x} + f(u)$$
$$u(0, x, y) = \chi^-(x), \qquad x \in R^1, \qquad y \in G, \qquad t > 0. \qquad (1.3.17)$$

Here the operators A_x and A_y are applied to $u(t, x, y)$ as a function of x and y respectively; $b(y)$ is a Lipschitz continuous function on G; $f(\cdot) \in F_1$. Define the process $\tilde{X}_t$ by the equality: $\tilde{X}_t = X_t + \int_0^t b(Y_s)\,ds$. The couple $(\tilde{X}_t, Y_t)$ is a strong Markov process on the phase space $R^1 \times G$.

Using the generalized Feynman–Kac formula one can write the following equation in the space of trajectories $(\tilde{X}_t, Y_t)$:

$$u(t, x, y) = E_{x,y}\chi^-(\tilde{X}_t) \exp\left\{\int_0^t c(u(t - s, \tilde{X}_s, Y_s))\,ds\right\}. \qquad (1.3.18)$$

Assuming that $c(u)$ is Lipschitz continuous we can derive from (1.3.18) the existence and uniqueness of the solution of problem (1.3.17) (compare Chap. 5 in [7]). We can expect that under certain assumptions an asymptotic speed will be established for problem (1.3.17).

One can derive from the Frobenius theorem that there exists an eigenvalue $\lambda = \lambda(\alpha)$ of the problem

$$A_y\varphi(y) + \alpha b(y)\varphi(y) = \lambda\varphi(y), \qquad y \in G,$$

corresponding to the positive eigenfunction. Under certain assumptions on ergodicity of the process Y_t this eigenvalue is simple. Then one can prove (see Chap. 7 in [11]) that the logarithmic asymptotic of the deviations of order t for $\int_0^t b(Y_s)\,ds$ as $t \to \infty$ is given by the action function $L(\beta) = \sup_\alpha (\alpha\beta - \lambda(\alpha))$:

$$\lim_{t\to\infty} t \ln P_y \left\{ \frac{1}{t} \int_0^t b(y)\,ds \in (a, a+da) \right\} = -L(a), \qquad a \in R^1.$$

Suppose that the logarithmic asymptotic of the deviations of order t for the process X_t is given by the action function $M(\beta)$:

$$\lim_{t\to\infty} t \ln P_0 \left\{ t^{-1} X_t \in (a, a+da) \right\} = -M(a). \qquad (1.3.19)$$

For example, if $X_t = W_t$ then $M(a) = a^2/2$. Since X_t and Y_t are independent the action function for $\tilde{X}_t = X_t + \int_0^t b(Y_s)\,ds$ is given by the expression

$$\mathcal{N}(\beta) = \min_a (M(a) + L(\beta - a)).$$

Taking into account that $f \in F_1$ we derive from (1.3.18) that

$$0 \leq u(t, \alpha t, y) \leq e^{ct} P_{\alpha t}\{\tilde{X}_t < 0\} \approx \exp\{t(c - \mathcal{N}(\alpha))\}, \qquad t \to \infty, \; c = f'(0). \tag{1.3.20}$$

Let α^* be the positive root of the equation $\mathcal{N}(\alpha) = c$. Then we can conclude from (1.3.20) that

$$\lim_{t\to\infty} u(t, \alpha t, y) = 0 \qquad (1.3.21)$$

if $\alpha > \alpha^*$. Now, using the large deviations principle, one can check that

$$\lim_{t\to\infty} \ln u(t, \alpha^* t, y) = 0 \qquad (1.3.22)$$

uniformly in $y \in G$. Taking into account the strong Markov property of the process $(\tilde{X}_t, Y_t)$, one can derive from (1.3.18) and (1.3.22) that

$$\lim_{t\to\infty} u(t, \alpha t, y) = 1 \qquad (1.3.23)$$

if $\alpha < \alpha^*$. The last relation is true uniformly in $\alpha \leq \alpha_0 < \alpha^*$. From (1.3.21) and (1.3.23) we see that α^* is the asymptotic speed.

We considered in Section 1.2 the Wiener process in a compact domain with normal reflection on the boundary as the process Y_t. Of course, any nondegenerate diffusion in the domain G with reflection on the boundary can be considered as a Y-process. But one can choose something else. For example, let $G = \{1, \dots, n\}$ and Y_t be a Markov chain with n states with continuous time and with positive transition intensities c_{ij}, $i \neq j$. Suppose that X_t is the Wiener process in R^1. Then equation (1.3.17) is equivalent to the system

$$\frac{\partial u_k(t, x)}{\partial t} = \frac{1}{2} \frac{\partial^2 u_k}{\partial x^2} + b_k \frac{\partial u_k}{\partial x} + \sum_{j=1}^{n} c_{kj} \cdot (u_j - u_k) + f(u_k)$$

$$u_k(0, x) = \chi^-(x), \qquad x \in R^1, \qquad t > 0, \qquad k = 1, \dots, n.$$

We will consider such systems and their generalizations later.

5. If the characteristics of the random process or of the nonlinear term depend on the coordinate x in an arbitrary way one cannot expect that an asymptotic speed of the wave front as $t \to \infty$ will be established. The medium should be asymptotically homogeneous in one or another sense. We can expect establishment of an asymptotic speed if the coefficients of the diffusion process and the nonlinear term are random fields invariant with respect to shifts along the tube (see [7], Chap. 7). An asymptotic speed will be established as well if the coefficients and the nonlinear terms are periodic in the variable x directed along the tube, or if those characteristics are close to periodic as $|x| \to \infty$.

Consider the periodic case:

$$\frac{\partial u(t,x,y)}{\partial t} = \frac{D}{2}\Delta_{x,y}u - b(x,y)\frac{\partial u}{\partial x} + c(x,y,u)u = \mathrm{L}u + c(x,y,u)u$$

$$t > 0, \quad x \in R^1, \quad y \in G \subset R^r, \quad \left.\frac{\partial u(t,x,y)}{\partial n}\right|_{t>0,x\in R^1,y\in\partial G} = 0,$$

$$u(0,x,y) = \chi^-(x). \tag{1.3.24}$$

We assume that the coefficients and the nonlinear term are smooth enough, $D > 0$, $f = c(x,y,u)u \in F_1$ for any $x \in R^1$, $y \in G \cup \partial G$. We assume also that $b(x,y)$ and $c(x,y)$ are periodic in x with period 1, and that $(1/|G|)\int_G b(x,y)\,dy = \bar{b} = $ const.

To formulate the result on the asymptotic speed let us introduce the family of operators L^z, $z \in R^1$:

$$L^z(x,y) = \frac{D}{2}\Delta_{x,y}\varphi - \tilde{b}(x,y)\frac{\partial\varphi}{\partial x} - z\frac{\partial\varphi}{\partial x} + \left(c(x,y) + \tilde{b}(x,y)z + \frac{Dz^2}{2}\right)\varphi$$

Here $c(x,y) = c(x,y,0)$, $\tilde{b}(x,y) = b(x,y) - \bar{b}$. Since all coefficients of L^z are periodic in x with period 1 we can consider L^z on the space of functions $\varphi(x,y)$ defined on the set $\Pi = S^1 \times (G \cup \partial G)$, where S^1 is the unit circle, with boundary condition $\partial\varphi/\partial n|_{S^1 \times \partial G} = 0$.

Denote by $\lambda(z)$ the eigenvalue, corresponding to the positive eigenfunction, of the following problem:

$$L^z\varphi(x,y) = \lambda(z)\varphi(x,y), \qquad (x,y) \in \Pi, \qquad \left.\frac{\partial\varphi}{\partial n}\right|_{S^1\times\partial G} = 0. \tag{1.3.25}$$

Such an eigenvalue exists. It is simple, convex, and differentiable in z. Denote by $H(\beta)$ the Legendre transformation of $\lambda(z)$: $H(\alpha) = \sup_\alpha[\alpha z - \lambda(z)]$.

One can prove that there exists a unique positive solution α^* of the equation $H(\alpha) = 0$. Then the asymptotic speed for problem (1.3.24) is equal to $\hat{\alpha} = \bar{b} + \alpha^*$.

The proof of the last statement is close to the proof of Theorem 7.3.1 in [7] and we omit it.

If the functions $b(x,y)$ and $c(x,y,u)$ in (1.3.24) are not periodic but there exist $b^1(x,y)$ and $c^1(x,y,u)$ periodic in x such that $\lim_{t\to\infty}[b(x,y) - b^1(x,y)] = 0$ and $\lim_{t\to\infty}(c(x,y) - c^1(x,y)) = 0$ uniformly in $y \in G \cup \partial G$, and the asymptotic speed $\hat{\alpha}$ calculated for these $c^1(x,y)$, $b^1(x,y)$ is positive then the asymptotic speed for (1.3.24) also equals $\hat{\alpha}$. It follows from a simple comparison theorem for equation

(1.3.24) and from the continuity of the dependence of the first eigenvalue of (1.3.25) on the coefficients $b(x, y)$ and $c(x, y)$.

1.4.   WAVE FRONT PROPAGATION IN ANISOTROPIC MEDIA

So far we considered isotropic diffusion. Now we examine wave front propagation in the tube when diffusion coefficients along the tube and across it have a different order. First, let us consider the case when the diffusivity in the y variables is much smaller than in the x-direction.

$$\frac{\partial u(t, x, y)}{\partial t} = \frac{\varepsilon}{2} \Delta_y u + \frac{D(y)}{2} \frac{\partial^2 u}{\partial x^2} + f(y, u), \qquad t > 0, \ x \in R^1, \ y \in G,$$

$$\left. \frac{\partial u(t, x, y)}{\partial n} \right|_{t>0, x \in R^1, y \in \partial G} = 0, \qquad u(0, x, y) = \chi^-(x). \tag{1.4.1}$$

Here $f(y, \cdot) \in F_1$ for $y \in G \cup \partial G$, $0 < D(y)$, $0 < \varepsilon \ll 1$. We put for brevity $b \equiv 0$.

If $t \to \infty$ but $\varepsilon t \to 0$ there is not enough time for mixing in the y-direction, and therefore, roughly speaking, an asymptotic speed as $t \to \infty$ will be established for each $y \in G \cup \partial G$ separately. If $\varepsilon^{-1} \approx t$ one can expect that a common asymptotic speed for all $y \in G \cup \partial G$ will be established. To consider this case let us rescale space and time. Denote by $u^\varepsilon(t, x, y) = u(t/\varepsilon, x/\varepsilon, y)$, where $u(t, x, y)$ is the solution of problem (1.4.1). The function $u^\varepsilon(t, x, y)$ is the solution of the following problem:

$$\frac{\partial u^\varepsilon(t, x, y)}{\partial t} = \frac{\varepsilon D(y)}{2} \frac{\partial^2 u^\varepsilon}{\partial x^2} + \frac{1}{2} \Delta_y u^\varepsilon + \frac{1}{\varepsilon} f(y, u^\varepsilon), \qquad t > 0, \ x \in R^1, \ y \in G,$$

$$\left. \frac{\partial u^\varepsilon}{\partial n}(t, x, y) \right|_{t>0, x \in R^1, y \in \partial G} = 0, \qquad u^\varepsilon(0, x, y) = \chi^-(x). \tag{1.4.2}$$

We shall examine the behavior of $u^\varepsilon(t, x, y)$ as $\varepsilon \to 0$.

If, as before, y_t is the Wiener process in G with normal reflection on ∂G, and

$$x_t^\varepsilon = x + W \left(\varepsilon \int_0^t D(y_s) \, ds \right),$$

where W_t is the Wiener process in R^1, then the Feynman–Kac formula gives the following equation for $u^\varepsilon(t, x, y)$:

$$u^\varepsilon(t, x, y) = E_{x,y} \chi^-(x_t^\varepsilon) \exp \left\{ \frac{1}{\varepsilon} \int_0^t c(y_s, u^\varepsilon(t - s, x_s^\varepsilon, y_s)) \, ds \right\}. \tag{1.4.3}$$

Since $f(y, \cdot) \in F_1$, $c(y) = c(y, 0) \geq c(y, u)$, and we conclude from (1.4.3):

$$0 \leq u^\varepsilon(t, x, y) \leq E_{x,y} \chi^-(x_t^\varepsilon) \exp \left\{ \frac{1}{\varepsilon} \int_0^t c(y_s) \, ds \right\}$$

$$= E_y \left[\exp \left\{ \frac{1}{\varepsilon} \int_0^t c(y_s) \, ds \right\} \right.$$

$$\left. \cdot P \left\{ x + W \left(\varepsilon \int_0^t D(y_s) \, ds \right) < 0 \mid Y_s, \ 0 \leq s \leq t \right\} \right]. \tag{1.4.4}$$

Since $W.$ and $y.$ are independent random processes, the conditional probability in the right side of (1.4.4) can be calculated explicitly. We need only the logarithmic asymptotic of this probability as $\varepsilon \downarrow 0$:

$$\lim_{\varepsilon \downarrow 0} \varepsilon \ln P \left\{ x + W \left(\varepsilon \int_0^t D(y_s)\, ds \right) < 0 \mid Y_s,\ 0 \le s \le t \right\}$$
$$= -x^2 \cdot \left(2 \int_0^t D(y_s)\, ds \right)^{-1}.$$

One can derive from the last equality, that the right side of (1.4.4) is logarithmically equivalent as $\varepsilon \downarrow 0$ to

$$E_y \exp \left\{ \frac{1}{\varepsilon} \left[\int_0^t c(y_s)\, ds - \frac{x^2}{2 \int_0^t D(y_s)\, ds} \right] \right\}$$
$$= E_y \exp \left\{ \frac{t}{\varepsilon} \left[\frac{1}{t} \int_0^t c(y_s)\, ds - \frac{(x/t)^2}{\frac{2}{t} \int_0^t D(y_s)\, ds} \right] \right\}. \qquad (1.4.5)$$

For any measurable set $\Gamma \subset G \cup \partial G$ denote by $\chi_\Gamma(y)$ the indicator of the set Γ and

$$\mu(\Gamma) = \mu_t(\Gamma;\ Y_s,\ 0 \le s \le t) = \frac{1}{t} \int_0^t \chi_\Gamma(y_s)\, ds;$$

$\mu(\Gamma)$ is a random probability measure. One can rewrite the right side of (1.4.5) in the form

$$E_y \exp \left\{ \frac{t}{\varepsilon} \left[\int_{G \cup \partial G} c(z)\mu(dz) - \frac{(x/t)^2}{2 \int_{G \cup \partial G} D(z)\mu(dz)} \right] \right\}. \qquad (1.4.6)$$

Since the support of the distribution of the process y_s, $0 \le s \le t$, is the whole space of continuous functions on $[0, t]$ with values in $G \cup \partial G$, starting at $y \in G \cup \partial G$, and since $c(z)$ and $D(z)$ are continuous functions on $G \cup \partial G$, it is easy to check that the expectation (1.4.6) is logarithmically equivalent as $\varepsilon \downarrow 0$ to

$$\exp \left\{ \frac{t}{\varepsilon} \sup_{\nu : \nu(G \cup \partial G) = 1} \left[\int_{G \cup \partial G} c(z)\nu(dz) - \frac{(x/t)^2}{2 \int_{G \cup \partial G} D(z)\nu(dz)} \right] \right\}. \qquad (1.4.7)$$

where the supremum is taken over all measures ν on $G \cup \partial G$ such that $\nu(G \cup \partial G) = 1$.

To calculate the supremum in (1.4.7) consider the mapping $T : G \cup \partial G \to R^2$ defined as follows

$$y \to T(y) = (c(y), D(y)),$$

and denote by $\mathcal{A}$ the image of $G \cup \partial G$:

$$\mathcal{A} = \{ z \in R^2 : z = T(y),\ y \in G \cup \partial G \}.$$

So, $\mathcal{A}$ is a compact connected set in the plane. Let $\mathcal{A}_{\mathrm{conv}}$ be the convex hull of the set $\mathcal{A}$. For any point $(z_1, z_2) \in \mathcal{A}_{\mathrm{conv}}$ one can choose a proper measure $\nu = \nu_{z_1, z_2}$ on $G \cup \partial G$, $\nu(G \cup \partial G) = 1$, such that

$$(z_1, z_2) = \left(\int_{G \cup \partial G} c(z)\nu(dz),\ \int_{G \cup \partial G} D(z)\nu(dz) \right).$$

Therefore the supremum in (1.4.7) is equal to

$$\max_{(z_1,z_2)\in\mathcal{A}_{\text{conv}}} \left[z_1 - \frac{(x/t)^2}{2z_2} \right] = \Gamma(x/t). \qquad (1.4.8)$$

Consider the function

$$\Gamma(\alpha) = \max_{(z_1,z_2)\in\mathcal{A}_{\text{conv}}} \left[z_1 - \frac{\alpha^2}{2z_2} \right].$$

This function is strictly decreasing for $\alpha > 0$,

$$\Gamma(0) = \max_{z\in G\cup\partial G} c(z) > 0, \qquad \lim_{\alpha\to\infty} \Gamma(\alpha) = -\infty.$$

Therefore the equation $\Gamma(\alpha) = 0$ has a unique positive root $\hat{\alpha}$.

Let us check that $\hat{\alpha} = \max_{(z_1,z_2)\in G\cup\partial G} \sqrt{2z_1 z_2}$: Define $z_1^* = z_1^*(\alpha)$, $z_2^* = z_2^*(\alpha)$ so that

$$\Gamma(\alpha) = \frac{2z_1^* z_2^* - \alpha^2}{2z_2^*}.$$

Then $\hat{\alpha} = \sqrt{2z_1^*(\hat{\alpha})z_2^*(\hat{\alpha})}$ and thus $\hat{\alpha} \le \max_{(z_1,z_2)\in\mathcal{A}_{\text{conv}}} \sqrt{2z_1 z_2}$.

Now, for any z_1, $z_2 \in \mathcal{A}_{\text{conv}}$

$$\frac{2z_1 z_2 - \hat{\alpha}^2}{2z_2} \le \max_{(z_1,z_2)\in\mathcal{A}_{\text{conv}}} \frac{2z_1 z_2 - \hat{\alpha}^2}{2z_2} = \Gamma(\hat{\alpha}) = 0,$$

and we conclude that $\hat{\alpha} \ge \max_{(z_1,z_2)\in\mathcal{A}_{\text{conv}}} \sqrt{2z_1 z_2}$. Thus

$$\hat{\alpha} = \max_{(z_1,z_2)\in\mathcal{A}_{\text{conv}}} \sqrt{2z_1 z_2}.$$

The function $\Gamma(\alpha)$ is negative when $\alpha > \hat{\alpha}$ and positive for $\alpha < \hat{\alpha}$. Combining (1.4.4)–(1.4.8) together, we conclude that

$$\lim_{\varepsilon\downarrow 0} u^\varepsilon(t,x,y) = 0 \qquad (1.4.9)$$

uniformly in any compact subset of $\{(t,x,y) : t \ge 0,\ y \in G\cup\partial G,\ x > \hat{\alpha}t\}$.

Let us show that

$$\lim_{\varepsilon\downarrow 0} u^\varepsilon(t,x,y) = 1 \qquad (1.4.10)$$

uniformly in $y \in G\cup\partial G$ if $x < \hat{\alpha}t$. To prove (1.4.10) let us check first that

$$\lim_{\varepsilon\downarrow 0} \varepsilon\ln u^\varepsilon(t,\hat{\alpha}t,y) = 0 \qquad (1.4.11)$$

uniformly in $y \in G\cup\partial\mathrm{G}$.

For any $\delta > 0$ small enough denote by $\hat{\varphi}_s^\delta$, $0 \le s \le t$, a function such that $\hat{\varphi}_t^\delta = -\delta$, $\hat{\varphi}_s^\delta = \hat{\alpha}t$ for $s \in [0,\delta]$, $\hat{\varphi}_s^\delta$ is linear on the interval $[t-\delta,t]$ and for $\delta \le s \le t - \delta$ (see Figure 5). The μ-neighborhood of the function $\hat{\varphi}^\delta$ in C_{0t} denote $\mathcal{E}_\mu(\hat{\varphi}^\delta) = \mathcal{E}_\mu$; $\chi_{\mathcal{E}_\mu}$ is the indicator of the set $\mathcal{E}_\mu \subset C_{0t}$.

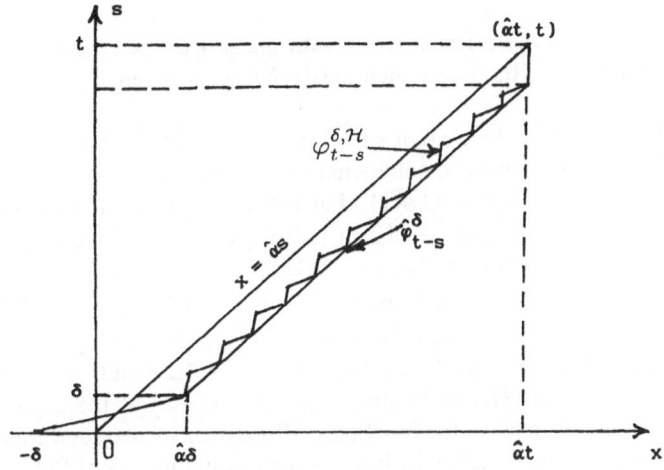

FIGURE 1.5.

Using (1.4.3) and (1.4.9) we can conclude that for any $h > 0$ one can choose so small positive δ and μ that the following lower bound is held:

$$u^\varepsilon(t, \hat{a}t, y) \geq E_{\hat{a}t, y} \chi_{\mathcal{E}_\mu}(x^\varepsilon_\cdot) \exp\left\{\frac{1}{\varepsilon}\left[\int_0^t c(y_s)\, ds - h\right]\right\} = e^{-h/\varepsilon} \cdot M^\varepsilon. \tag{1.4.12}$$

The trajectory $x^\varepsilon_\cdot$ starting from $\hat{a}t$ can be written as follows:

$$x^\varepsilon_s = at + W\left(\varepsilon \int_0^t D(y_{s_1})\, ds_1\right).$$

Now we can calculate the logarithmic asymptotic of the right side of (1.4.12). The action functional for the family x^ε_s for fixed y_s, $0 \leq s \leq t$, in C_{0t} on a smooth function $\psi_\cdot \in C_{0t}$ has the form

$$\frac{1}{\varepsilon} S^{y_0}_{0,t}(\psi) = \frac{1}{2\varepsilon}\int_0^t \frac{\dot{\psi}_s^2\, ds}{D(y_s)}.$$

Taking it into account we conclude that

$$M^\varepsilon = E_y E_{\hat{a}t}\left[\chi_{\mathcal{E}_\mu}(x^\varepsilon) \exp\left\{\frac{1}{\varepsilon}\int_0^t c(y_s)\, ds\right\}\,\Big|\, y_s,\ 0 \leq s \leq t\right]$$

$$\approx E_y \exp\left\{\frac{1}{\varepsilon} \sup_{\psi \in \mathcal{E}_\mu}\left[\int_0^t c(y_s)\, ds - \frac{1}{2}\int_0^t \frac{\dot{\psi}_s^2\, ds}{D(y_s)}\right]\right\}, \quad \text{as } \varepsilon \downarrow 0; \tag{1.4.13}$$

here "$\approx$" is the sign of logarithmic equivalency.

Since the support of the measure in the functional space for the process y_s, $0 \leq s \leq t$, coincides with the space $C_{G,y}$ of continuous functions $[0, t] \to G \cup \partial G$, $y_0 = y$, the right side of (1.4.13) is logarithmically equivalent as $\varepsilon \downarrow 0$ to

$$\exp\left\{\frac{1}{\varepsilon} \sup_{y_\cdot \in C_{G,y}} \sup_{\psi \in \mathcal{E}_\mu}\left[\int_0^t c(y_s)\, ds - \frac{1}{2}\int_0^t \frac{\dot{\psi}_s^2\, ds}{D(y_s)}\right]\right\}. \tag{1.4.14}$$

Now we will choose a function $\hat{\psi}^\delta$ with the properties described above in such a way that (1.4.14) will be bigger than $\exp\{-h/\varepsilon\}$ for given $h > 0$ and $\varepsilon > 0$ small enough.

Let $(z_1^*, z_2^*) \in \mathcal{A}_{\text{conv}}$ be a point such that $\hat{\alpha} = \sqrt{2z_1^* z_2^*}$. The point (z_1^*, z_2^*), like any point of $\mathcal{A}_{\text{conv}}$, is a linear combination of at most two points of $\mathcal{A} = \{(z_1, z_2) : z_1 = c(y), \ z_2 = D(y), \ y \in G \cup \partial G\}$. Let $(z_1^*, z_2^*) = P_1 \cdot (c_1, D_1) + P_2 \cdot (c_2, D_2)$, $P_1, P_2 \geq 0$, $P_1 + P_2 = 1$. In particular, if $(z_1^*, z_2^*) \in \mathcal{A}$ one of the weights is equal to zero. Choose small $\mathcal{H} > 0$ such that $(t - 2\delta)\mathcal{H}^{-1}$ is an integer.

Define $\varphi_s^{\delta, \mathcal{H}}$, $0 \leq s \leq t$, such that: $\varphi_s^{\delta, \mathcal{H}} = \hat{\alpha}t$ for $0 \leq s \leq \delta$, $\varphi_t^{\delta, \mathcal{H}} = -\delta < 0$, $\varphi_{t-\delta}^{\delta, \mathcal{H}} = \hat{\alpha} \cdot (t - 2\delta)$, $\varphi_s^{\delta, \mathcal{H}}$ is linear for $t - \delta \leq s \leq t$. Define $\varphi_s^{\delta, \mathcal{H}}$ for $s \in [\delta, t - \delta]$ as the saw-tooth function which has the vertices at the points $s_k = t - k\mathcal{H} - \delta$, $k = 0, 1, 2, \ldots, (t - 2\delta)/\mathcal{H}$, and equal to $\hat{\alpha} \cdot (t - s - \delta)$ at these points. Between $t - (k+1)\mathcal{H} - \delta$ and $t - k\mathcal{H} - \delta$ the function $\varphi_s^{\delta, \mathcal{H}}$ consists of two segments such that for $s \in (t - (k + P_1)\mathcal{H} - \delta, \ t - k\mathcal{H} - \delta)$ it has a slope equal to $\hat{\alpha}D_1/(P_1 D_1 + P_2 D_2)$ and equal to $-\alpha D_2/(P_1 D_1 + P_2 D_2)$ for the rest of the interval $[t - (k+1)\mathcal{H} - \delta, \ t - k\mathcal{H} - \delta]$, $k = 0, 1, \ldots, (t - 2\delta - \mathcal{H})/\mathcal{H}$. It is easy to see that $\varphi_s^{\delta, \mathcal{H}} \in \mathcal{E}_\mu$ if $\mathcal{H}$ is small enough, and $\varphi_s^{\delta, \mathcal{H}}$ for $s \in [\delta, t - \delta]$ belongs to the region where $u^\varepsilon(t, x, y)$ tends to 0 as $\varepsilon \downarrow 0$.

Taking into account that $\hat{\alpha}$ is the positive root of the equation $\Gamma(\alpha) = 0$ one can check that for any $h > 0$

$$\sup_{y. \in C_{G,y}} \left[\int_0^t c(y_s) \, ds - \frac{1}{2} \int_0^t \frac{|\dot{\psi}_s^{\delta, \mathcal{H}}|^2}{D(y_s)} ds \right] > h \tag{1.4.15}$$

if δ is small enough and $\mathcal{H} = \mathcal{H}(\delta)$ is small enough.

Equality (1.4.11) follows from (1.4.12)–(1.4.15) and the large deviation principle, if we take into account that $\sup_{t, x, y} u^\varepsilon(t, x, y) = 1$. The last inequality results from the negativity of $c(y, u)$ for $u > 1$ since $u(0, x, y) = \chi^-(x) \leq 1$.

Using the integral equation (1.4.3) and (1.4.11) one can prove equality (1.4.10) in the standard way (compare with [7], Chap. 6).

Combining (1.4.9) and (1.4.10) we come to the following result.

Theorem 1.5. *Let $\mathcal{A}_{\text{conv}}$ be the convex hull of the set $\mathcal{A} = \{(z_1, z_2) \in R^2 : z_1 = c(y), \ z_2 = D(y), \ y \in G \cup \partial G\}$, and $\hat{\alpha} = \max\{\sqrt{2z_1 z_2} : (z_1, z_2) \in \mathcal{A}_{\text{conv}}\}$.*

If $u^\varepsilon(t, x, y)$ is the solution of the problem (1.4.2) then

$$\lim_{\varepsilon \downarrow 0} u^\varepsilon(t, x, y) = 0 \qquad \text{for } t > 0, \ x > \hat{\alpha}t, \ y \in G \cup \partial G;$$

$$\lim_{\varepsilon \downarrow 0} u^\varepsilon(t, x, y) = 1 \qquad \text{for } t > 0, \ x < \hat{\alpha}t, \ y \in G \cup \partial G.$$

Remark. It follows actually from our consideration that $u^\varepsilon(t, x, y)$ tends to the limit as $\varepsilon \downarrow 0$ uniformly in any compact subset of $\{(t, x, y) : t > 0, \ x \neq \alpha t, \ y \in G \cup \partial G\}$.

The constant $\hat{\alpha}$ for equation (1.4.2) is an asymptotic speed. Two cases are possible: $\alpha^* = \max_{y \in G \cup \partial G} \sqrt{2c(y)D(y)} = \hat{\alpha}$ or $\alpha^* < \hat{\alpha}$. In the first case there exists $y^* \in G \cup \partial G$ such that $\sqrt{2c(y^*)D(y^*)} = \hat{\alpha}$. It means that the speed $\hat{\alpha}$ is equal to the asymptotic speed which will be established along the straight line

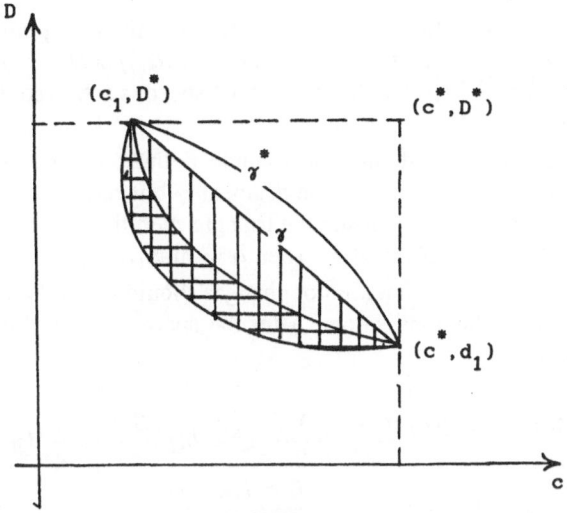

FIGURE 1.6.

$\{(x, y^*), \ -\infty < x < \infty\}$. The speed along other lines $\{(x, y), \ -\infty < x < \infty\}$, in the absence of mixing in y-directions is less than or equal to the speed for $y = y^*$. The mixing helps to establish the maximal speed $\hat{\alpha} = \alpha^*$ of the front for all $y \in G \cup \partial G$.

In the case $\alpha^* < \hat{\alpha}$ a common speed will be established in the tube which is bigger than the asymptotic speed for each fixed $y \in G \cup \partial G$. In this case, roughly speaking the particles defining motion of the front, use some points of the section $G \cup \partial G$ for multiplication (for chemical reaction) and other points of the section for the motion in the x-direction.

Let, for example, $\max_{y \in G \cup \partial G} c(y) = c^*$, $\max_{y \in G \cup \partial G} D(y) = D^*$, $d_1 = \max_{y:c(y)=c^*} D(y)$, $c_1 = \max_{y:D(y)=D^*} c(y)$ (Figure 6).

Denote by γ the segment in the plane (C, D) connecting the points (c_1, D^*) and $(c*, d_1)$. Suppose that the set $\mathcal{A} = \{(c(y), D(y)), \ y \in G \cup \partial G\}$ lies below the straight line defined by γ. The set $\mathcal{A}$ in Figure 6 is shaded by horizontal and vertical lines. The convex hull $\mathcal{A}_{\mathrm{conv}}$ of $\mathcal{A}$ is shaded by vertical lines, $\gamma \subset \mathcal{A}_{\mathrm{conv}}$.

The function $\sqrt{2z_1 z_2}$ increases in both variables z_1 and z_2, therefore

$$\hat{\alpha} = \max_{(z_1, z_2) \in \mathcal{A}_{\mathrm{conv}}} \sqrt{2z_1 z_2} = \max_{(z_1, z_2) \in \gamma} \sqrt{2z_1 z_2}.$$

The last maximum can be calculated explicitly:

$$z_2 = -\frac{D^* - d_1}{c^* - c_1} z_1 + \frac{c^* D^* - c_1 d_1}{c^* - c_1}, \qquad c_1 \le z_1 \le c^*,$$

for $(z_1, z_2) \in \gamma$; taking it into account we conclude that

$$\hat{\alpha} = \max_{c_1 \le z_1 \le c^*} \sqrt{2z_1 z_2} = \begin{cases} \sqrt{2c_1 D^*}, & \text{if } \dfrac{c^* D^* - c_1 d_1}{2(D^* - d_1)} \le c_1, \\[3mm] \dfrac{c^* D^* - c_1 d_1}{\sqrt{2(c^* - c_1)(D^* - d_1)}}, & \text{if } c_1 < \dfrac{c^* D^* - c_1 d_1}{2(D^* - d_1)} < c^*, \\[3mm] \sqrt{2c^* d_1}, & \text{if } \dfrac{c^* D^* - c_1 d_1}{2(D^* - d_1)} \ge c^*. \end{cases}$$

If $c_1 = c^*$ or $d_1 = D^*$ then $\hat{\alpha} = \sqrt{2c^*D^*}$. If only two points (c_1, D^*) and (c^*, d_1) of $\mathcal{A}$ belong to γ and $c_1 < (c^*D^* - c_1d_1)/2(D^* - d_1) < c^*$, then $\hat{\alpha} > \max_{y \in G \cup \partial G} \sqrt{2c(y)D(y)}$ and the asymptotic speed is achieved due to the mixing in the y-directions.

If the intersection $\hat{\mathcal{A}}$ of $\mathcal{A}$ and the triangle defined by the vertices (c_1, D^*), (c^*, D^*), (c^*, d_1) consists not only of the segment γ but has some other points, one can consider the convex curve γ^* inside the triangle such that the convex hull of $\hat{\mathcal{A}}$ is the area between γ^* and the segment γ. Then $\hat{\alpha} = \max_{(z_1, z_2) \in \gamma^*} \sqrt{2z_1z_2}$.

Now we consider briefly what kind of changes should be made if equation (1.4.1) includes a drift term in the x-direction. After the space and time rescaling we have the following equation:

$$\frac{\partial u^\varepsilon(t, x)}{\partial t} = \frac{\varepsilon D(y)}{2} \frac{\partial^2 u^\varepsilon}{\partial x^2} + \frac{1}{2}\Delta_y u^\varepsilon - b(y)\frac{\partial u^\varepsilon}{\partial x} + \frac{1}{\varepsilon}f(y, u^\varepsilon)$$

$$t > 0, \quad x \in R^1, \quad y \in G, \quad \frac{\partial u^\varepsilon(t, x, y)}{\partial n}\bigg|_{t>0, x\in R^1, y\in \partial G} = 0,$$

$$u^\varepsilon(0, x, y) = \chi^-(x). \tag{1.4.16}$$

Here $D(y) > 0$, $f(y, \cdot) \in F_1$, $b(y)$ is smooth enough. If D, b, and $f = c(u)u$ independent of y the asymptotic speed will be equal to $\alpha^* = b + \sqrt{2cD}$, $c = c(0)$. In the absence of the diffusion in the y-direction we can speak of an asymptotic speed $\alpha^*(y)$ for each $y \in G \cup \partial G$: $\alpha^*(y) = b(y) + \sqrt{2c(y)D(y)}$. It turns out that owing to the diffusion in the y-direction the asymptotic speed can increase. This effect is similar to the situation described in Theorem 5. To formulate the result let us consider the mapping $\tilde{T} : G \cup \partial G \to R^3 : \tilde{T}(y) = (c(y), D(y), b(y))$. Denote by $\tilde{\mathcal{A}} = \{(z_1, z_2, z_3) \in R^3 : z_1 = c(y), z_2 = D(y), z_3 = b(y) : y \in G \cup \partial G\}$, $\tilde{\mathcal{A}}_{\text{conv}}$ is the convex hull of $\tilde{\mathcal{A}}$ in R^3. The asymptotic speed for equation (1.4.16) is equal to

$$\tilde{\alpha} = \max_{(z_1, z_2, z_3) \in \tilde{\mathcal{A}}_{\text{conv}}} \left[z_3 + \sqrt{2z_1z_2}\right].$$

That means that $\lim_{\varepsilon \downarrow 0} u^\varepsilon(t, x, y) = 0$ if $x > \tilde{\alpha}t$ and $y \in G \cup \partial G$, and $\lim_{\varepsilon \downarrow 0} u^\varepsilon(t, x, y) = 1$ if $x < \tilde{\alpha}t$, $y \in G \cup \partial G$, $t > 0$. The proof of this statement is similar to the proof of Theorem 5.

So far we considered the case when the diffusion coefficients in the y-direction were small in comparison to the coefficient in the x-direction. Now let the diffusivity in the x-direction be small and the diffusivity in the y-direction be of order 1:

$$\frac{\partial u(t, x, y)}{\partial t} = \frac{1}{2}\Delta_y u + \frac{\varepsilon^2}{2}a(y)\frac{\partial^2 u}{\partial x^2} + f(y, u), \qquad t > 0, \ x \in R^1, \ y \in G$$

$$u(0, x, y) = \chi^-(x), \quad \frac{\partial u(t, x, y)}{\partial n}\bigg|_{t>0, x\in R^1, y\in \partial G} = 0, \quad 0 < \varepsilon \ll 1. \tag{1.4.17}$$

It is easy to see that the speed of propagation of the region, where $u(t, x, y)$ is close to 1, is of order ε as $\varepsilon \downarrow 0$. Therefore it is useful to rescale the time: denote

$u^\varepsilon(t, x, y) = u(t/\varepsilon, x, y)$. The new function $u^\varepsilon(t, x, y)$ satisfies the equation

$$\frac{\partial u^\varepsilon(t, x, y)}{\partial t} = \frac{1}{2\varepsilon} \Delta_y u + \frac{\varepsilon}{2} a(y) \frac{\partial^2 u^\varepsilon}{\partial x^2} + \frac{1}{\varepsilon} f(y, u^\varepsilon),$$

$$\left.\frac{\partial u^\varepsilon}{\partial n}\right|_{\partial G} = 0, \qquad u^\varepsilon(0, x, y) = \chi^-(x), \qquad f = c(y, u)u \in F_1. \quad (1.4.18)$$

Let, as before, y_t be the Wiener process in G with reflection on the boundary and

$$x_t^\varepsilon = x + \sqrt{\varepsilon}\, W\left(\int_0^t a(y_{s/\varepsilon})\, ds\right),$$

where W_t is the Wiener process in R^1 independent of y_t. Then we have

$$u^\varepsilon(t, x, y) = E_{x,y}\chi(x_t^\varepsilon) \exp\left\{\frac{1}{\varepsilon}\int_0^t c(y_{s/\varepsilon}, u^\varepsilon(t-s, x_s^\varepsilon, y_{s/\varepsilon}))\, ds\right\}$$

$$\leq E_{x,y}\chi(x_t^\varepsilon) \exp\left\{\frac{1}{\varepsilon}\int_0^t c(y_{s/\varepsilon})\, ds\right\}$$

$$\approx E_y \exp\left\{\frac{1}{\varepsilon}\left[\int_0^t c(y_{s/\varepsilon})\, ds - \frac{x^2}{2\int_0^t a(y_{s/\varepsilon})\, ds}\right]\right\}, \qquad \varepsilon \downarrow 0. \quad (1.4.19)$$

To calculate the asymptotic of (1.4.19) as $\varepsilon \downarrow 0$ consider the eigenvalue problem

$$\frac{1}{2}\Delta\varphi(y) + [\alpha_1 c(y) + \alpha_2 a(y)]\varphi(y) = \lambda \cdot \varphi(y), \qquad y \in G, \quad \left.\frac{\partial\varphi}{\partial n}\right|_{\partial G} = 0.$$

Let $\lambda = \lambda(\alpha_1, \alpha_2)$ be the first eigenvalue of this problem. We assume that $c(y)$ and $a(y)$ are smooth enough. Then $\lambda(\alpha_1, \alpha_2)$ is a differentiable convex function. Denote by $L(\beta_1, \beta_2)$ the Legendre transformation of $\lambda(\alpha_1, \alpha_2)$. The function $\varepsilon^{-1}tL(\beta_1, \beta_2)$ is the action function for deviations of order 1 as $\varepsilon^{-1}t \to \infty$ for the family of random variables $\left(\frac{\varepsilon}{t}\int_0^t c(y_s)\, ds, \frac{\varepsilon}{t}\int_0^t a(y_s)\, ds\right)$. Thus the right side of (1.4.19) is logarithmically equivalent to

$$\exp\left\{\frac{t}{\varepsilon} \cdot \sup_{z_1, z_2 > 0}\left[z_1 - \frac{(x/t)^2}{2z_2} - L(z_1, z_2)\right]\right\}, \qquad t\varepsilon^{-1} \to \infty. \quad (1.4.20)$$

Denote

$$\Lambda(\alpha) = \sup_{z_1, z_2 > 0}\left[z_1 - \frac{\alpha^2}{2z_2} - L(z_1, z_2)\right], \qquad \alpha > 0,$$

and let $\hat{\alpha}$ be the positive root of the equation $\Lambda(\alpha) = 0$. It is easy to check that such a root exists and is unique. One can conclude from (1.4.19) and (1.4.20) that

$$\lim_{\varepsilon \downarrow 0} u^\varepsilon(t, x, y) = 0, \qquad \text{if } x > \hat{\alpha}t, \ t > 0.$$

Using the last equality, the strong Markov property of the process $(x_t^\varepsilon, y_{t/\varepsilon})$, and the upper bound of the large deviation principle one can check that

$$\lim_{\varepsilon \downarrow 0} u^\varepsilon(t, x, y) = 1, \qquad \text{if } x < \hat{\alpha}t, \ t > 0.$$

That means that $\hat{\alpha}$ is the asymptotic speed of the wave front for problem (1.4.18).

This problem was briefly considered in [6]. Its generalization was considered in [2].

1.5.   WAVE FRONT PROPAGATION IN A TUBE WITH ABSORBING WALLS

So far we considered wave front propagation in tubes with reflecting boundary conditions. Now we examine a problem with absorbing conditions on the walls. We restrict ourselves to the case $y \in [-r, r] \subset R^1$. More general equations can be studied in a similar way. Consider the problem

$$\frac{\partial u(t, x, y)}{\partial t} = \frac{D}{2} \Delta_{x,y} u + f(u), \qquad t > 0, \; x \in R^1, \; |y| \leq r,$$

$$u(t, x, \pm r) = 0, \qquad u(0, x, y) = g((y)\chi^-(x). \tag{1.5.1}$$

Assume that $f = c(u)u \in F_1$ and, besides, let $c'(u) < 0$ for $u \in [0, 1]$; $g(y)$ is a continuously differentiable function positive inside $(-r, r)$, and $g(\pm r) = 0$.

Let $-\lambda$ be the eigenvalue of the problem

$$\frac{D}{2} \frac{d^2 \varphi(y)}{dy^2} = -\lambda \varphi(y), \qquad |y| < r, \; \varphi(\pm r) = 0,$$

corresponding to the positive eigenfunction $\varphi(y)$. It is easy to calculate that $\lambda = \pi^2/8r^2$ and $\varphi(y) = \frac{1}{2} \cos(\pi y/2r)$.

Let y_t be the Markov process corresponding to the operator $(D/2)d^2/dy^2$ in R^1, $\tau = \tau_{[-r,r]} = \min\{t : y_t \notin [-r, r]\}$.

Lemma 1.1. *For any $y \in (-r, r)$*

$$\lim_{t \to \infty} e^{-\lambda t} P_y\{\tau > t\} = \varphi(y) \cdot \left(\int_{-r}^{r} \varphi(z) \, dz \right).$$

Proof. The function $v((t, y) = P_y\{\tau > t\}$ satisfies the following equation and boundary conditions:

$$\frac{\partial v}{\partial t} = \frac{D}{2} \frac{\partial^2 v}{\partial y^2}, \qquad |y| < r, \; v(t, \pm r) = 0, \; v(0, y) = 1.$$

Solving this problem by Fourier method we have

$$v(t, x) = \sum_k c_k e^{-\lambda_k t} \varphi_k(y), \qquad c_k = \int_{-r}^{r} \varphi_k(y) \, dy,$$

where $\{\lambda_k\}$ is the sequence of eigenvalues and $\{\varphi_k(y)\}$ are the corresponding eigenfunctions. The statement of the Lemma follows from this expansion.

For any $\Gamma = (a, b) \subset R^1, 0 < a < b < 1$, let

$$\theta_\Gamma(t) = \int_0^t \chi_\Gamma(y_s) \, ds,$$

where χ_Γ is the indicator of the set Γ.

Lemma 1.2. *For any $A, M > 0$ there exists $\delta > 0$ such that if $|b - a| \leq 2\delta$ then*

$$\varlimsup_{t \to \infty} \frac{1}{t} \ln P_y \left\{ \int_0^{t \wedge \tau} X_\Gamma(y_s)\, ds > At \right\} < -M$$

for any $y \in [-r, r]$.

Proof. It is sufficient to prove that

$$\varlimsup_{t \to \infty} \frac{1}{t} \ln P_y \left\{ \int_0^\tau X_\Gamma(y_s)\, ds > At \right\} < -M \tag{1.5.2}$$

for $\Gamma = (-\delta, \delta)$ and $\delta > 0$ small enough.

Let us introduce Markov times $\beta_0, \beta_1, \dots$ and $\sigma_0, \sigma_1, \sigma_2, \dots$ as follows

$$\sigma_0 = \min\{t : |y_t| = \delta \quad \text{or} \quad |y_t| = r\},$$
$$\beta_0 = \min\{t : |y_t| = 2\delta\}, \qquad \sigma_1 = \min\{t > \beta_0 : |y_t| = \delta \quad \text{or} \quad |y_t| = r\}$$
$$\beta_k = \min\{t > \sigma_k : |y_t| = 2\delta\}, \qquad k = 1, 2, \dots,$$
$$\sigma_k = \min\{t > \beta_{k-1}, |y_t| = \delta \quad \text{or} \quad |y_t| = r\}, \qquad \kappa_k = \beta_k - \sigma_k.$$

Then we have

$$P_y \left\{ \int_0^\tau \chi_{-\delta,\delta}(y_s)\, ds > At \right\} \leq P_y \left\{ \sum_{l=0}^{\nu-1} \beta_l > At \right\}, \tag{1.5.3}$$

where $\nu = \min\{n : |y_{\sigma_n}| = r\}$. The random variables β_l, $l = 0, 1, 2, \dots$ and ν are independent. One can calculate (see, for example, [7]), that

$$E_y e^{k\beta_l} = \frac{\cos y\sqrt{2kD^{-1}}}{\cos \dfrac{2\delta\sqrt{2k}}{\sqrt{D}}} \leq \frac{1}{\cos \dfrac{2\delta\sqrt{2k}}{\sqrt{D}}} \sim e^{4\delta^2 k/D} \quad \text{as } \delta \downarrow 0. \tag{1.5.4}$$

Taking into account that $P_y\{y_{\sigma_0} = r\}$ is a linear function of y on $[\delta, r]$ equal to 0 at $y = \delta$ and to 1 at $y = r$, we conclude that

$$P_\delta\{\nu = l\} \approx \frac{\delta}{r}\left(1 - \frac{\delta}{r}\right)^l \quad \text{as } \delta \downarrow 0. \tag{1.5.5}$$

Using the exponential Chebyshev inequality we derive from (1.5.3)–(1.5.5)

$$P_\delta \left\{ \int_0^\tau \chi_{-\delta,\delta}(y_s)\, ds > At \right\} \leq e^{-kAt} E_\delta \left(E_\delta e^{k\beta_0}\right)^\nu \leq N e^{-kAt} \tag{1.5.6}$$

where

$$N = E_\delta \exp\left\{ \frac{4\delta^2 k\nu}{D} \right\} \approx \sum_{l=0}^\infty \exp\left\{ l\left(\frac{4\delta^2 k}{D} - \frac{\delta}{r}\right) \right\} < \infty \quad \text{if } 0 < \delta < \frac{D}{4kr}.$$

Choosing k large enough and then $\delta = \delta(k)$ small enough we derive (1.5.2) from (1.5.6).

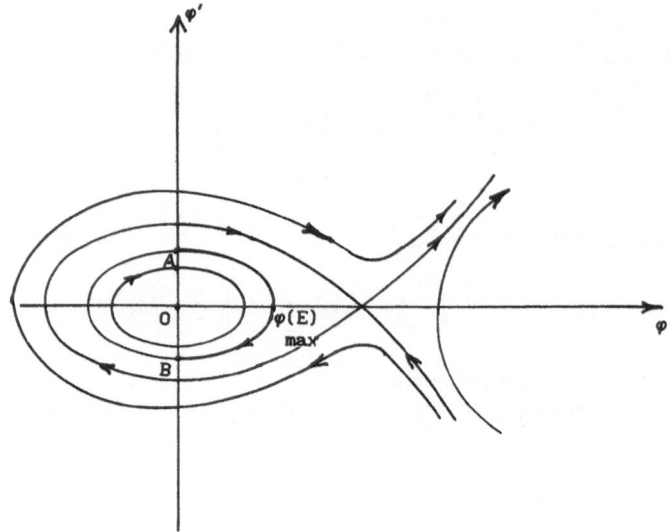

FIGURE 1.7.

Consider the following boundary problem:

$$\frac{D}{2}\frac{d^2\varphi(y)}{dy^2} + f(\varphi(y)) = 0, \qquad \varphi(y) > 0,\ |y| < r,\ \varphi(\pm r) = 0. \qquad (1.5.7)$$

Under our assumptions on f the phase picture of the equation (1.5.7) is as in Figure 7. As is well known, equation (1.5.7) has a first integral:

$$\frac{D}{4}(\varphi'(y))^2 + F(\varphi(y)) = E, \qquad F(\varphi) = \int_0^\varphi f(z)\,dz.$$

The solution of problem (1.5.7) is represented on Figure 7 by the part of periodic solution connecting the points A, $\varphi_{\max}$, and B. The "amplitude" $\varphi = \varphi_{\max}(E)$ is defined by the following condition:

$$\int_0^{\varphi_{\max}(E)} \frac{\sqrt{D}\,d\varphi}{\sqrt{2(E - F(\varphi))}} = r. \qquad (1.5.8)$$

The integral in (1.5.8) approaches $\pi\sqrt{D}/2\sqrt{2c}$ as $E \to 0$, $c = f'(0)$. (It is quarter of the period of the linear oscillations with potential $c\varphi^2/2D$.) One can see from (1.5.8) that the integral tends to infinity when $\varphi_{\max}$ approaches 1. It means that equation (1.5.7) has a positive solution if

$$\pi^2 D < 8cr^2.$$

Thus (1.5.7) is solvable if the width of the strip $r > \pi\sqrt{D}/2\sqrt{2c} = r^*$. Denote by $\varphi^*(y) = \varphi_r^*(y)$ a solution of (1.5.7). It is clear that $|\varphi_r^*(y)| < 1$, and $|(d\varphi_r^*/dy)(y)|$ is bounded uniformly in r and y, and there exists a constant A such that

$$A^{-1}\cos\frac{\pi y}{2r} \le \varphi_r^*(y) \le A\cos\frac{\pi y}{2r}. \qquad (1.5.9)$$

In general, problem (1.5.7) can have more than one solution. We shall assume for simplicity that there exists exactly one solution of problem (1.5.7) If it is not the case one should consider the maximal solution.

Theorem 1.6. *Let $u(t, x, y)$ be the solution of problem (1.5.1). Then*

$$\lim_{t \to \infty} u(t, x, y) = 0$$

uniformly in $x \in R^1$, $y \in [-r, r]$ if $r < \pi\sqrt{D}/2\sqrt{2c}$.

If $r > \pi\sqrt{D}/2\sqrt{2c}$ and $\varphi_r^*(y) = \varphi^*(y)$ is the unique solution of problem (1.5.7), then for any $h > 0$

$$\lim_{t \to \infty} \sup_{x > (\alpha^* + h)t} u(t, x, y) = 0, \tag{1.5.10}$$

$$\lim_{t \to \infty} \sup_{x < (\alpha^* - h)t} u(t, x, y) = \varphi_r^*(y) \tag{1.5.11}$$

uniformly on $y \in [-r, r]$, where $\alpha^* = \sqrt{2D(c - \lambda)}$, $c = f'(0)$, $\lambda = \pi^2 D/8r^2$.

Proof. Let (x_t, y_t) be the Markov process in R^2 governed by the operator $(D/2)\Delta_{x,y}$; $\tau = \min\{t : |y_t| = r\}$ is the first exit time from the strip $R^1 \times [-r, r]$; $\chi_{\tau > t}$ is the indicator function of the event $\{\tau > t\}$.

Using the Feynman–Kac formula we have

$$0 \le u(t, x, y) = E_{x,y}\chi^-(x_t)\chi_{\tau > t}g(y_t) \exp\left\{\int_0^t c(u(t - s, x_s, y_s))\,ds\right\}$$

$$\le E_{x,y}\chi^-(x_t)\chi_{\tau > t} \cdot \max_{-r \le y \le r} g(y)e^{ct}$$

$$\le \max_{|y| \le r} g(y)e^{ct}P_x\{x_t < 0\}P_y\{\tau > t\}$$

$$\approx \varphi(y)\exp\left\{-\frac{x_+^2}{2tD} + ct - \lambda t\right\}. \tag{1.5.12}$$

We used the notation $x_+ = x$ for $x > 0$ and $x_+ = 0$ for $x < 0$. Lemma 1 was used in (1.5.12).

If $r < \pi\sqrt{D}/2\sqrt{2c}$ then $\lambda > c$ and the right side of (1.5.12) exponentially fast tends to zero as $t \to \infty$ uniformly in $x \in R^1$ and $y \in [-r, r]$.

If $r > \pi\sqrt{D}/2\sqrt{2c}$ and $x > (\alpha^* + h)t$ where $\alpha^* = 2\sqrt{2D(c - \lambda)}$ one can derive equality (1.5.10) from (1.5.12).

Let us prove equality (1.5.11). First of all let us check that for any $\delta > 0$ there exists $t_0 > 0$ such that

$$u(t, \alpha^* t, y) \ge \varphi_r^*(y)e^{-\delta t} \tag{1.5.13}$$

for $t > t_0$. Choose small $\kappa > 0$. Let g_s, $0 \le s \le t$, be the function such that $g_s = \alpha^* t$ for $0 \le s < \kappa t$; $g_s = \alpha^* s + \alpha^* \kappa t$ for $\kappa t \le s \le t(1 - \kappa)$; g_s is linear for

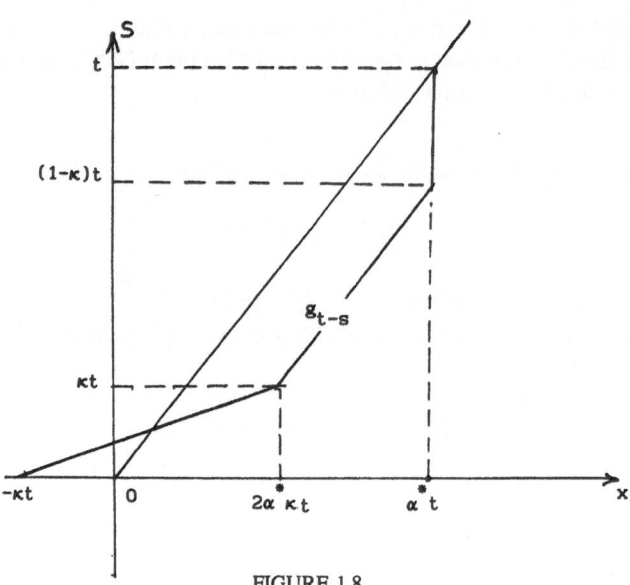

FIGURE 1.8.

$t(1 - \kappa) \le s \le t$, $g_t = -\kappa t$, $g_{t(1-\kappa)} = 2\alpha^* \kappa t$ (see Figure 8). Let $\mathcal{E}_\kappa$ be the set of trajectories x_s, $0 \le s \le t$, such that

$$\sup_{0 \le s \le t} |x_s - g_s| < \frac{\kappa}{4} t.$$

The functions in $\mathcal{E}_\kappa$ for $s \in [\kappa t, (1 - \kappa) t]$ have values from the domain $\{(s, x) : x > \alpha^*(t - s)\} \subset R^2$. In this domain $u(s, x, y)$ is close to zero for large s. The last statement follows from (1.5.10). Thus there exists a constant A_1 independent of κ and t such that

$$\int_0^t c(u(t - s, x_s \cdot y_s) \, ds > ct(1 - A_1 \kappa),$$

if $x. \in \mathcal{E}_\kappa$ and κ is small enough.

It is easy to check that

$$\lim_{t \to \infty} t \ln P_{\alpha^* t} \{x. \in \mathcal{E}_\kappa\} > -\frac{\alpha^{*2}}{2} (1 - A_2 t \kappa).$$

The last two inequalities, (1.5.9) and (1.5.12), give us bound (1.5.13):

$$
\begin{aligned}
u(t, \alpha^* t, y) &\ge E_{\alpha^* t, y} \chi_{\mathcal{E}_\kappa} \chi^-(x_t) \chi_{\tau > t} g(y_t) \exp \left\{ \int_0^t c(u(t - s, x_s, y_s)) \, ds \right\} \\
&\ge \exp\{ct - A_3 t \kappa\} E_{\alpha^* t} \chi_{\mathcal{E}_\kappa} \chi^-(x_t) \cdot E_y \chi_{\tau > t} g(y_t) \\
&\ge \exp\{t(c - A_3 \kappa)\} \exp \left\{ -\frac{\alpha^{*2}}{2} (1 - A_2 t \kappa) \right\} \cdot A_4 \varphi(y) \exp\{-\lambda t\} \\
&\ge \exp \left\{ t \left(c - \frac{\alpha^{*2}}{2} - \lambda - A_5 \kappa \right) \right\} \cdot \varphi_r^*(y)
\end{aligned}
$$

for t large enough. We used here that $v(t, y) = E_y g(y_t)\chi_{\tau > t} \geq A_4 \varphi(y)\exp\{-\lambda t\}$, $t \gg 1$. This bound can be proved by using the Fourier method to solve the initial-boundary problem for $v(t, y)$.

Now let us prove that for any $h > 0$

$$\lim_{t \to \infty} u(t, x, y) \geq \varphi_r^*(y) \tag{1.5.14}$$

uniformly on $y \in [-r, r]$ and $x < (\alpha^* - h)t$.

Let us assume that the converse is true: For any n there exists $t > n$, $z < \alpha^* - h < \alpha^*$, and $\hat{y} \in [-r, r]$ such that

$$u(t, zt, \hat{y}) < \varphi^*(\hat{y})(1 - d), \qquad d > 0.$$

Denote by $G_1 = \{(t, x, y) : u(t, x, y) < \varphi^*(y)(1 - d/2)\}$, $G_2 = \{(t, x, y) : x < \alpha^* t, \ t > 0, \ y \in [-r, r]\}$, $G_3 = \{(t, x, y) : t > 0, \ |y| \leq r, \ x \in R^1\}$. Let τ_k be the first exit time from G_k, $k = 1, 2, 3$, for the heat process $z_s = (t - s, x_s, y_s)$ and $\tau = \min(\tau_1, \tau_2, \tau_3)$. Using strong Markov property we have from (1.5.12):

$$u(t, zt, \hat{y}) = E_{zt, \hat{y}} u(z_\tau) \exp\left\{\int_0^\tau c(u(z_s))\, ds\right\}$$

$$= \sum_{k=1}^3 E_{zt, \hat{y}}\chi_{\tau = \tau_k} u(z_{\tau_k}) \exp\left\{\int_0^{\tau_k} c(u(z_s))\, ds\right\} \tag{1.5.15}$$

Since $u < (1 - d/2)\varphi^*$ in G_1 and $u = (1 - d/2)\varphi^*$ on ∂G_1, taking into account that $c(u)$ is a decreasing function, we conclude that

$$E_{zt, \hat{y}}\chi_{\tau = \tau_1} u(z_\tau) \exp\left\{\int_0^\tau c(u(z_s))\, ds\right\}$$

$$\geq E_{zt, \hat{y}}\chi_{\tau = \tau_1}\left(1 - \frac{d}{2}\right)\varphi^*(y_{\tau_1}) \exp\left\{\int_0^{\tau_1} c(\varphi^*(y_s))\, ds\right\}. \tag{1.5.16}$$

Denote by $\theta_\mu(t)$ the time that the trajectory y_s, $0 \leq s \leq (\tau_3 \wedge t)$, spends in the μ-neighborhood of the points $y = \pm r$. Using Lemma 2 one can choose $\mu > 0$ such that

$$P_{\hat{y}}\left\{\theta_\mu(t) > \frac{t}{2}\right\} \leq \exp\{-2ct\} \tag{1.5.17}$$

for t large enough.

Let $\alpha_1 > 0$ be so small that

$$P_{zt, \hat{y}}\{\tau_2 < \alpha_1 t\} \leq \exp\{-2ct\}. \tag{1.5.18}$$

Denote by $\hat{\chi}_{\tau = \tau_2, \mu, \alpha_1} = \hat{\chi}_{\tau = \tau_2}$ the indicator function of the set $\{\tau = \tau_2, \ \theta_\mu(t) < t/2, \ \tau_2 > \alpha_1 t\}$. It follows from (1.5.17) and (1.5.18) that

$$E_{zt, \hat{y}}\chi_{\tau = \tau_2} u(z_\tau) \exp\left\{\int_0^\tau c(u(z_s))\, ds\right\}$$

$$= (1 + o_t(1)) E_{zt, \hat{y}}\hat{\chi}_{\tau = \tau_2} u(z_\tau) \exp\left\{\int_0^\tau c(u(z_s))\, ds\right\}, \qquad t \to \infty. \tag{1.5.19}$$

Denote $-\bar{c} = \max_{0 \le u \le 1} c'(u) < 0$. We derive from (1.5.9), (1.5.13), (1.5.16), (1.5.19) that

$$
E_{zt,\hat{y}} u(z_\tau) \hat{\chi}_{\tau=\tau_2} \exp \left\{ \int_0^\tau c(u(z_s)) \, ds \right\}
$$

$$
\ge E_{zt,\hat{y}} \hat{\chi}_{\tau=\tau_2} A \varphi^*(y_\tau) e^{-\delta t} \exp \left\{ \int_0^\tau c(\varphi^*(y_s)) \, ds \right\} (1 - o_t(1))
$$

$$
\times \exp \left\{ \int_0^\tau [c(u(z_s)) - c(\varphi^*(y_s))] \, ds \right\}
$$

$$
\ge (1 - o_t(1)) E_{zt,\hat{y}} \hat{\chi}_{\tau=\tau_2} \varphi^*(y_\tau) \exp \left\{ \int_0^\tau c(\varphi^*(y_s)) \, ds \right\}
$$

$$
A \exp \left\{ -\delta t + \frac{d}{2} \int_0^\tau \bar{c} \varphi^*(y_s) \, ds \right\}. \tag{1.5.20}
$$

Since $\tau_2 > \tau \alpha_1$ for trajectories y_s for which $\hat{\chi}_{\tau=\tau_2} = 1$ and since

$$
\min_{-r+\mu \le y \le r-\mu} \varphi^*(y) = \gamma > 0,
$$

we conclude that the right side of (1.5.20) is bigger than

$$
E_{zt,\hat{y}} \hat{\chi}_{\tau=\tau_2} \varphi^*(y_\tau) \exp \left\{ \int_0^\tau c(\varphi^*(y_s)) \, ds \right\}
$$

$$
\times \exp\{-\delta t + \gamma \alpha_1 t \bar{c}\} \ge E_{zt,\hat{y}} \hat{\chi}_{\tau=\tau_2} \varphi^*(y_\tau) \exp \left\{ \int_0^\tau c(\varphi^*(y_s)) \, ds \right\} \tag{1.5.21}
$$

if t is large enough and $\delta < \alpha_1 \gamma \bar{c}$.

From (1.5.15), (1.5.16), (1.5.20), and (1.5.21) one can derive

$$
u(t, zt, \hat{y}) \ge \left(1 - \frac{d}{2} \right) E_{zt,\hat{y}} \chi_{\tau < \tau_3} \varphi^*(y_\tau) \exp \left\{ \int_0^\tau c(\varphi^*(y_s)) \, ds \right\}
$$

$$
= \left(1 - \frac{d}{2} \right) \varphi^*(\hat{y}). \tag{1.5.22}
$$

We used in (1.5.22) the equality

$$
\varphi^*(y) = E_y \chi_{\beta \le \tau_3} \varphi(y_\beta) \exp \left\{ \int_0^\beta c(\varphi^*(y_s)) \, ds \right\},
$$

which holds for any Markov time β. Inequality (1.5.22) contradicts our assumption that $u(t, zt, \hat{y}) < (1 - d)\varphi^*(\hat{y})$. Thus (1.5.14) is true.

Let us show now that

$$
\overline{\lim_{t \to \infty}} \, u(t, x, y) \le \varphi^*(y) \tag{1.5.23}
$$

uniformly in $x \in R^1$, $y \in [-r, r]$. Using the Feynman–Kac formula we have for any Markov time $\beta \le t$:

$$
u(t, x, y) = E_{x,y} u(t - \beta, x_\beta, y_\beta) \exp \left\{ \int_0^\beta c(u(t - s, x_s y_s)) \, ds \right\} \chi_{\tau_3 > \beta}. \tag{1.5.24}
$$

Let $u(t, x_0, y_0) \geq (1 + d)\varphi^*(y_0)$ for some x_0 and y_0, $|y_0| \neq z$, and for some arbitrarily large t. Use the notation

$$G_t = \left\{ (x, y) : x \in R^1, \ |y| \leq r, \ u(t, x, y) > \left(1 + \frac{d}{2} \right) \varphi^*(y) \right\},$$

$$\theta_t = \min\{s : (x_s, y_s) \neq G_{t-s}\} \wedge t.$$

Using (1.5.24) for $\beta = \theta_t$ one can write

$$\varphi^*(y_0)(1 + d) \leq u(t, x_0, y_0)$$

$$= E_{x_0, y_0} \chi_{\tau_3 > \theta_t} u(t - \theta_t, x_{\theta_t}, y_{\theta_t}) \exp \left\{ \int_0^{\theta_t} c(u(t - s, x_s y_s)) \, ds \right\}$$

$$\leq \left(1 + \frac{d}{2} \right) E \varphi^*(y_{\theta_t}) \chi_{\theta_t < t} \chi_{\tau_3 > \theta_t} \exp \left\{ \int_0^{\theta_t} c(\varphi^*(y_s)) \, ds \right\}$$

$$+ \left(1 + \frac{d}{2} \right) E_y \varphi^*(y_{\theta_t}) \chi_{\tau_3 > \theta_t = t} \exp \left\{ \int_0^{\theta_t} c(\varphi^*(y_s)) \, ds \right\}$$

$$\cdot \left[\sup_{|y| < r} \frac{g(y)}{\varphi^*(y)} \exp \left\{ \int_0^t \frac{d\tilde{c}}{2} \varphi^*(y_s) \, ds \right\} \right]. \tag{1.5.25}$$

Here $\tilde{c} = -\min_{0 \leq u \leq 1} |c'(u)| < 0$. Taking into account Lemma 2 we can conclude that the expression in square brackets is less than 1 for t large enough. Thus it follows from (1.5.25) that

$$(1 + d)\varphi^*(y_0) \leq \left(1 + \frac{d}{2} \right) E_{y_0} \varphi^*(y_{\theta_t}) \chi_{\tau_3 > \theta_t} \exp \left\{ \int_0^{\theta_t} c(\varphi^*(y_s)) \, ds \right\}$$

$$= \left(1 + \frac{d}{2} \right) \varphi^*(y_0).$$

This contradiction proves (1.5.23) and Theorem 6.

1.6. WAVE FRONT PROPAGATION IN SLOWLY NONHOMOGENEOUS MEDIA

Consider the following differential equation:

$$\frac{\partial u(t, x, y)}{\partial t} = \frac{1}{2} \sum_{i,j=1}^r a^{ij}(\varepsilon x, y) \frac{\partial^2 u}{\partial x^i \partial x^j} + \delta D(\varepsilon x, y) \frac{\partial u}{\partial y^2} + f(\varepsilon x, y, u),$$

$$t > 0, \quad x \in R^r, \quad y \in [-1, 1], \quad \frac{\partial u}{\partial y}(t, x, y) \Big|_{y = \pm 1} = 0. \tag{1.6.1}$$

Here $(a^{ij}(x, y))$ is a positively defined symmetric matrix, $D(x, y) > 0$, $f(x, y, \cdot) \in F_1$ for each $x \in R^r$, $y \in [-1, 1]$. All coefficients and the nonlinear term are assumed to be continuously differentiable and bounded together with their first derivatives. The parameters ε and δ supposed to be positive, $\varepsilon \ll 1$. Smallness of ε means that the coefficients change slowly in x.

To study wave front propagation for equation (1.6.1) let us rescale space and time: denote by $u^{\varepsilon,\delta}(t,x,y) = u(t/\varepsilon, x/\varepsilon, y)$. If $u(t,x,y)$ satisfies (1.6.1), the new function $u^{\varepsilon,\delta}(t,x,y)$ satisfies the equation

$$\frac{\partial u^{\varepsilon,\delta}(t,x,y)}{\partial t} = \frac{\varepsilon}{2} \sum_{i,j=1}^{r} a^{ij}(x,y)\frac{\partial^2 u^{\varepsilon,\delta}}{\partial x^i \partial x^j} + \frac{\delta}{2\varepsilon}D(x,y)\frac{\partial^2 u^\varepsilon}{\partial y^2} + \frac{1}{\varepsilon}f(x,y,u^{\varepsilon,\delta}),$$

$$t > 0, \quad x \in R^r, \quad |y| < 1, \quad \left.\frac{\partial u^{\varepsilon,\delta}(t,x,y)}{\partial y}\right|_{y=\pm 1} = 0. \qquad (1.6.2)$$

We supplement (1.6.2) with the initial condition

$$u^{\varepsilon,\delta}(0,x,y) = g(x,y) \geq 0, \qquad \sup_{x \in R^1, |y| \leq 1} g(x,y) < \infty. \qquad (1.6.3)$$

Denote $G_0 = \text{support } g$ and assume that the closure $[G_0]$ of G_0 coincides with the closure of the interior (G_0) of G_0. Let $\bar{G}_0 \subset R^r$ be the projection of G_0 on the space R^r.

It turns out that the function $u^{\varepsilon,\delta}(t,x,y)$ for each $t > 0$ tends to 1 in a domain $G_t \subset R^r \times [-1,1]$ and tends to zero in the interior points of the complement of the domain G_t as $\varepsilon \downarrow 0$. The domain G_t expands as t increases, and the law of expansion depends on the relation between ε and δ.

One can prove that if $\varepsilon \downarrow 0$ and $\delta \to \infty$ then $G_t = \bar{G}_t \times [-1,1]$, $\bar{G}_t \subset R^r$. The sets $\bar{G}_t$ are defined as follows: $\bar{G}_t = \{x \in R^r : \lim_{\varepsilon \downarrow 0} v^\varepsilon(t,x) = 1\}$, where $v(t,x)$ is the solution of the problem

$$\frac{\partial v^\varepsilon(t,x)}{\partial t} = \frac{\varepsilon}{2} \sum_{i,j=1}^{r} \bar{a}^{ij}(x)\frac{\partial^2 v^\varepsilon}{\partial x^i \partial x^j} + \frac{1}{\varepsilon}\bar{f}(x,v^\varepsilon),$$

$$v^\varepsilon(0,x) = \left\{ \begin{array}{ll} 1, & x \in \bar{G}_0, \\ 0, & x \notin \bar{G}_0; \end{array} \right.$$

$$\bar{a}^{ij}(x) = \int_{-1}^{1} \frac{a^{ij}(x,y)}{D(x,y)}dy \left(\int_{-1}^{1} \frac{dy}{D(x,y)}\right)^{-1},$$

$$\bar{f}(x,v) = \int_{-1}^{1} \frac{f(x,y,v)\,dy}{D(x,y)} \left(\int_{-1}^{1} \frac{dy}{D(x,y)}\right)^{-1}. \qquad (1.6.4)$$

This statement can be easily proved if one takes into account that

$$D^{-1}(x,y)\left(\int_{-1}^{1} dy/D(x,y)\right)^{-1}$$

is the invariant density of the diffusion process in $[-1,1]$ governed by the operator $(\delta D(x,y)/2\varepsilon)d^2/dy^2$, x is fixed, with reflection in the ends of the interval, and that the probabilities of deviations of order 1 (as $\varepsilon \downarrow 0$, $\delta \to \infty$) of the normalized occupation time from the invariant measure are of order $\exp\{(\delta/\varepsilon) \cdot \text{const}\}$ as $\delta/\varepsilon \to \infty$.

The problem (1.6.4) was studied in detail in [6], [7], [8], and [3]. To formulate the general result use the notation

$$V_1(t,x) = \sup_{\varphi \in C_{0,t}, \varphi_0 = x, \varphi_t \in \bar{G}_0} \min_{0 \le a \le t} \int_0^a \left[c(\varphi_s) - \frac{1}{2} \sum_{i,j=1}^r \bar{a}_{ij}(\varphi_s) \dot{\varphi}_s^i \dot{\varphi}_s^j \right] ds,$$

$$c(x) = \left. \frac{\partial \bar{f}(x,u)}{\partial u} \right|_{u=0},$$

$$(\bar{a}_{ij}(x)) = (\bar{a}^{ij}(x))^{-1}.$$

It is easy to check that $V_1(t,x)$ is Lipschitz continuous and $V_1(t,x) \le 0$. The limiting behavior of $v^\varepsilon(t,x)$ and thus of $u^{\varepsilon,\delta}(t,x,y)$ can be described as follows:

$$\lim_{\varepsilon \downarrow 0, \delta \to \infty} u^{\varepsilon,\delta}(t,x,y) = 0, \qquad \text{if } V_1(t,x) < 0, \ |y| \le 1;$$

$$\lim_{\varepsilon \downarrow 0, \delta \to \infty} u^{\varepsilon,\delta}(t,x,y) = 1, \qquad \text{if } (t,x) \text{ is an interior point of}$$

$$\{(s,z) : V_1(s,z) = 0\}.$$

In particular one can derive from the last statement, that if $c(x) = \partial \bar{f}(x,u)/\partial u|_{u=0} = c = \text{const}$ then the propagation of the wave front is governed by a Huygens principle. The Huygens principle has the simplest form in the Riemannian metric $ds^2 = \sum_{i,j=1}^r \bar{a}_{ij}(x)\, dx^i\, dx^j$: in this metric the corresponding velocity field is homogeneous and isotropic and the magnitude of the velocity is equal to $\sqrt{2c}$. If $c(x)$ depends on x the front can have jumps and the propagation actually cannot be described by a universal Huygens principle: corresponding velocity field depends on the initial conditions (see [6], [7]).

Now let $\delta \to a \in (0,\infty)$. In this case the probabilities of deviations of order 1 for the process with small diffusion and the deviations for the occupation time are of the same order. Both of them will influence the speed of the front. To describe the motion of the wave front, we assume, for brevity, that the diffusion coefficients $a^{ij}(x)$ are independent of y.

Denote by $\lambda = \lambda(x)$ the eigenvalue of the problem

$$\frac{aD(x,y)}{2} \varphi_x''(y) + c(x,y)\varphi_x(y) = \lambda(x)\varphi_x(y), \qquad |y| < 1,$$

$$\frac{\partial \varphi_x}{\partial y}(\pm 1) = 0, \qquad c(x,y) = \left. \frac{\partial f(x,y,u)}{\partial u} \right|_{u=0},$$

$$x \in R^r \text{ is a parameter,} \tag{1.6.5}$$

corresponding to the positive eigenfunction $\varphi_x(y)$, $|y| \le 1$.

Let $(x_t^\varepsilon, y_t^\varepsilon)$ be the diffusion process in the layer $R^r \times [-1,1]$ governed by the operator

$$\frac{\varepsilon}{2} \sum_{i,j=1}^r a^{ij}(x) \frac{\partial^2}{\partial x^i \partial x^j} + \frac{aD(x,y)}{2\varepsilon} \frac{\partial^2}{\partial y^2}$$

with reflection on the boundary. Then we can write

$$u^\varepsilon(t,x,y) = u^{\varepsilon,a}(t,x,y)$$

$$= E_{x,y} g(x_t^\varepsilon, y_t^\varepsilon) \exp\left\{ \int_0^t c(x_s^\varepsilon, y_s^\varepsilon, u^\varepsilon(t-s, x_s^\varepsilon, y_s^\varepsilon))\, ds \right\}$$

$$\leq E_{x,y} g(x_t^\varepsilon, y_t^\varepsilon) \exp\left\{ \frac{1}{\varepsilon} \int_0^t c(x_s^\varepsilon, y_s^\varepsilon)\, ds \right\}. \tag{1.6.6}$$

Using the results of ([11], Chaps. 3 and 7) one can check that the right side of (1.6.6) is logarithmically equivalent to

$$\exp\left\{ \frac{1}{\varepsilon} \sup \left[\int_0^t \left(\lambda(\varphi_s) - \frac{1}{2} \sum_{i,j=1}^r a_{ij}(\varphi_s)\dot{\varphi}_s^i \dot{\varphi}_s^j \right) ds : \right. \right.$$

$$\left. \left. \varphi \in C_{0t};\ \varphi_0 = x,\ \varphi_t \in \bar{G}_0 \right] \right\}.$$

Let, for a moment, $\lambda(x) = \lambda$ be independent of x. Then one can check that the supremum in the last expression is equal to

$$\lambda t - \frac{\rho^2(x, G_0)}{2t},$$

where $\rho(\cdot, \cdot)$ is the Riemannian metric corresponding to the form

$$ds^2 = \sum_{i,j=1}^r a_{ij}(x)\, dx^i\, dx^j.$$

We derive from (1.6.6) that

$$\lim_{\varepsilon \downarrow 0} u^\varepsilon(t,x,y) = 0, \qquad \text{if } \rho(x, G_0) > t\sqrt{2\lambda},\ |y| \leq 1.$$

In the same way as it was proved in [7], Chap. 6, one can prove that

$$\lim_{\varepsilon \downarrow 0} u^\varepsilon(t,x,y) = 1, \qquad \text{if } \rho(x, G_0) < t\sqrt{2\lambda},\ |y| \leq 1.$$

Thus in this case the front propagates according to the Huygens principle and the corresponding velocity field is homogeneous and isotropic in the metric ρ and the magnitude of the velocity is equal to $\sqrt{2\lambda}$.

If λ depends on x we should introduce the function

$$V_2(t,x) = \sup_{\varphi_0 = x, \varphi_t \in G_0} \inf_{0 \leq a \leq t} \int_0^a \left[\lambda(\varphi_s) - \frac{1}{2} \sum_{i,j=1}^r a_{ij}(\varphi_s)\dot{\varphi}_s^i \dot{\varphi}_s^j \right] ds.$$

The function $V_2(t,x)$ is not positive. One can prove that

$$\lim_{\varepsilon \downarrow 0, \delta \to a \in (0,\infty)} u^{\varepsilon,\delta}(t,x,y) = 0, \qquad \text{if } V_2(t,x) < 0;$$

$$\lim_{\varepsilon \downarrow 0, \delta \to a \in (0,\infty)} u^{\varepsilon,\delta}(t,x,y) = 1, \qquad \text{if } (t,x) \text{ belongs to the interior of}$$

$$\{(s,z) : V_2(s,z) = 0\},\ |y| \leq 1.$$

A similar description of the motion of the front can be given in the general case when $a_{ij}(x, y)$ depend on x and y.

If $\delta = o(\varepsilon^2)$ as $\varepsilon \downarrow 0$ the mixing in the y-direction is very slow and the wave fronts for different y's are, in a sense, independent. Use the notation

$$V_3(t, x, y) = \sup_{\substack{\varphi : \varphi_0 = x \\ g(\varphi_t, y)}} \left\{ \min_{0 \leq a \leq t} \int_0^a \left[c(\varphi_s, y) - \frac{1}{2} \sum_{i,j=1}^r a_{ij}(\varphi_s, y) \dot{\varphi}_s^i \dot{\varphi}_s^j \right] ds \right\},$$

$y \in [-1, 1]$ is a parameter.

Then

$$\lim_{\varepsilon \downarrow 0, \delta = 0(\varepsilon^2)} u^{\varepsilon, \delta}(t, x, y) = 0$$

if $V(t, x, y) < 0$, and

$$\lim_{\varepsilon \downarrow 0, \delta = 0(\varepsilon^2)} u^{\varepsilon, \delta}(t, x, y) = 1$$

if (t, x) belongs to the interior of the set $\{(s, z) : V_3(s, z, y) = 0\}$.

If $\delta \downarrow 0$ and $\delta/\varepsilon^2 \to b \in (0, \infty)$ the position of the front will be the same as for the problem

$$\frac{\partial \tilde{u}(t, x, y)}{\partial t} = \frac{\varepsilon}{2} \left(\sum_{i,j=1}^r a^{ij}(x, y) \frac{\partial^2 \tilde{u}^\varepsilon}{\partial x^i \partial x^j} + b D(x, y) \frac{\partial^2 \tilde{u}^\varepsilon}{\partial y^2} \right) + \frac{1}{\varepsilon} f(x, y, \tilde{u}^\varepsilon)$$

$$\tilde{u}^\varepsilon(0, x, y) = g(x, y), \qquad \frac{\partial \tilde{u}}{\partial y}(t, x, y) \bigg|_{y = \pm 1} = 0. \tag{1.6.7}$$

The motion of the front for problem (1.6.7) is described by the function

$$V_4(t, x, y) = \sup_{\substack{\varphi_0 = x, \psi_0 = y \\ (\varphi_t, \psi_t) \in G_0}} \left[\min_{0 \leq a \leq ??} \left[\int_0^{ta} (c(\varphi_s, \psi_s) \right. \right.$$

$$\left. \left. - \frac{1}{2} \sum a_{ij}(\varphi_s, \psi_s) \dot{\varphi}_s^i \dot{\varphi}_s^j \right) - \frac{\dot{\psi}_s^2}{bD(\varphi_s, \varphi_s)} \right] ds \right] ;$$

here $\varphi. : [0, t] \to R^r$, $\psi. : [0, t] \to R^1$ (see [7], Chap. 6, [8], [9]).

Now let $\varepsilon, \delta \downarrow 0$ and $\delta/\varepsilon^2 \to \infty$. That means that $\varepsilon \ll \delta/c \ll 1/\varepsilon$, the probabilities of deviations of order 1 as $\varepsilon, \delta \downarrow 0$, of the normalized occupation time for the process $y_t^{\varepsilon, \delta}$ from its typical behavior for such values of the parameters tend to zero more slowly than the probabilities of the deviations of order 1 for $x_t^{\varepsilon, \delta}$. Using this fact propagation will be the same for all kinds of behavior of the parameters ε and δ such that $\varepsilon, \delta \downarrow 0$, $\varepsilon \ll \delta/\varepsilon \ll \varepsilon^{-1}$. We shall consider in more detail the case $\delta = \varepsilon$. Then the problem (1.6.2)-(1.6.3) can be rewritten as follows:

$$\frac{\partial u^\varepsilon(t, x, y)}{\partial t} = \frac{\varepsilon}{2} \sum_{i,j=1}^r a^{ij}(x, y) \frac{\partial^2 u^\varepsilon}{\partial x^i \partial x^j} + \frac{D(x, y)}{2} \frac{\partial^2 u^\varepsilon}{\partial y^2} + \frac{1}{\varepsilon} f(x, y, u^\varepsilon),$$

$$\frac{\partial u^\varepsilon(t, x, y)}{\partial y} \bigg|_{y = \pm 1} = 0, \qquad u^\varepsilon(0, x, y) = g(x, y). \tag{1.6.8}$$

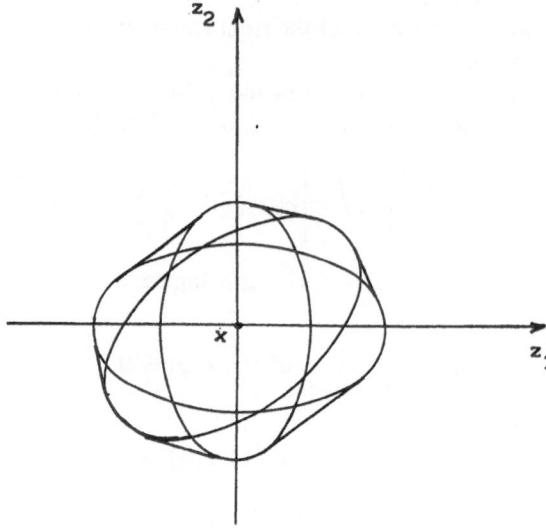

FIGURE 1.9.

Denote $\rho_y(\cdot,\cdot)$ the Riemannian metric in R^r corresponding to the form $ds^2 = \sum_{i,j=1}^r a_{ij}(x,y)\,dx^i\,dx^j$, $(a_{ij}(x,y)) = (a^{ij}(x,y))^{-1}$, $y \in [-1,1]$ is a parameter. Let $S_y(x) = \left\{z \in R^r : \sum_{i,j=1}^r a_i(x,y)z^i z^j = 1\right\}$ be the unit sphere in the "tangent space" at point x. Denote by $\hat{S}(x)$ the convex hull of all $S_y(x)$, $|y| \le 1$; $\hat{S}(x)$ is a convex symmetric set (Figure 9). We define a Finsler metric $\hat{\rho}(\cdot,\cdot)$ such that $\hat{S}(x)$ will be the unit sphere in the "tangent space" at point x in this metric: Let $h(x,z)$ be the function such that $h(x,z) = 1$ for $x \in R^r$, $z \in \hat{S}(x)$ and $h(x,\alpha z) = |\alpha|h(x,z)$ for any real α.

Then

$$\hat{\rho}(a,b) = \inf_{\varphi_0=a,\varphi_T=b} \int_0^T h(\varphi_s,\dot{\varphi}_s)\,ds; \qquad a,b \in R^r,\ T > 0.$$

The infimum is taken over all curves connecting the points a and b. It follows from the properties of $h(x,z)$ that $\hat{\rho}(a,b)$ is actually independent of T and is a metric. Note that any convex symmetric set with smooth enough boundary can be considered as the convex hull of a proper family of ellipsoids.

Theorem 1.7. *Let $u^\varepsilon(t,x,y)$ be the solution of problem (1.6.8), $f(x,y,\cdot) \in F_1$ for any $x \in R^r$, $|y| \le 1$, and $\partial f(x,y,u)/\partial u|_{u=0} = c = \text{const.}$ Then*

$$\lim_{\varepsilon \downarrow 0} u^\varepsilon(t,x,y) = \begin{cases} 1, & \text{if } \hat{\rho}(x,\bar{G}_0) < t\sqrt{2c} \\ 0, & \text{if } \hat{\rho}(x,\bar{G}_0) > t\sqrt{2c} \end{cases}$$

That means that the wave front (the interface between the regions in $R^r \times [-1,1]$ where $u^\varepsilon(t,x,y)$ is close to 0 and to 1 as $\varepsilon \downarrow 0$) moves according to the Huygens principle and the velocity field is homogeneous and isotropic and its absolute value is equal to $\sqrt{2c}$ if calculated in the metric $\hat{\rho}(\cdot,\cdot)$.

To prove this theorem we consider the Markov process $(x_t^\varepsilon, y_t^\varepsilon)$ in $R^r \times [-1, 1]$ corresponding to the operator

$$(\varepsilon/2) \sum_{i,j=1}^r a^{ij}(x,y)(\partial^2/\partial x^i \partial x^j) + (D(x,y)/2)\partial^2/\partial y^2$$

with normal reflection on the boundary. Since $f \in F_1$ we have from the Feynman–Kac formula that for some constant $A > 0$

$$0 \le u^\varepsilon(t,x,y) = E_{x,y} g(x_t^\varepsilon, y_t^\varepsilon) \exp\left\{\frac{1}{\varepsilon} \int_0^t c(x_s^\varepsilon, y_s^\varepsilon, u^\varepsilon(t-s, x_s^\varepsilon, y_s^\varepsilon))\, ds\right\}$$

$$\le A e^{ct/\varepsilon} P_{x,y}\{(x_t^\varepsilon, y_t^\varepsilon) \in [G_0]\}. \tag{1.6.9}$$

We can calculate the logarithmic asymptotic of the probability in the right side of (1.6.9) as $\varepsilon \downarrow 0$:

$$\lim_{\varepsilon \downarrow 0} \varepsilon \ln P_{x,y}\{(x_t^\varepsilon, y_t^\varepsilon) \in [G_0]\}$$

$$= \lim_{\varepsilon \downarrow 0} \varepsilon \ln P_{x,y}\{x_t^\varepsilon \in [\bar{G}_0]\}$$

$$= \lim_{\varepsilon \downarrow 0} \varepsilon \ln E_{x,y} P_x\{x_t^\varepsilon \in [\bar{G}_0] \mid y_s^\varepsilon, \ 0 \le s \le t\}$$

$$= -\frac{1}{2} \sup_\psi \sup_{\substack{\varphi \in c_{0t}, \\ \varphi_0 = x, \\ \varphi_t \in [\bar{G}_0]}} \left\{ \int_0^t \sum_{i,j=1}^r a_{ij}(\varphi_s, \psi_s)\dot{\varphi}_s^i \dot{\varphi}_s^j \right\}. \tag{1.6.10}$$

The first supremum on the right side of (1.6.10) is taken over all continuous functions ψ_s, $0 \le s \le t$, with values in $[-1, 1]$. Equality (1.6.10) is the implication of the large deviation principle and the arguments used in Section 1.4.

Using the same reasoning as in the proof of Theorem 3 in [8] one can conclude that the right side in (1.6.10) is equal to

$$-\frac{\hat{\rho}^2(x, G_0)}{2t}.$$

This together with (1.6.9) implies that $\lim_{\varepsilon \downarrow 0} u^\varepsilon(t,x) = 0$ if $\hat{\rho}(x, \bar{G}_0) > t\sqrt{2c}$.

The proof of the statement

$$\lim u^\varepsilon(t,x) = 1 \qquad \text{if } \hat{\rho}(x, G_0) < t\sqrt{2c},$$

is also similar to the proof of corresponding statement in Theorem 3 of [8], and we omit it.

To describe the motion of the wave front in the case of a general nonlinear term $f(x, y, \cdot) \in F_1$ in equation (1.6.2) one should introduce the function

$$V_5(t,x) = \sup\left\{\min_{0 \le a \le t} \int_0^a \left[c(\varphi_s, \psi_s) - \frac{1}{2}\sum_{i,j=1}^r a_{ij}(\varphi_s, \psi_s)\dot{\varphi}_s^i \dot{\varphi}_s^j\right] ds : \right.$$

$$\left. \varphi \in C_{0t}, \quad \varphi_0 = x, \quad \varphi_t \in [\bar{G}_0], \quad \psi: [0,t] \to [-1,1]\right\}.$$

It is clear that $V_5(t, x) \leq 0$. One can prove that $\lim u^\varepsilon(t, x, y) = 0$ if $V_5(t, x) < 0$, and that $\lim_{\varepsilon \downarrow 0} u^\varepsilon(t, x, y) = 1$ if (t, x) is an interior point of the set $\{(s, z) : V_5(s, z) = 0\}$. The proof of this statement is similar to the proof of Theorem 8.3 in [9].

In the case of a variable $c(x, y) = \partial f(x, y, u)/\partial u|_{u=o}$ the motion of the wave front cannot be described by the universal Huygens principle. The evolution of the front may have a non-Markovian character: The position of the front at time $t > s$, given the position of the front at time s, may depend on the behavior of the front before time s (compare with example 2 from [6]).

One can consider a slowly nonhomogeneous equation in a tube $R^1 \times G$. After a proper rescaling this equation is as follows:

$$\frac{\partial u^\varepsilon(t, x, y)}{\partial t} = \frac{\varepsilon}{2} a(x, y) \frac{\partial^2 u^\varepsilon}{\partial x^2} + \frac{1}{2} D(x, y) \Delta_y u^\varepsilon + \frac{1}{\varepsilon} c(x, y, u^\varepsilon) u^\varepsilon,$$

$$t > 0, \qquad x \in R^1, \qquad y \in G, \qquad \left.\frac{\partial u^\varepsilon}{\partial n}\right|_{\partial G} = 0,$$

$$u^\varepsilon(0, x, y) = g(x, y) \geq 0. \tag{1.6.11}$$

I will not consider the general case, and will mention only that if either $a(x, y)$ or $c(x, y, 0)$ does not depend on y the wave front motion problem for (1.6.11) can be reduced to the one-dimensional case. For example, if $a(x, y) = a(x)$ then the motion of the wave front will be the same as for the problem

$$\frac{\partial \bar{u}^\varepsilon(t, x)}{\partial t} = \frac{\varepsilon}{2} a(x) \frac{\partial^2 \bar{u}^\varepsilon}{\partial x^2} + \frac{1}{\varepsilon} \max_{y \in G \cup \partial G} c(x, y) \bar{u}^\varepsilon (1 - \bar{u}^\varepsilon),$$

$$\bar{u}^\varepsilon(0, x) = \begin{cases} 1, & \text{if } \max_y g(x, y) > 0, \\ 0, & \text{if } g(x, y) = 0 \text{ for } y \in G \cup \partial G. \end{cases}$$

The latter is considered in [6], [7], [8].

I will mention now some interesting effects which appear as a result of the nonhomogeneous drift term. A one-dimensional equation with variable drift can be used as a model of wave front propagation in a tube of variable diameter. Because of the conservation law for the flow the velocity of the flow is bigger in the narrow parts of the tube and less in thick parts. We have seen in Section 1.2 that the wave front can move against the flow if the speed of the flow is not too large. If the tube anywhere becomes very thin (and flow is directed against the direction of the propagation of the wave front), the speed in these narrow parts can be very large. The front cannot move immediately against such a strong flow. But if after the narrow parts there is a part of the tube where the tube is wide enough (thus the speed of the flow on such parts is small), the wave front can "jump" over the narrow parts and the excitation can come back to the narrow parts along the flow.

To demonstrate such behavior let us consider the following equation:

$$\frac{\partial u^\varepsilon(t, x)}{\partial t} = \frac{\varepsilon}{2} \frac{\partial^2 u^\varepsilon}{\partial x^2} + b(x) \frac{\partial u^\varepsilon}{\partial x} + \frac{1}{\varepsilon} c(u^\varepsilon) \bar{u}^\varepsilon, \qquad u^\varepsilon(0, x) = \chi^-(x). \tag{1.6.12}$$

We assume that $c(u)u \in F_1$ and thus $c = c(0) = \max c(u)$. Let $b(x) = b_1 > 0$ for $x \notin [h_1, h_2)$, $b(x) = b_2 > b_1$ for $h_1 \leq x < h_2$, $h_1 > 0$. (The positive sign of $b(x)$ means that the real flow moves to the left, since we consider (1.5.8) as a backward

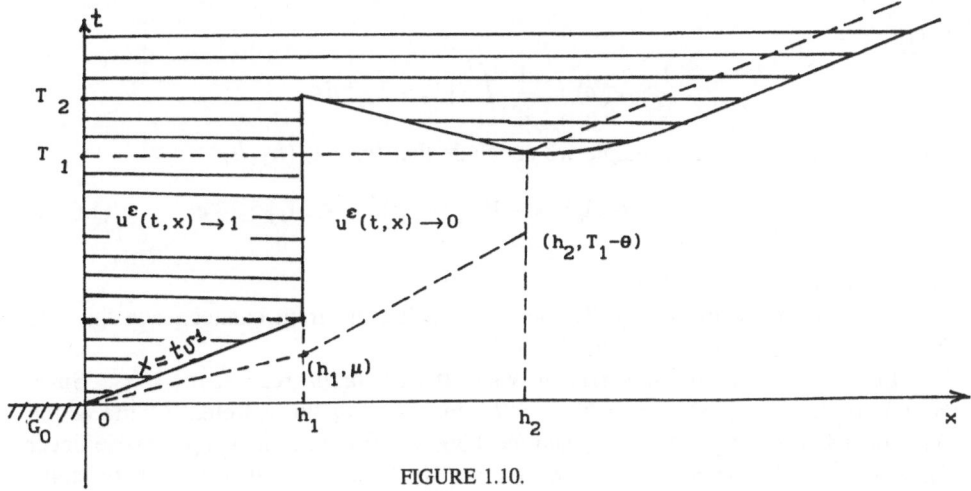

FIGURE 1.10.

equation.) Assume that $b_1 < \sqrt{2c}$, but $b_2 > \sqrt{2c}$. Recall that $\sqrt{2c}$ is the Kolmogorov–Petrovskii–Piskunov speed for the equation (1.6.12) without drift. For such drift $b(x)$ the front as $\varepsilon \downarrow 0$ moves to the right with the speed $v_1 = \sqrt{2c} - b_1$ until it hits the point $x = h_1$ (see Figure 10). Then the front stops at the point $x = h_1$ and will stay up to a time T_2. At time $T_1 < T_2$ "a new source" switches on at the point $x = h_2$. From this source the front will move in both directions. It will move to the left with the speed $v_2 = b_2 + \sqrt{2c}$ and will hit point $x = h_1$ at time $T_2 = T_1 + (h_2 - h_1)/v_2$. Thus the front moves continuously to the right with the constant speed v_1 until it hits point $x = h_1$. The front stays there and at time T_1 jumps to $x = h_2$. Then at time T_2 the excitation will come to $x = h_1$ from the right. In the shaded area $\lim_{\varepsilon \downarrow 0} u^\varepsilon(t, x) = 1$, and $\lim_{\varepsilon \downarrow 0} u^\varepsilon(t, x) = 0$ in the unshaded area.

To explain such behavior and calculate T_1 let us consider the diffusion process X_t^ε connected with (1.6.12):

$$X_t^\varepsilon = x + \int_0^t b(X_s^\varepsilon)\, ds + \sqrt{\varepsilon}\, W_t,$$

W_t is the Wiener process in R^1. We can write

$$u^\varepsilon(t, x) = E_x \chi^-(X_t^\varepsilon) \exp\left\{ \frac{1}{\varepsilon} \int_0^t c(u^\varepsilon(t - s, X_s^\varepsilon))\, ds \right\} \le e^{ct/\varepsilon} P_x\{X_t^\varepsilon \le 0\}. \tag{1.6.13}$$

From (1.6.13), using the standard arguments (see [7], Chap. 6) we can derive that the front moves to the right with speed $v_1 = \sqrt{2c} - b_1$ up to the point $x = h_1$.

Denote by $\mathcal{A}_t$, $t > T_0 = h_1/v_1$, the following subset in C_{0t}:

$$\mathcal{A}_t = \{\varphi_s,\ 0 \le s \le t,\ \varphi_0 = x \in (h_1, h_2),\ \varphi_t \le 0,$$
$$\max_{0 \le s \le t} \varphi_s < h_2,\ \varphi_s > h_1 \quad \text{for } 0 \le s < t - T_0\}.$$

The action functional $(1/\varepsilon)S_{0T}(\varphi)$ for the family X_t^ε, $0 \le t \le T$, in C_{0T} as $\varepsilon \downarrow 0$ for absolutely continuous φ_s, $0 \le s \le T$, such that $h_1 < \varphi_s < h_2$, has the form (see

[11]):

$$\frac{1}{\varepsilon} S_{0T}(\varphi) = \frac{1}{2\varepsilon} \int_0^T |\dot{\varphi}_s - b_2|^2 \, ds.$$

From the large deviation principle we conclude that for $x \in (h_1, h_2)$

$$\lim_{\varepsilon \downarrow 0} \varepsilon \ln P_x \{X_{\cdot}^\varepsilon \in \mathcal{A}_t\} = -\inf\{S_{0t}(\varphi) : \varphi \in \mathcal{A}_t\} \leq -ct. \qquad (1.6.14)$$

We used here that $b_2 > \sqrt{2c}$.

One can derive from (1.6.13) and (1.6.14) that the front cannot reach the area $h_1 < x < h_2$ before it reaches the area $x \geq h_2$.

Let us consider now the moment when the excitation reaches $x = h_2$. Since $b(x)$ is piecewise constant and $b(x) = b_1$ for $x \geq h_2$ the extremal of the action functional starting at point $x = h_2$ and reaching $x = 0$ at time T is a piecewise linear function $\varphi = \varphi^{\theta,\mu}$ such that $\varphi_s = h_2$ for $s \in [0, \theta]$, $\varphi_{T-\mu} = h_1$, $\varphi_T = 0$ for some $0 \leq \theta \leq T - \mu \leq T$; on the intervals $[0, \theta]$, $[\theta, T - \mu]$ and $[T - \mu, T]$ the function $\varphi^{\theta,\mu}$ is linear. The constants θ and μ should be found from the minimality condition

$$\min_{\substack{0 \leq \theta \leq T, \mu < h_1/v_1 \\ 0 \leq \mu \leq T-\theta}} S_{0T}(\varphi^{\theta,\mu})$$

$$= \min_{\substack{0 \leq \theta \leq T, \mu < h_1/v_1 \\ 0 \leq \mu \leq T-\theta}} \left[\theta b_1^2 + (t - \theta - \mu) \left(\frac{h_2 - h_1}{T - \theta - \mu} + b_2 \right)^2 + \mu \left(\frac{h_1}{\mu} + b_1 \right)^2 \right]$$

$$= A(T).$$

Let $\hat{\varphi} = \varphi^{\theta,\mu}$ have the optimal θ and μ, and T_1 be the solution of the equation

$$A(T_1) = 2c.$$

Then taking into account that $\hat{\varphi}_s$ for $s \in (0, T_1)$ belongs to the area where $u^\varepsilon(t, x) \to 0$ as $\varepsilon \downarrow 0$, one can derive from the equality (1.6.13) that $\lim_{\varepsilon \downarrow 0} u^\varepsilon(t, x) = 1$ for $t > T_1$, $x = h_2$.

Now, if the excitation has already come to the point $x = h_2$, it can go to the left along the flow up to the point $x = h_1$ with speed $-v_2 = b_2 + \sqrt{2c}$ and will reach the point $x = h_1$ at the time $T_2 = T_1 + (h_2 - h_1)/v_2$.

Note that if $b(x) = \hat{b} < -\sqrt{2c}$ for $x > h_2$, the front can jump over the interval (h_1, h_2) and the excitation will never reach this interval.

Finally, I will consider one more effect where we can observe jumps of the wave front. The one-dimensional case is considered for brevity. Let $f(x, u) = c(x, u) \cdot u$ and assume that $c(x, u) < 0$ for $x \in R^1 \setminus (\cup_k \{x \in R^1 : |x - k| < \delta\})$, $u \geq 0$, $0 < \delta < 1$, k runs through all integers. Suppose that a continuous function $\gamma(x) > 0$, $x \in R_1$, exists such that $c(x, u) > 0$ for $x \in \mathcal{E}_\delta = \cup_k \{x \in R^1, |x - k| < \delta\}$, $u < \gamma(x)$, and $c(x, u) < 0$ for $u > \gamma(x)$, $x \in R^1$. Let $c(x) = c(x, 0) = \max_{u \geq 0} c(x, u)$ for $x \in \mathcal{E}_\delta$ (Figure 11). Consider the Cauchy problem

$$\frac{\partial u^{\varepsilon,\delta}(t, x)}{\partial t} = \frac{\varepsilon}{2} \frac{\partial^2 u^{\varepsilon,\delta}}{\partial x^2} + \frac{1}{\varepsilon} c(x, u^{\varepsilon,\delta}) u^{\varepsilon,\delta}, \qquad u^{\varepsilon,\delta}(0, x) = g(x) \geq 0.$$

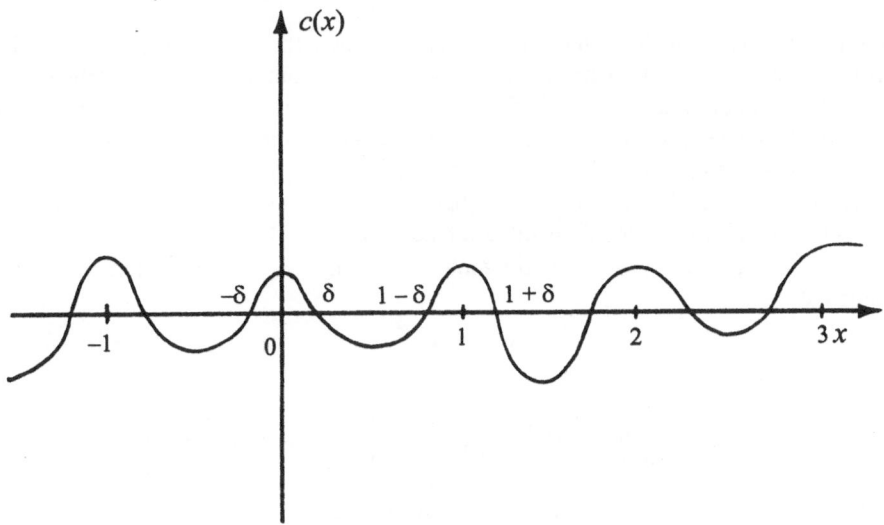

FIGURE 1.11.

Assume that $G_0 = [-\delta, \delta]$ is the support of $g(x)$. It is not difficult to check that for $t > 0$ small enough $u^{\varepsilon,\delta}(t, x) \to \gamma(x)$ for $|x| < \delta$ and $u^{\varepsilon,\delta}(t, x) \to 0$ for $|x| > \delta$ as $\varepsilon \downarrow 0$. What will be the behavior of $u^{\varepsilon,\delta}(t, x)$ for $0 < \varepsilon \ll 1$ as t increases? Will the area, where $u^\varepsilon(t, x)$ has a positive limit, expand with time, and if so what is the law of expansion? It turns out that one can calculate a sequence $t_1, t_2, \ldots, t_n, \ldots$ such that $\lim_{\varepsilon \downarrow 0} u^{\varepsilon,\delta}(t, x) = 0$ for $|x - k| < \delta$, $t < t_k$, and $\lim_{\varepsilon \downarrow 0} u^{\varepsilon,\delta}(t, x) = \gamma(x)$ for $t > t_k$ at least for some points of the interval $|x - k| < \delta$. Then as time increases the excitation will expand on the whole interval $\{x : |x - k| < \delta\}$. Thus the front jumps through the area where $c(x) < 0$. Using a counterpart of the V-functions introduced before in this section one can describe the behavior of the front. I will consider the case when $\delta \downarrow 0$ together with $\varepsilon \downarrow 0$ so that $\overline{\lim}_{\varepsilon, \delta \downarrow 0}(\varepsilon/\delta^2) = \infty$. The result in this case is more explicit. Suppose there are a smooth function $c(x) \leq 0$, $x \in R^1$, and a sequence $c_k > 0$, $k = 0, \pm 1, \pm 2, \ldots$. Let the nonlinear term $f^\delta(x, u) = c^\delta(x, u) \cdot u$ in our equation be such that $c^\delta(x, 0) = c(x)$ outside the set $\{x : |x - k| < \delta$ for some $k = 0, \pm 1, \pm 2, \ldots\}$, $c^\delta(x, 0) = c_k$ for $|x - k| < \delta/2$ and $c^\delta(x, 0)$ is continuous.

Use the notations

$$V_n(t) = \inf_{\varphi_0 = 0, \varphi_t = n} \int_0^t \left[|c(\varphi_s)| + \frac{\dot\varphi_s^2}{2} \right] ds$$

$$W_n(t) = \sup_{0 < \alpha \leq t} \left[\alpha \cdot \max_{1 \leq k \leq n} c_k - V_n(t - \alpha) \right].$$

The function $W_n(t)$ increases as t increases, changing from $-\infty$ as $t \to 0$ to $+\infty$ as $t \to \infty$. Thus there exists the unique $t^* = t^*(n)$ such that $W_n(t^*(n)) = 0$. Let $\hat{t} = \min\{t^*(k) : k = 1, \ldots, n, \ldots\}$. It is easy to check that $\hat{t} = t^*(\hat{k}) > 0$, if c_k does not increase too fast as $|k| \to \infty$. For example, $\hat{t} > 0$ if $c_k < \bar{c} < \infty$ for all $k = 0, 1, 2, \ldots$. Note that the $\min\{t^*(k) : k = 1, \ldots, n\}$ can be achieved on some set $\mathcal{K}$ consisting of more than one point. One can prove that $u^{\varepsilon,\delta}(t, x) \to 0$ outside $[G_0] = \{x : |x| \leq \delta\}$ for $t < \hat{t}$, and $u^{\varepsilon,\delta}(t, x) \to \gamma(x)$ if $t > \hat{t}$, for a point $x = x_k$

such that $|k - x_k| < \delta$, for some $k \in \mathcal{K}$. Thus at time $\hat{t}$ the new sources of excitation will be switched on. Note that, in general, several new sources can be switched on at time $\hat{t}$, and the new sources may be different from $k = \pm 1$: the excitation can "jump" over closest integer points if the constants c_k at those points is relatively small. In the same way one can describe where and when after the time $\hat{t}$ the next source will appear and so on. If the sequence c_k changes slowly enough the excitation will move first to the neighbors of already excited integer points.

Let, for example, $c(x) \equiv 0$ and denote $c^{(n)} = \max(c_1, \ldots, c_n)$. Then it is easy to calculate that

$$V_n(t) = \frac{n^2}{2t},$$

$$W_n(t) = \sup_{0 < \alpha < t} \left[\alpha c^{(n)} - \frac{n^2}{2(t - \alpha)} \right] = c^{(n)}t - n\sqrt{2c^{(n)}},$$

$$t^*(n) = \frac{n\sqrt{2}}{\sqrt{c^{(n)}}}.$$

If $kc_1 > c_k$ for any $k = 2, 3, \ldots$, then $\hat{t} = t^*(1)$ and the excitation will jump to $k = 1$ before it comes to any other point $k > 0$. If $c_k > kc_1$ for some $k \geq 2$ the excitation can jump from G_0 to the neighborhood of k before it reaches the neighborhood of the point 1. One can describe the next jumps in the same way.

1.7.   RDE SYSTEMS OF KPP TYPE

The use of probabilistic methods for studying wave front propagation for semilinear PDEs is based on the probabilistic representations of the solutions of corresponding linear problems. Formulas which serve as representations of the solutions in the linear case give integral equations in the functional space in the case of semilinear equations. These integral equations turn out to be convenient for studying asymptotic behavior of the solutions of the semilinear problems. To study semilinear PDE systems we need probabilistic representation for solutions of linear systems.

Consider the following linear system:

$$\frac{\partial u_k(t, x)}{\partial f} = L_k u_k + c_k(t, x)u_k + \sum_{j=1}^{n} c_{kj}(x)(u_j - u_k),$$

$$u_k(0, X) = g_k(x), \quad x \in R^r, \quad t > 0, \quad k = 1, \ldots, n. \tag{1.7.1}$$

Here $L_k = \frac{1}{2} \sum_{i,j=1}^{r} a_k^{ij}(X)\partial^2/\partial x^i \partial x^j + \sum_{i=1}^{r} b_k^i(x)\partial/\partial x^i$ are elliptic operators with smooth enough coefficients, $c_{kj}(t, x) \geq 0$ are continuous functions.

A Markov process (x_t, ν_t) in $R^r \times \{1, \ldots, n\}$ can be connected with system (1.7.1). The process is defined by the stochastic differential equations

$$dx_t = \sigma_{\nu_t}(x_t)\, dW_t + b_{\nu_t}(x_t)\, dt, \quad x_0 = x \in R^r,\ \nu_0 = k \in \{1, \ldots, n\},$$

$\sigma_l(x) = (a_l^{ij}(x))^{1/2}$, $b_l(x) = (b_l^1(x), \ldots, b_l^r(x))$, W_t is the Wiener process in R^r, and by the equality

$$P\{\nu_{t+\Delta} = k | \nu_t = l, \ x_t = x; \ \nu_s, x_s, \ 0 \le s \le t\} = c_{lk}(x)\Delta + o(\Delta), \qquad \Delta \downarrow 0.$$

It is not difficult to prove that the solution of problem (1.7.1) can be written as follows:

$$u_k(t,x) = E_{x,k}g_{\nu_t}(x_t)\exp\left\{\int_0^t c_{\nu_s}(t-s,x_s)\,ds\right\}. \qquad (1.7.2)$$

Representation (1.7.2) is a version of the Feynman–Kac formula (see [8]). Similar representations can be given for the solutions of natural mixed problems and for solutions of corresponding stationary problems.

Now consider some semilinear PDE systems. Let us start with weakly coupled systems (see, for example, [8]). In the simplest space-homogeneous case the problem in the tube $R^1 \times (G \cup \partial G)$ can be reduced to a problem with a one-dimensional space variable. After proper rescaling of space and time we come to the following problem:

$$\frac{\partial u_k^\varepsilon(t,x)}{\partial t} = \varepsilon\frac{D_k}{2}\frac{\partial^2 u_k^\varepsilon}{\partial x^2} - b_k\frac{\partial u_k^\varepsilon}{\partial x} + \frac{1}{\varepsilon}f_k(u_k^\varepsilon) + \sum_{j=1}^n c_{kj}\cdot(u_j^\varepsilon - u_k^\varepsilon)$$

$$t > 0, \quad x \in R^1, \quad u_k^\varepsilon(0,x) = g_k(x) \ge 0, \quad k = 1,\ldots,n,$$

$$f_k \in F_1, \quad c_{kj} > 0. \qquad (1.7.3)$$

For brevity, let the union of the supports of the initial functions be equal to $G_0 = (x \in R^1, x \le 0)$.

To describe the asymptotic behavior of problem (1.7.3) let us introduce the following set of points in R^3:

$$M = \{(b_k, c_k, D_k), \ k = 1,\ldots,n\},$$

where $c_k = df_k(u)/du|_{u=0}$. The convex hull of the set M is denoted M_{conv}.

Theorem 1.8. *Let $(u_1^\varepsilon(t,x),\ldots,u_n^\varepsilon(t,x))$ be the solution of problem (1.7.3). Use the notation*

$$\hat{\alpha} = \max\{z_1 + \sqrt{2z_2z_3} : (z_1, z_2, z_3) \in M_{\text{conv}}\}.$$

Then

$$\lim_{\varepsilon\downarrow 0} u_k^\varepsilon(t,k) = \begin{cases} 1, & \text{if } x < \hat{\alpha}t, \\ 0, & \text{if } x > \hat{\alpha}t. \end{cases} \qquad (1.7.4)$$

Proof. To prove this theorem we introduce the family of Markov processes (x_t^ε, ν_t), $\varepsilon > 0$, corresponding to the linear part of system (1.7.3). Using (1.7.2) we can write the following integral equation in the space of trajectories for $u_k^\varepsilon(t,x)$, $k = 1,\ldots,n$:

$$u_k^\varepsilon(t,x) = E_{x,k}g_{\nu_t}(x_t^\varepsilon)\exp\left\{\frac{1}{\varepsilon}\int_0^t c_{\nu_s}(u_{\nu_s}^\varepsilon(t-s,x_s^\varepsilon))\,ds\right\},$$

$$c_k(u_k) = f_k(u_k)u_k^{-1}. \qquad (1.7.5)$$

Using this formula we can derive (1.7.4) in a way similar to the proof of Theorem 5. The component ν_t^ε plays here the same part as the y-component in Section 1.4. We omit the detailed proof.

Remark. The maximum in the definition of $\hat{\alpha}$ is reached on the boundary of M_{conv}. It can be bigger that the speed of the wave front in each separated equation.

The asymptotic speed for problem (1.7.3) can be larger than in each separated equation owing to "almost free" transitions of the component ν_t: The probability that ν_t, $0 \le t \le T < \infty$, is close to a given function $\psi_t : [0, T] \to \{1, \dots, n\}$ with finite number of jumps is independent of ε.

Consider now a system of equations coupled only on the boundary of the tube $R^1 \times G$:

$$\frac{\partial u^\varepsilon(t, x, y)}{\partial t} = \frac{\varepsilon D_k}{2} \Delta u_k^\varepsilon - b_k \frac{\partial u_k^\varepsilon}{\partial x} + \frac{1}{\varepsilon} c_k(u_k^\varepsilon) u_k^\varepsilon,$$

$$t > 0, \ x \in R^1, \ y \in G \subset R^r,$$

$$\left. \frac{\partial u_k^\varepsilon}{\partial n} - \sum_{j=1}^n c_{kj} \cdot (u_j^\varepsilon - u_k^\varepsilon) \right|_{t>0, x \in R^1, y \in \partial G} = 0,$$

$$u_k^\varepsilon(0, x, y) = g_k(x) \ge 0; \ k = 1, 2, \dots, n. \tag{1.7.6}$$

One can write an integral equation of the type (1.7.5) for the solution of problem (1.7.6). It turns out that in this case the transitions of the jumping component are also "almost free"; and thus a counterpart of Theorem 8 is true for the solution of problem (1.7.6).

Consider now the following weakly coupled system that is nonhomogeneous in space:

$$\frac{\partial u^\varepsilon(t, x)}{\partial t} = \frac{\varepsilon}{2} \sum_{i,j=1}^r a_k^{ij}(x) \frac{\partial^2 u_k^\varepsilon}{\partial x^i \partial x^j} + \frac{1}{\varepsilon} c_k(x, u_k^\varepsilon) u_k^\varepsilon$$

$$+ \sum_{j=1}^n c_{kj}(x)(u_j^\varepsilon - u_k^\varepsilon),$$

$$u_k^\varepsilon(0, x) = g_k(x) \ge 0, \quad t > 0, \quad x \in R^r, \quad k = 1, \dots, n. \tag{1.7.7}$$

We assume that the coefficients are smooth enough, the forms $\sum_{i,j=1}^n a_k^{ij}(x) \lambda_i \lambda_j$, $k = 1, \dots, n$, are positive definite, $c_{kj}(x) > 0$, the closures of the supports $G_0^{(k)}$ of $g_k(\cdot)$ coincide with the closures of their interiors. Let $G_0 = \cup_{k=1}^n G_0^{(k)}$.

Suppose that $c_k(x, 0) = c$ independent of k and of x. Then the motion of the wave front can be described by the Huygens principle and in a proper Finsler metric $\hat{\rho}$ the corresponding velocity field $v(e)$, $e \in R^r$, will be homogeneous and isotropic. The unit ball in the tangent space at point x for this metric $\hat{\rho}$ is equal to the convex hull of the Riemannian balls $S_k = \left\{ z \in R^r : \sum_{i,j=1}^r a_{ij,k}(x) z^i z^j < 1 \right\}$, $(a_{ij,k}(x)) = (a_k^{ij}(x))^{-1}$. In this metric $|v(e)| = \sqrt{2c}$. The proof of this statement is given in [8]. The general case is also considered in that paper.

I will mention one result concerning the diffusion-transmutation process (X_t, ν_t)

corresponding to the system

$$\frac{\partial u_k}{\partial t} = \frac{1}{2} \sum_{i,j=1}^{r} a_k^{ij}(x) \frac{\partial^2 u}{\partial x^i \partial x^j} + \sum_{j=1}^{n} c_{kj}(x)(u_j - u_k),$$

$$x \in R^r, \quad k = 1, \ldots, n; \quad c_{kj}(X) > 0. \tag{1.7.8}$$

In the case of one equation there is a well known result of S .R. S. Varadhan [16] concerning the behavior of the transition density as $t \downarrow 0$:

$$\lim_{t \to \infty} 2t \ln p(t, x, y) = -\rho^2(x, y),$$

where $\rho(\cdot, \cdot)$ is the Riemannian metric corresponding to the form

$$\sum_{i,j=1}^{r} a_{ij}(x) \, dx^i \, dx^j = ds^2,$$

$(a_{ij}(x)) = (a^{ij}(x))^{-1}$. Consider the transition density $p(t, (x, k), (y, l))$ of the process (X_t, ν_t) in the general case. One can derive from the results of [8], that

$$\lim_{t \to \infty} 2t \ln p(t, (x, k), (y, l)) = -\hat{\rho}^2(x, y),$$

where $\hat{\rho}(x, y)$ is the Finsler metric, which was described earlier in this section as the convex hull of the Riemannian metrics corresponding to the forms $ds_k^2 = \sum_{i,j=1}^{r} a_{ij,k}(x) \, dx^i \, dx^j$, $k = 1, 2, \ldots, n$. Calculation of the preexponential term in the asymptotic of $p(t, (x, k), (y, l))$ as $t \downarrow 0$ is an interesting problem. As is well known, this asymptotic is connected with the spectral asymptotic for problem (1.7.8).

One can consider a system of coupled equations when the transmutation coefficients have the same order as the other terms of the equations:

$$\frac{\partial u_k(t, x)}{\partial t} = \frac{D_k}{2} \Delta u_k + f_k(u_k) + \sum_{j=1}^{n} c_{kj} \cdot (u_j - u_k),$$

$$u_k(0, x) = g_k(x) \geq 0, \quad x \in R^r, \quad k = 1, 2, \ldots, n.$$

Here $f_k \in F_1$, $c_{kj} > 0$. Such systems were considered in [6], [7], [8]. Systems with nonlinear terms of more general form were studied in [1] by analytical methods. The most general result is presented in [10]. In particular, the equations with slowly changing coefficients are considered there.

The systems without diffusion can also be of interest:

$$\frac{\partial u_k(t, x)}{\partial t} = -\sum_{i=1}^{r} b_k^i \frac{\partial u_k}{\partial x^i} + \sum_{j=1}^{n} c_{kj} \cdot (u_j - u_k) + f_k(u_k),$$

$$t > 0, \quad x \in R^r, \quad u_k(0, k) = g_k(x) \geq 0, \quad k = 1, 2, \ldots, n. \tag{1.7.9}$$

We assume that $c_{kj} > 0$; $k, j \in \{1, \ldots, n\}$, and that the function $\sum_{k=1}^{n} g_k(x)$ has compact support G_0 coinciding with the closure of its interior.

The Markov process (x_t, ν_t) in $R^r \times \{1, \dots, n\}$ is defined as follows:

$$x_t = x - \int_0^t b_{\nu_s}\, ds, \qquad b_k = (b_k^1, \dots, b_k^r), \ k = 1, \dots, n,$$

$$P\{\nu_{t+\Delta} = j \mid \nu_t = i, \ x_t = y\} = c_{ij}(y)\Delta + O(\Delta), \qquad \Delta \downarrow 0, \ i \neq j.$$

Then we can write

$$u_k(t, x) = E_{x,k} g_{\nu_t}(x_t) \exp\left\{ \int_0^t c_{\nu_s}(u(t - s, x_s))\, ds \right\}, \tag{1.7.10}$$

where $c_k(u) = f_k(u)u^{-1}$. We conclude from (1.7.10) that

$$0 \leq u_k(t, x) \leq E_{x,k} g(X_t) \exp\left\{ \int_0^t c_{\nu_s}\, ds \right\}, c_k = c_k(0). \tag{1.7.11}$$

Denote by $\lambda(\beta_0, \beta)$, $\beta_0 \in R^1$, $\beta \in R^r$, the first eigenvalue of the matrix $Q^{\beta_0, \beta} = (q_{ij}^{\beta_0, \beta})_1^n$, $\beta_0 \in R^1$, $\beta \in R^r$, with the entries $q_{ij}^{\beta_0, \beta} = c_{ij}$ for $i \neq j$ and $q_{ii}^{\beta_0, \beta} = \beta_0 c_i - (\beta, b_i) - \sum_{j:j\neq i} c_{ij}$. Let $L(z_0, z)$, $z_0 \in R^1$, $z \in R^r$, be the Legendre transformation of $\lambda(\beta_0, \beta)$: $L(z_0, z) = \sup_{\beta_0, \beta}[\beta_0 z_0 + (z, \beta) - \lambda(\beta_0, \beta)]$. As is shown in Section 1.7.4 of [11], the function $tL(z_0, z)$ is the action function for the large deviations for the family of random variables $\left(t^{-1} \int_0^t c_{\nu_s}\, ds, \ t^{-1} \int_0^t b_{\nu_s}\, ds \right)$ as $t \to \infty$. Using this fact one can derive from (1.7.10) and (1.7.11) that the asymptotic speed $\alpha^*(e)$ for problem (1.7.9) in the direction of a unit vector $e \in R^r$ is equal to the solution of the equation

$$\max_{z_0}[z_0 - L(z_0, \alpha^* e)] = 0.$$

That means that for any $h > 0$

$$\lim_{t \to \infty} \sup_{\substack{|x| > t \cdot (\alpha^*(x/|x|) + h) \\ k=1,\dots,n}} u_k(t, x) = 0,$$

$$\lim_{t \to \infty} \sup_{\substack{|x| < t \cdot (\alpha^*(x/|x|) - h) \\ k=1,\dots,n}} u_k(t, x) = 1.$$

Note that, in general, for some $e \in R^r$, $|e| = 1$, $\alpha^*(e) = 0$. That means that the excitation does not propagate in the direction e. One can formulate conditions providing positivity of $\alpha^*(e)$ for each $e \in R^r$: the origin should be an interior point of the convex hull of the vectors $b_1, \dots, b_n$ in R^r. In particular, if $n < r + 1$ the excitation governed by (1.7.9) cannot propagate with positive asymptotic speed in all directions. In general, the excitation as $t \to \infty$ will occupy the set $H \subset R^r$, which can be described as follows: Let K be the cone induced by the convex hull of the vectors $b_1, \dots, b_n$. Denote by $K + z$ the shift of K on vector $z \in R^r$. Then $H = \cup_{z \in G_0}(K + z)$.

Consider now a version of (1.7.9) with a small parameter

$$\frac{\partial u_k^\varepsilon(t, x)}{\partial t} = -\varepsilon^{-\kappa}(b_k, \nabla u_k^\varepsilon) + \varepsilon^{-1} \sum_{j=1}^n c_{kj} \cdot (u_j^\varepsilon - u_k^\varepsilon) + \varepsilon^{2\kappa - 1} f_k(u_k^\varepsilon)$$

$$u_k^\varepsilon(0, x) = g_k(x), \quad x \in R^r, \quad k = 1, \dots, n, \quad 0 < \varepsilon \ll 1. \tag{1.7.12}$$

If $\kappa = 0$, then system (1.7.12) is equivalent to (1.7.9). The interesting case is when $0 < \kappa < 1/2$. Assume that $\bar{b} = \sum_{k=1}^{n} \gamma_k b_k = 0$, where $(\gamma_1, \ldots, \gamma_n)$ is the invariant distribution of the Markov process $\nu(t)$ in $\{1, 2, \ldots, n\}$ with transition intensities c_{kj} (we assume that $c_{kj} > 0$), and that the convex hull of the vectors $b_1, \ldots, b_n \in R^r$ is an open set in R^r. Then $\int_0^t b_{\nu(s/\varepsilon)} ds \rightarrow 0$ as $\varepsilon \downarrow 0$. A version of the central limit theorem holds here. There exists a matrix B such that $\varepsilon^{-1/2} \int_0^t b_{\nu(s/\varepsilon)} ds$ converges as $\varepsilon \downarrow 0$ in the distribution sense to the Gaussian distribution in R^r with mean zero and covariation matrix tB. The proof of this statement and expression for B can be found in Chap. 7 of [11].

To describe the asymptotic behavior of the solution of (1.7.12) as $\varepsilon \downarrow 0$ we need the logarithmic asymptotics of the deviations of order ε^κ for $\int_0^t b_{\nu(s/\varepsilon)} ds$. As follows from Section 1.7.7 of [11], the probabilities of such deviations for $0 < \kappa < 1/2$ are logarithmically equivalent to the probabilities of deviations of order $\varepsilon^{1/2-\kappa}$ for the limit Gaussian distribution of $\varepsilon^{-1/2} \int_0^t b_{\nu(s/\varepsilon)} ds$ as $\varepsilon \downarrow 0$. Thus for any set $G \subset R^r$ such that $[G] = [(G)]$:

$$P\left\{\frac{1}{\varepsilon^\kappa} \int_0^t b_{\nu(s/\varepsilon)} ds \in G\right\} \approx \exp\left\{-\inf_{zt \in G} \frac{t(B^{-1}z, z)}{2\varepsilon^{1-2\kappa}}\right\}, \qquad \varepsilon \downarrow 0. \quad (1.7.13)$$

Suppose that $c = f_k'(0)$ independent of k. Then from (1.7.10) and (1.7.13) one can derive that the asymptotic speed $\alpha^*(e)$ in a direction $e \in R^r$, $|e| = 1$ is given by the formula

$$\alpha^*(e) = \sqrt{\frac{2c}{(B^{-1}e, e)}}, \qquad e \in R^r, |e| = 1. \quad (1.7.14)$$

If $c_k = f_k'(0)$ depends on k we should replace c in (1.7.14) by $\bar{c} = \sum_{k=1}^{n} \gamma_k c_k$.

1.8. RDE'S IN NARROW BRANCHING CHANNELS

1. Consider a reaction-diffusion equation in a narrow branching channel. Let us have a connected graph Γ in R^r consisting of segments $\gamma_1, \ldots, \gamma_n$ connecting the vertices $O_1, \ldots, O_m$ (Figure 12).

Suppose that near each segment γ_k a tube $\mathcal{E}_k \subset R^r$ of radius εr_k is taken. And let G_ε be the union of all such tubes (Figure 13). We assume that ε is small enough so that no points $O_1 \cdots O_m$ belong to any tube $\mathcal{E}_k$ besides the ends of the segment γ_k. Consider in G_ε a reaction-diffusion equation

$$\frac{D}{2}\Delta u^\varepsilon(t, x) + f(u^\varepsilon(t, x)) = \frac{\partial u^\varepsilon}{\partial t}, \qquad t > 0, \ x \in G_\varepsilon$$
$$u^\varepsilon(0, x) = g(x), \qquad x \in G_\varepsilon. \quad (1.8.1)$$

We should add boundary conditions.

$$\left.\frac{\partial u^\varepsilon(t, x)}{\partial n}\right|_{x \in \partial G_\varepsilon \backslash A_\varepsilon} = 0, \quad (1.8.2)$$

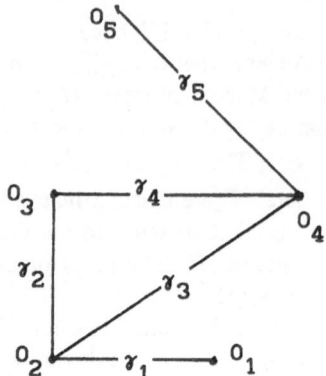

FIGURE 1.12.

where A_ε is the set of points where ∂G_ε is not smooth. The set A_ε is thin enough so that problem (1.8.1)–(1.8.2) has a unique solution.

One can expect that the solution of the problem (1.8.1)–(1.8.2) approaches a solution of a problem on the limiting graph Γ as $\varepsilon \downarrow 0$. Let X_t^ε be the process in G_ε corresponding to the operator $(D/2)\Delta$ with normal reflection on the boundary. Denote by $Y_t^\varepsilon = y(X_t^\varepsilon)$ the projection of X_t^ε on the graph Γ. One should be more careful defining the projection in the small neighborhoods of the vertices O_k, but it can be done (see [12]). It is proved in the latter article that the family Y_t^ε, $0 \le t \le T < \infty$, weakly converges in the space of continuous functions as $\varepsilon \downarrow 0$ to a process Y_t on Γ, which can be described as follows. The process Y_t is a Markov process on Γ, which is governed by the operator $(D/2)d^2/dx^2$ in the interior points of the segment $\gamma_k \subset \Gamma$ and at the vertices governed by the gluing conditions

$$\sum_{i=1}^{l_k} r_i^{r-1} \frac{du(O_k)}{dy^{ik}} = O, \qquad k = 1, \dots, m. \tag{1.8.3}$$

where l_k is the number of segments γ_i connected with the point O_k, r_i^{r-1} is, up to a constant factor depending only on the dimension, the volume of the cross-section of the tube $\mathcal{E}_i$ near γ_i and y_i is the coordinate on γ_i directed from O_k inside γ_i. If a

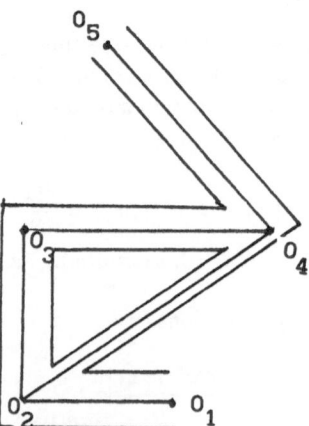

FIGURE 1.13.

point O_k is the end of only one segment γ_i, then (1.8.3) will be a simple reflection condition.

The last result together with some a priori bounds of the derivatives of the solution of problem (1.8.1)-(1.8.2) implies the following result.

Theorem 1.9. *Assume that $c(u) = u^{-1}f(u)$ in (1.8.1) is Lipschitz continuous and $g(x)$ is continuous. Then $\lim_{\varepsilon \downarrow 0} u^\varepsilon(t,x) = \tilde{u}(t, y(x))$, where $\tilde{u}(t,y)$ is the unique solution of the following problem:*

$$\frac{\partial \tilde{u}(t,y)}{\partial t} = \frac{D}{2}\frac{\partial^2 \tilde{u}}{\partial y^2} + f(\tilde{u}), \qquad t > 0, \ y \in \Gamma \setminus \cup\{O_k\}$$

$$\sum_{i=1}^{l_k} r_i^{r-1}\frac{\partial u}{\partial y_i}(O_k) = 0, \qquad \tilde{u}(t,0) = g(y), \ k = 1,\dots,m. \tag{1.8.4}$$

A similar result holds for stationary problems. Note that the stationary problem on the graph can be solved explicitly.

One can consider a linear equation on a graph with nonlinear gluing conditions

$$\frac{\partial v(t,y)}{\partial t} = \frac{D}{2}\frac{\partial^2 v}{\partial y^2}, \qquad t > 0, \ y \in \Gamma \setminus (\cup_{i=1}^m \{O_i\});$$

$$\sum_{i=1}^{l_k} r_i\frac{\partial v}{\partial y_i}(O_k) = f_k(v), \qquad k = 1,\dots,m, \ v(0,y) = g(y). \tag{1.8.5}$$

The notations in (1.8.5) are the same as in (1.8.6). Problems of (1.8.5) type can arise when "the chemical reaction" occurs in the neighborhoods of the vertices of a system of narrow branching channels. If, for example, $D = \varepsilon D$, $f_k = (1/\varepsilon)\hat{f}_k$, $\hat{f}_k \in F_1$ for $k = 1,\dots,m$, and the initial function is nonnegative and $g(O_1) > 0$, one can, using the approach developed earlier in this article, describe the propagation of the excitation in the set of vertices.

Systems of reaction-diffusion equations in a domain consisting of narrow branching channels can be reduced to a RDE-system on the corresponding graph as well.

I will mention briefly one more effect concerning the narrow tubes. Suppose the tube where we consider the equation is narrow and has changing cross-section: $G_\varepsilon = \{(x, r, \varphi), \ x \in R^1, \ 0 \leq \varphi \leq 2\pi, \ 0 \leq r \leq \varepsilon f(x, \varphi)\}$. We assume that $f(x, \varphi)$ is smooth enough and positive. Consider the problem

$$\frac{\partial u^\varepsilon(t,x,r,\varphi)}{\partial t} = \frac{1}{2}\Delta u^\varepsilon + f(u), \qquad t > 0, \ (x, r_i\varphi) \in G_\varepsilon$$

$$\left.\frac{\partial u^\varepsilon}{\partial n^\varepsilon}\right|_{t>0,\partial G_\varepsilon} = 0, \qquad u^\varepsilon(\varphi, x, r, \varphi) = \chi^-(x). \tag{1.8.6}$$

Of course, as one can expect, $u^\varepsilon(t,x,r,\varphi)$ converges to a function $u(t,x)$ as $\varepsilon \downarrow 0$. The limiting function satisfies the same initial conditions. The variability of the cross-section of the tube leads to an additional term in the equation for limiting function:

$$\frac{\partial u(t,x)}{\partial t} = \frac{1}{2}\frac{\partial^2 u}{\partial x^2} + b(x)\frac{\partial u}{\partial x} + f(u), \qquad x \in R^1, \ t > 0, \ u(0,x) = \chi^-(x).$$

Using the results of Section 1.2.5 from [7], we can calculate $b(x)$:

$$b(x) = \frac{1}{2} \frac{d}{dx} \ln \int_0^{2\pi} f(x, \varphi) \, d\varphi.$$

The term $\frac{1}{2} \int_0^{2\pi} f(x, \varphi) \, d\varphi$ is actually the area of the cross-section at point x. There are a number of interesting effects concerning the wave front propagation caused by this drift. We will consider them in detail elsewhere.

REFERENCES

[1] G. Barles, L. C. Evans, and P. E. Souganidis, Wave front propagation for reaction-diffusion systems of PDE, *Duke Math. J.* **62**, 835–838 (1990).

[2] S. Carmona, An asymptotic problem for a reaction-diffusion equation with a fast diffusion component, *Stochastic and Stoch. Reports* **52**, 43–80 (1995).

[3] L. C. Evans and P. E. Souganidis, A PDE approach to geometric optics for certain semilinear parabolic equations, *Indiana U. Math. J.* **38**, 141–172 (1989).

[4] M. I. Freidlin, Propagation of concentration waves by a random motion connected with growth, *Dokl. Akad. Nauk SSSR* **246**, 544–548 (1979).

[5] M. I. Freidlin, On wave front propagation in periodic media, In *Stochastic Analysis and Applications* (M. A. Pinsky, ed.), Marcel Decker, New York, pp. 147–166, 1984.

[6] M. I. Freidlin, Limit theorems for large deviations and reaction-diffusion equations, *Ann. Probab.* **13**, 639–675 (1985).

[7] M. I. Freidlin, *Functional Integration and Partial Differential Equations*, Princeton U. Press, 1985.

[8] M. I. Freidlin, Coupled reaction-diffusion equations, *Ann. Probab.* **19**(1), 29–57 (1990).

[9] M. I. Freidlin, Semi-linear PDEs and limit theorems for large deviations, pp. 1–107, *Lectures in the Summer School in Probability,* Saint Flour 1990, Lecture Notes in Mathematics 1527, Springer-Verlag, New York.

[10] M. I. Freidlin and T.-Y. Lee, Wave front propagation and large deviations for diffusion-transmutation processes, submitted to *Theory Probab. Appl.*

[11] M. I. Freidlin and A. D. Wentzell, *Random Perturbations of Dynamical Systems,* Springer-Verlag, New York, 1984.

[12] M. I. Freidlin and A. D. Wentzell, Diffusion processes on graph and averaging principle, *Ann. Probab.* **21**, 4, 2215–2245 (1993).

[13] J. Gärtner and M. I. Freidlin, On the propagation of concentration waves in periodic and random media, *Dokl. Akad. Nauk SSSR* **249**, 521–525 (1979).

[14] T. Kato, *Perturbation Theory for Linear Operators,* Springer-Verlag, New York, 1976.

[15] A. Kolmogorov, I. Petrovskii, and N. Piskunov, Étude de l'équation de la diffusion avec croissence de la matière at son application a un probleme biologique, *Moscow Univ. Bull. Math.* **1**, 1–24 (1937).

[16] S. R. S. Varadhan, On the behavior of the fundamental solution of the heat equation with variable coefficients, *Commun. Pure Appl. Math.* **20**(2), 431–455, (1967).

[17] S. R. S. Varadhan, *Large Deviations and Applications,* Society for Industrial and Applied Mathematics, Philadelphia, 1984.

2

Particle and Wave Transmission in One-Dimensional Disordered Systems

Sergey A. Gredeskul, Andrew V. Marchenko, and
Leonid A. Pastur

2.1. INTRODUCTION

Localization is one of the fundamental quantum-mechanical phenomena in disordered media. It reflects the most general properties of disordered systems and stems from the wave nature of elementary excitations and the random character of scattering media.

Localization was introduced in the pioneering paper by Anderson [6], in which he formulated the problem, made the link between localization and transport, and gave the first quantitative estimate for the critical disorder for the transition between the delocalized (diffusive) and the localized (nondiffusive) regimes.

One may also mention the paper [97], in which some features of localization were anticipated. Authors considered the distribution of states in the part of the spectrum that is now called the fluctuation region (see, e.g., [105]) and interpreted their

SERGEY A. GREDESKUL • Department of Physics, Ben-Gurion University of the Negev, Beer-Sheva, 84105 Israel, Internet: sergeyg@bgumail.bgu.ac.il. ANDREW V. MARCHENKO • Moscow State University of Transportation, 15 Obraztsova Str., Moscow, Russia. LEONID A. PASTUR • B. Verkin Institute for Low Temperature Physics and Engineering, Ukrainian Academy of Sciences, 47 Lenin Ave., Kharkov, 310164, Ukraine.

Surveys in Applied Mathematics, Volume 2. Edited by Mark Freidlin, Sergey A. Gredeskul, John K. Hunter, Andrew V. Marchenko, and Leonid A. Pastur, Plenum Press, New York, 1995.

numerical results "in terms of an approximation in which each electron is confined to its own group of adjacent identical atoms." Furthermore, they practically formulated the definition of the Anderson dielectric: "In a disordered alloy what can be expected if the Fermi level occurs in an almost forbidden range [the fluctuation region!—Authors] ... the wave functions at the Fermi level are localized to the large ordered inclusions, and therefore cannot carry a current."

In subsequent decades, localization became the principal notion in the physics of disordered systems. It was followed by such important notions as the Anderson dielectric, the Anderson transition, the scaling theory of localization, the weak localization, etc. that are now quite common in textbooks and physical encyclopedias. Several monographs on the subject were published [114], [136], [16], [105], [41]. The number of international meetings on various aspects of localization and related problems is now extremely large.

Recently interest has been rapidly increasing in the localization of a wide variety of other types of waves and elementary excitations in all kinds of media without spatial homogeneity (or periodicity). Intensive studies are carried out on the localization of sound waves [119], [71], [74], [11], [82], [128], [32], of electromagnetic waves in solids and plasma [42], [128], of gravity waves in a shallow water channel with a rough bottom [66], [35], [14], of the third and fourth sounds in a liquid helium film on a randomly inhomogeneous substrate [30], [31], [90], of surface waves in metals [44], [110], [111], etc. The interest in transmission of short irregular pulses through homogeneous media has given the impetus to the study of localization properties of Dirac-type equations [18], [62], describing one-particle excitations in superconductors and semiconductors [19], [63], [39], and the wave transmission through a layered structure similar to those used in x-ray mirrors [52].

In this article we deal mostly with localization and associated phenomena in a random media supposing that its parameters depend on a single coordinate only. We mention three reasons justifying the study of such one-dimensional systems. The first (and the most trivial) one is that owing to their technical simplicity one-dimensional systems often lead to exactly solvable or rather feasible problems. The second reason is based on a considerable practical interest since such layered "one-dimensional" structures are widely used in optics, radiophysics, and acoustics as models of propagation media. The last (and maybe the most important) reason is that in such media the interference effects being the main mechanism of field formation are most pronounced. Therefore the associated localization manifests itself most completely: for whatever small disorder in a one-dimensional random medium all the states are localized (Mott [115]).

The complete localization in a one-dimensional disordered system significantly influences its scattering characteristics, in particular transmission and reflection coefficients, which in turn determine directly various kinetic properties of the system.These problems constitute the main subject of the present review.

In the first introductory section we discuss the connections between transmission characteristics and kinetic quantities and describe the models that will be used further on. We also discuss here the main statistical properties of the random potential and some physical quantities.

Section 2.2 is devoted to the one-dimensional localization theory. We discuss here the localization criteria and introduce such important quantities as the Lyapunov exponent and the localization length. The final subsection contains the brief review of the modern state of the localization theory.

Section 2.3 deals with transmission characteristics. Some formulas are derived for the transmission coefficient, the exponential decay of the average transmission coefficient and of the transmission coefficient on realizations is proved, as well as self-averaging of the logarithmic decrement of the transmission coefficient on a realization. We also construct a number of expansions of the decrement with respect to the concentration of scatterers, energy, etc.

Properties of the reflection coefficient and resonance effects are considered in Section 2.4. The behavior of the reflection coefficient on a realization is qualitatively described. The notion of a resonant (and nonresonant) transmission is introduced and the existence of transparent energies and realizations is discussed.

Explicit formulas for the transmittivity decrements in some limit cases are obtained in Section 2.5.

Section 2.6 begins with the discussion of a connection between the calculated averaged values and the observed ones. The asymptotics of the decrement of the observed transparency of a disordered layer are obtained. This result is applied to the surplus tunnel currents theory and associated mesoscopic effects.

In Section 2.7 the problem under consideration is the point source field in a randomly stratified layer. The exact expressions for the energy fluxes are obtained and their mesoscopic properties are analyzed. The existence of quasihomogeneous waves in the layer is predicted and the theory of a fluctuational waveguide is constructed.

Some recent results concerning the "interaction" between the nonlinearity and disorder are contained in Section 2.8. They concern various important physical problems: nonlinear sound amplification in superconductors, the generation of dark solitons in nonlinear optical fibers by the random input pulse, transmission of the plane wave through the nonlinear disordered segment, and soliton propagation in the nonlinear medium with weak random point scatterers.

2.1.1. Wave Transmission and Kinetics

Quantum mechanical scattering problem for the one-dimensional Schrödinger equation (SE)

$$-\psi'' + v(x)\psi = E\psi \qquad (2.1.1.1)$$

with the scattering potential $v(x) \to 0$ as $|x| \to \infty$ and the energy $E = k^2 > 0$ deals with the solutions $\psi_{1,2}(x)$ of this equation satisfying the boundary conditions

$$\psi_1(x) \approx e^{ikx} + r_+(k)e^{-ikx}, \qquad x \to -\infty;$$
$$\psi_1(x) \approx t_+(k)e^{ikx}, \qquad x \to +\infty \qquad (2.1.1.2a)$$

$$\psi_2(x) \approx t_-(k)e^{-ikx}, \qquad x \to -\infty;$$
$$\psi_2(x) \approx e^{-ikx} + r_-(k)e^{ikx}, \qquad x \to +\infty \qquad (2.1.1.2b)$$

For the scalar wave equation describing the propagation of the plane wave through an inhomogeneity of the medium

$$v(x) = -\frac{\omega^2}{c^2}\delta\varepsilon(x), \qquad k^2 = \frac{\omega^2}{c^2}\varepsilon_0,$$

where ω and c are the frequency and the velocity of wave in the homogeneous medium and $\varepsilon(x) = \varepsilon_0 + \delta\varepsilon(x)$ is the fluctuating characteristic of the medium (dielectric permittivity, density, etc.).

In this article we consider a scattering potential $v(x)$, that is zero if $x < 0$ or $x > L$ and is the restriction of a random stationary process if $0 < x < L$. It models a one-dimensional disordered medium with spatially homogeneous disorder. We call this potential a random barrier of the length L.

For random barriers the amplitudes $r_\pm$ and $t_+ = t_- \equiv t$ are random functions of the barrier length L and the energy $E = k^2$ of the incident wave. We are interested in the statistical behavior of $r_\pm$ and $t_\pm$ for large L. In particular, the important quantities are the reflection and transmission coefficients (reflectivity and transmittivity)

$$R = |r_\pm|^2, \qquad T = |t|^2. \tag{2.1.1.3}$$

Being physically observable characteristics of the scattering, they enter into a variety of other formulas. For example the intensity I of the wave passing through a disordered segment $[0, L]$ measured at its inner point x is [87]

$$I(x) = I(0) \cdot \frac{|1 + r_-(x)|^2}{1 - |r_-(x)|^2} \tag{2.1.1.4}$$

where $r_-(x)$ is the reflection amplitude of the segment $[0, x]$ the wave incident from the right.

The radiation flux density j of a point source in a one-dimensional problem is expressed by the similar formula [58]:

$$j = \frac{2}{\sqrt{E}} \cdot \frac{1 - |r_+(E)|^2}{|1 - r_+(E)|^2} \tag{2.1.1.5}$$

(here E is the energy of the incident wave).

The field G of a source at a point $\vec{R}_0$ of a randomly layered medium is [21]:

$$G(\vec{R}, \vec{R}_0) = \frac{i\sqrt{E_0}}{2} \int \frac{1 + r_-(x_0)}{1 - r_+(x_0)r_-(x_0)} H_0^{(1)}(k\rho\sin\theta)$$
$$\cdot f_{1,2}(\theta, x\,\text{sign}(x - x_0))\sin\theta\,d\theta,$$
$$(x \neq x_0) \tag{2.1.1.6}$$

where reflection amplitudes themselves $r_\pm(x_0)$ are present.

Owing to the wave nature of quantum particles and elementary excitations in the condensed matter (quasiparticles) many important physical quantities can be expressed via the scattering characteristics of the sample. For instance, according to the Landauer formula [96] the electric conductance G of a one-dimensional disordered segment embedded into an infinite ideal circuit is

$$G = \frac{e^2 T}{hR} \qquad (2.1.1.7)$$

where e is the electron charge, h is the Planck constant, and T and R are the transmission and reflection coefficients of the segment.

There are also other similar formulas. We mention the mean thermal flux through a disordered segment (see the book [105] and references therein), the power absorbed by a superconductor from the field of a sound signal [17], etc.

All these formulas contain the reflection and transmission coefficients or amplitudes that are subjects to the localization effects as we shall see below.

The localization of waves and particles itself results from the randomness of coefficients (the dielectric constant, the refractive index, the potential, etc.) of the corresponding dynamic equations considered in the infinite region. This property should naturally manifest itself in the case of randomness in a sufficiently long but finite region, e.g., in scattering problems for long random barriers. Although these problems deal with open systems (finite disordered segment in an infinite free space), still the important physical quantities can be expressed in terms of solutions related to a closed system and therefore both the scattering states and the transmission and reflection characteristics "feel" the localization effects. And these as we have already seen directly influence the kinetic properties of the sample.

2.1.2. Models

We shall discuss now the basic equations considered in this article.

The most important physical problems considered below lead as a rule to the Schrödinger or Dirac-type equations. The disorder is modelled by a random character of the coefficients of the equations.

We use the stationary Schrödinger equation (SE) in the form (2.1.1.1). This form of the SE corresponds to the choice of units where $\hbar = 2m = 1$ ($\hbar$ is the Planck constant and m is the mass of the particle).

Systems that have an initial two-band spectrum (e.g., superconductors or semiconductors) are described by the Dirac-type equation (DE). The most general form of the (two-component) DE in an appropriate system of units is

$$-i\sigma_3 \frac{d\vec{\psi}}{dx} + \Delta\sigma_1\vec{\psi} + \sum_{i=0}^{3} v_i(x)\sigma_i\vec{\psi} = E\vec{\psi} \qquad (2.1.2.1)$$

$$\vec{\psi} = \begin{pmatrix} \psi_1 \\ \psi_2 \end{pmatrix} \qquad (2.1.2.2)$$

Here $\sigma_{1,2,3}$ are the Pauli matrices, $\sigma_0 = \begin{pmatrix} 1 & 0 \\ 0 & 1 \end{pmatrix}$, $v_i(x)$ are random potentials, Δ is the super- (semi-) conducting gap.

Note that the dimensionality of the spectral parameter E in the DE is the inverse length while in the SE it is the square of the inverse length.

The problem of the absorption of a sound pulse by the superconductor can be reduced to the scattering problem for the modified Dirac-type equation (MDE) [17]

$$-i(\sigma_3 - \beta)\frac{d\vec{\psi}}{dx} + \Delta\sigma_1\vec{\psi} + v(x)\vec{\psi} = E\vec{\psi} \qquad (2.1.2.3)$$

The parameter β is the ratio of the sound velocity to the projection of the Fermi velocity on the direction of the sound propagation.

Finally almost all the (nonlinear) problems considered in Section 2.8 include the stationary nonlinear Schrödinger equation (NLSE), that we write in the form

$$-\psi'' + v(x)\psi - \alpha|\psi|^2\psi = E\psi. \qquad (2.1.2.4)$$

The nonlinearity parameter α may have any sign (+ or −).

2.1.3. Statistics and Self-Averaging

The equations mentioned in the previous section have random coefficients. In what follows we shall always assume that all of them possess two properties: the translation invariance in the mean and the vanishing of correlations at infinitely distant points [122], [105].

The first property reflects the natural assumption that a macroscopically large disordered system is translation invariant in the mean. In microscopic terms it can be expressed, for instance, as the requirement that all the correlations, i.e., averages of the type

$$\langle v(\vec{r}_1)v(\vec{r}_2)\cdots v(\vec{r}_n)\rangle \qquad (2.1.3.1)$$

are invariant under translations of all the $\vec{r}_i$ by the same vector $\vec{a}$.

The second property, that is, the vanishing of statistical correlations, means that macroscopically distant parts of the system are statistically independent in some sense. More precisely it can be formulated in microscopical terms as the condition of asymptotic factorization of any correlation:

$$\lim_{|\vec{a}|\to\infty} \langle v(\vec{r}_1)\cdots v(\vec{r}_n)v(\vec{r}_{n+1}+\vec{a})\cdots v(\vec{r}_{n+m}+\vec{a})\rangle$$
$$= \langle v(\vec{r}_1)\cdots v(\vec{r}_n)\rangle\langle v(\vec{r}_{n+1})\cdots v(\vec{r}_{n+m})\rangle \qquad (2.1.3.2)$$

In mathematical terms these two properties mean that our coefficients are stationary random processes with good enough mixing properties.

All the widely used theoretical models (the white noise, random walk or Poisson potentials, the telegraph process, etc.) possess both these properties. To guarantee the validity of these properties in the experiment one should perform generally speaking special testing and even use special techniques while preparing solid samples. This may include neutron doping or annealing of radiation-induced defects with subsequent slow cooling that results in a statistically uniform distribution of defects in the sample [146], [133].

The important consequence of the general properties of disordered systems formulated above is the existence of self-averaging quantities (Lifshits [102]), which play a fundamental role in physics of disordered systems. The quantity f_V depending on the volume V of the disordered system is called self-averaging if

$$\lim_{V \to \infty} f_V = f \tag{2.1.3.3}$$

where f is a nonrandom quantity and convergence is to be understood in some conventional probabilistic sense, i.e., in probability, in variance, with probability 1, etc.

Examples of such quantities are provided, e.g., by densities of all extensive (asymptotically proportional to the volume of the system) physical quantities, in particular the density of one-particle states in a disordered system, the free energy density and therefore all the thermodynamic characteristics [102], as well as various kinetic ones (the electric conductivity [120], the decrement of the transmittivity of a one-dimensional system [124], [7], etc. The most obvious property of self-averaging quantities is the coincidence of their values at individual realizations with the mean value, which is the case for the infinite system. This means that calculated average values equal to those really observed on a sample.

The property of self-averaging is an analog of ergodicity (in particular the law of large numbers) in probability theory. Proceeding with this analogy one can show that for disordered systems with sufficiently short correlated randomness the fluctuations of self-averaging quantities have a Gaussian distribution centered at $\langle f_V \rangle$ and have a variance of the form

$$\langle f_V^2 \rangle - \langle f_V \rangle^2 = \frac{V_f}{V} \langle f_V \rangle^2 \tag{2.1.3.4}$$

where V_f is some characteristic (for a given quantity and the region of parameters) volume. Thus when $V \gg V_f$ the system becomes macroscopic with respect to the quantity f_V, while for $1 \ll V \ll V_f$ fluctuations of f_V that are sensitive to the individual realization of the system remain comparable to f_V. In many cases V_f is a microscopic volume.

However, in recent years an interesting concept has clearly emerged in the theoretical and experimental studies of quantum transport—the non-self-averaging of the conductance of samples in the mesoscopic size range. These are sizes large in the atomic scale but small enough for the electron wave function to have a well-defined phase over the entire sample. Because of this quantum coherence, statistical fluctuations of many important physical quantities become strongly correlated and do not average to zero on the mesoscopic (submicron at millikelvin temperatures) scale (see, e.g., books [64], [3]). It turns out that these fluctuations show remarkable universality: the variance of f_V is independent of the sample size and amount of disorder. This universality was discovered by perturbative calculations [2], [98] and supported by nonperturbative arguments, based on the multichannel analog of the Landauer formula (2.1.1.2) and a plausible hypothesis on the properties of random matrices of large order [72], [112].

For non-self-averaging quantities (as a rule dealt with in the theory of wave propagation in a random medium) their average values provide poor information for the description of their values on individual realizations. To relate these averages to experiments (that are usually carried out on individual realizations) one needs some special procedures (see, e.g., Section 2.6 below).

Some of these quantities have the relative fluctuations that increase unboundedly when the volume tends to infinity. The reason for such behavior is that the main contribution to the mean value of such a quantity f_V is made by exponentially few (in V) so-called representative realizations of the system. Each of them provides the value of f_V, which is exponentially large (in V) compared to that of typical (most probable) realizations. The balance of these two exponents defines the behavior of the quantity. In such a situation some additional information on the behavior of the random function for typical (but not representative) realizations may be obtained by studying their dependence on the self-averaging quantities. This approach starts from establishing the connection of the type $f_V = f(\gamma_V, V)$ where γ_V is a self-averaging quantity. When V increases to infinity γ_V tends to a nonrandom limit γ. Therefore the function $f(\gamma, V)$ ($\neq \langle f(\gamma_V, V) \rangle$) with large V characterizes at least qualitatively the behavior of f_V. This approach will be used further to study the reflectivity and transmittivity of a disordered segment (Sections 2.3 and 2.4) and the distribution of energy fluxes from the point source in a randomly layered medium (Section 2.7).

2.2. LOCALIZATION

2.2.1. The Localization Criteria

The localization in disordered systems means the existence of a macroscopically large number of localized states belonging to the point part of the spectrum of the corresponding random operator. In a system with a spatial cutoff the whole spectrum is pure point (discrete) and strictly speaking the notion of the localization is meaningless. It appears in a thermodynamic limit when V tends to infinity. Thus in order to discover the localization in the infinite system one should construct some quantities that are able to discriminate between the states condensing to the continuous part of the spectrum and those remaining discrete (or more precisely point) in the limit. We shall now show how it can be done restricting ourselves for the sake of simplicity by the multidimensional SE.

We note first that when $V \to \infty$ the states $\psi_{n,V}$ of the system normalized to the unit L^2 norm behave differently depending on their energy, namely, whether this energy belongs to the continuous or point part of the spectrum of the infinite system. The states of the first type tend to the localized states of the infinite system, while the others are of the order $V^{-1/2}$. One can use this difference to construct a quantity that detects the presence of a point spectrum in the neighborhood of a given energy E.

Consider for example the function

$$\sum_n \delta(E - E_{nV})|\psi_{nV}(\vec{r})|^\alpha \qquad (2.2.1.1)$$

where E_{nV} is the nth energy level of the system confined to a finite volume V, and α is a by now arbitrary exponent. The reader can easily see that if $\alpha > 2$ the states whose energy belongs to a continuous spectrum of the infinite system contribute nothing to the $V \to \infty$ limit of (2.2.1.1), since the respective terms are of the order $V^{-\alpha/2} = o(V^{-1})$, which is insufficient to form the integral sum. But if $\alpha = 2$ formula (2.2.1.1) is an expression for the local density of states at a point E which includes the sum over the discrete spectrum and the integral over the continuous one (when $V \to \infty$). Thus if (2.2.1.1) is positive for $\alpha > 2$ in the macroscopic limit then in the neighborhood of the energy E the point spectrum (hence localized states) is present. Otherwise the spectrum near E is continuous (or absent).

To calculate (2.2.1.1) with $\alpha > 2$ choose $\alpha = 4$ and consider the quantity

$$p(\vec{r}, \vec{r}'; E) = \sum_n \delta(E - E_n)|\psi_n(\vec{r})\psi_n(\vec{r}')|^2 \qquad (2.2.1.2)$$

where the summation is over the point part of the spectrum of an infinite system. (2.2.1.2) can be expressed in terms of the Green function $G = (E - H)^{-1}$ (H is the Hamiltonian defined by the l.h.s. of SE):

$$p(\vec{r}, \vec{r}'; E) = \lim_{\delta \to 0} \frac{\delta}{\pi}|G(\vec{r}, \vec{r}'; E - i\delta)|^2 \qquad (2.2.1.3)$$

Another convenient quantity is the integral

$$p(\vec{r}, \vec{r}') = \int_{-\infty}^{\infty} p(\vec{r}, \vec{r}'; E)\, dE = \sum_n |\psi_n(\vec{r})\psi_n(\vec{r}')|^2 \qquad (2.2.1.4)$$

representing the whole point spectrum of the system. It can also be expressed in terms of the (nonstationary) Green function $G = \exp(-itH)$ of the SE:

$$p(\vec{r}, \vec{r}') = \lim_{T \to \infty} \frac{1}{T} \int_0^T |G(\vec{r}, \vec{r}'; t)|^2\, dt. \qquad (2.2.1.5)$$

Formula (2.2.1.5) shows that $p(\vec{r}, \vec{r}')$ is the quantum mechanical probability for the particle starting from a point $\vec{r}$ at $t = 0$ to be found in the vicinity of the point $\vec{r}'$ in an infinite time. In particular, $p(\vec{r}, \vec{r})$ is the quantum mechanical probability for the particle to return to the initial point. It is clear that this quantity is closely related to the localization properties of a particle [6], [29].

The criteria of the localization formulated above deal with a single realization of the system. However, the quantities $p(\vec{r}, \vec{r}'; E)$ and $p(\vec{r}, \vec{r}')$ have an important property [121]: their averages as well as almost all individual realizations are simultaneously equal to or greater than zero. This property enables us to formulate the localization criteria describing the nature of the states in all typical realizations of the system, which is especially interesting. One such criterion states that the localization in the neighborhood of the point E exists if and only if the function (density-density correlator)

$$p(\vec{r}, ; E) = \langle p(\vec{r} + \vec{r}', \vec{r}'; E)\rangle = \left\langle \sum_n \delta(E - E_n)|\psi_n(\vec{r})\psi_n(0)|^2 \right\rangle \qquad (2.2.1.6)$$

is positive.

For other criteria for the existence of localized states (point spectrum) and the complete localization (pure point spectrum), see e.g., the review [109] and books [27], [125].

2.2.2. One-Dimensional Systems

This section contains some notation and technicalities used below.

Consider the system described by the SE with a random potential $v(x)$ in a region of high (compared with v) energy. Distinguishing the rapidly fluctuating exponents from the slowly varying amplitudes we can write the wave function in the form

$$\psi = \psi_+(x)e^{ikx} + \psi_-(x)e^{-ikx} \qquad (2.2.2.1)$$

where $k^2 = E$ is the particle momentum. To fix the factors $\psi_\pm$ we impose the following condition:

$$\psi'_+(x)e^{ikx} + \psi'_-(x)e^{-ikx} = 0 \qquad (2.2.2.2)$$

The coefficients $\psi_\pm$ may be considered as amplitudes of the forward and backward movement of the particle. Substituting (2.2.2.1)–(2.2.2.2) in the SE one obtains the Dirac type equation for the two-component vector $\vec{\psi} = \left(\begin{smallmatrix}\psi_+\\\psi_-\end{smallmatrix}\right)$ (cf. (2.1.2.2)):

$$-i\sigma_3\vec{\psi}' + \frac{v}{2k}\vec{\psi} + \frac{v}{2k}(\cos 2kx \cdot \sigma_1 + \sin 2kx \cdot \sigma_2)\vec{\psi} = 0 \qquad (2.2.2.3)$$

Note that only the two last terms proportional to Pauli matrices $\sigma_{1,2}$ mix the amplitudes $\psi_\pm$ and therefore are responsible for the scattering and localization.

Sometimes it is convenient to work with the complex mixture of $\psi_\pm$:

$$\psi_{1,2}(x) = \frac{1}{2ik}[ik\psi(x) \pm \psi'(x)] \qquad (2.2.2.4)$$

and study the vector $\vec{\psi} = \left(\begin{smallmatrix}\psi_1\\\psi_2\end{smallmatrix}\right)$ satisfying the equation

$$\vec{\psi}' = \hat{\mathbf{M}}\vec{\psi} \qquad (2.2.2.5)$$

where

$$\hat{\mathbf{M}} = \left(ik + \frac{v(x)}{2ik}\right)\sigma_3 + \frac{1}{2k}\sigma_2 \qquad (2.2.2.6)$$

Multiplying (2.2.2.5) by $-i\sigma_3$ and using (2.2.2.6) one can transform it to the form

$$-i\sigma_3\vec{\psi}' + \frac{1}{2k}\sigma_1\vec{\psi} + \frac{v}{2k}\vec{\psi} = k\vec{\psi} \qquad (2.2.2.7)$$

very similar to the DE (2.1.2.2) with $\Delta = 1/2k$, $v_0(x) = v/2k$, $E = k$, and $v_{1,2,3} \equiv 0$. The unique but important difference is that in (2.2.2.7) both the gap and the random potential explicitly depend upon the spectral parameter.

The current density j for the SE has the form

$$j = 2|\psi|^2 \, \mathrm{Im}(\psi'/\psi) \qquad (2.2.2.8)$$

and satisfies the current conservation law $j \equiv$ const.

If the system is closed all the states must satisfy a self-adjoint boundary condition that leads to the absence of the current ($j \equiv 0$). In this case it is natural to introduce the polar coordinates in the (ψ', ψ) plane as follows:

$$\frac{\psi'}{\psi} = z = \mathrm{ctg}\,\alpha \qquad (2.2.2.9)$$

with

$$\psi = e^{\Xi} \sin\alpha$$
$$\psi' = e^{\Xi} \cos\alpha \qquad (2.2.2.10)$$

These variables satisfy the equations

$$\alpha' = 1 - (1 + v - E)\sin^2\alpha$$
$$\Xi' = \frac{1}{2}(1 + v - E)\sin 2\alpha \qquad (2.2.2.11)$$

All the above transformations may by applied with appropriate modifications to the other models mentioned in Section 2.1. In the case of DE we should represent the wave function as the sum

$$\vec{\psi}(x) = A_+(x)\vec{\psi}_+^{(0)}(x) + A_-(x)\vec{\psi}_-^{(0)}(x), \qquad (2.2.2.12)$$

where $\vec{\psi}_\pm^{(0)}(x) \propto e^{\pm ipx}$ are the solutions of DE in free space ($v_i \equiv 0$) and $A_\pm(x)$ are slowly varying amplitudes. The corresponding equation for the pair $A_\pm$ is also the DE and only the terms proportional to $\sigma_{1,2}$ are responsible for the localization.

In the case of the MDE with $\beta > 1$ the free particle cannot move in both directions. Thus the localization may exist only if $0 < \beta < 1$.

The current density for NLSE coincides with that for the SE (2.2.2.8), and for the rest of equations has the form

$$\text{(DE)} \qquad j = (\vec{\psi}, \sigma_3\vec{\psi}), \qquad (2.2.2.13)$$

and

$$\text{(MDE)} \qquad j = (\vec{\psi}, (\sigma_3 - \beta)\vec{\psi}) \qquad (2.2.2.14)$$

The analog of to (2.2.2.9) for MDE has the form

$$\vec{\psi} = e^{\Xi - i\Psi} \begin{pmatrix} \frac{1}{\sqrt{1-\beta}} \\ \frac{e^{e\Phi}}{\sqrt{1+\beta}} \end{pmatrix} \qquad (2.2.2.15)$$

and that for the DE coincides with (2.2.2.15) with $\beta = 0$.

The variables Ξ and Φ obey the equations following from MDE:

$$(1 - \beta^2)\Xi'(x) = E_0 \sin\Phi(x)$$
$$\frac{1 - \beta^2}{2}\Phi'(x) = v(x) - E + E_0 \cos\Phi(x) \qquad (2.2.2.16)$$

where $E_0 = \Delta\sqrt{1-\beta^2}$. For the DE we have

$$\Xi' = (v_1 + \Delta)\sin\Phi - v_2\cos\Phi,$$

$$\frac{1}{2}\Phi' = v_0 - E + (\Delta + v_1)\cos\Phi + \sin\Phi. \qquad (2.2.2.17)$$

2.2.3. The Lyapunov Exponent. Definition and Positivity

The variables Ξ and α (Φ for the DE) introduced in the previous subsection describe the growth (decay) of the solutions and their oscillations, respectively. Since the localization is connected with the decay of the solution we are mainly interested in Ξ. Under the assumptions made in Section 2.1 (see (2.1.3.1), (2.1.3.2)) on the random potential one can prove that for any fixed energy with probability 1 there exists a limit

$$\gamma(E) = \lim_{x\to\infty}\frac{\Xi(x)}{x} = \lim_{x\to\infty}\frac{\langle\Xi(x)\rangle}{x} \qquad (2.2.3.1)$$

that does not depend on an individual realization of the system (that is of the random potential). This limit is known as the Lyapunov exponent, see, e.g., [105] or [125]. Recalling the notion of self-averaging (see Section 2.1.3) we can state that the ratio Ξ/x has this property.

It follows from the current conservation law that $\gamma(E) \geq 0$. Our goal in this section is to prove that if the random potential is the ergodic Markov process then $\gamma(E) > 0$ for all E.

This fact is closely connected to the localization phenomena. To make it clear consider the closed disordered system of a finite length and the two solutions of a corresponding equation (any of (2.1.2.1)–(2.1.2.3)) with the same value of the spectral parameter E but satisfying the boundary conditions on the ends of the interval occupied by the system. The positivity of $\gamma(E)$ leads to the exponential growth of both the solutions while moving from the endpoints to the middle of the interval. The value E belongs to the spectrum if these solutions match at some inner point of the interval and in that case we obtain an exponentially localized state.

To proceed we must obtain some expression for $\gamma(E)$. Usually the joint probability density of the pair $(\alpha(x), v(x))$ tends to x-independent (stationary) limit as x tends to infinity. Using this and the dynamic equation (2.2.2.11) by (2.2.2.10) one can derive the formula

$$\gamma(E) = \frac{1}{2}\langle(1 + v - E)\sin 2\alpha\rangle_{\text{st}} \qquad (2.2.3.2)$$

were $\langle\cdot\rangle_{\text{st}}$ denotes the averaging over the joint stationary distribution of α and v.

The analog of this relation for the DE has the form

$$\gamma(E) = \langle(\Delta + v_1)\sin\Phi - v_2\cos\Phi\rangle_{\text{st}} \qquad (2.2.3.3)$$

(Note that only $v_{1,2}$ are present!)

To prove the positivity of the Lyapunov exponent we need more convenient expressions than (2.2.3.2) and (2.2.3.3). One can obtain them under the assumption that

$v(x)$ ($v_{1,2}(x)$ in the DE case) is a function of a Markov process. In this case one can write the Fokker–Planck-type equation

$$\frac{\partial}{\partial \alpha}\{[1 - (1 + v - E)\sin^2 \alpha]p\} + L_v p = 0 \qquad (2.2.3.4)$$

for the joint stationary probability density $p(\alpha, v)$. The operator L_v here is the infinitesimal operator of the Markov process $v(x)$ and acts on the v-variable only. The Fourier transform of p with respect to α

$$p_m(v) = \int_0^\pi p(\alpha, v)e^{-2im\alpha}\,d\alpha \qquad (2.2.3.5)$$

satisfies the equation

$$(1 - v + E)p_m + \frac{1}{2}(1 + v - E)(p_{m-1} + p_{m+1}) - \frac{i}{m}L_v p_m = 0, \quad (2.2.3.6)$$

and in view of (2.2.3.2) the Lyapunov exponent in terms of p_m is

$$\gamma(E) = \frac{1}{2}\int (1 + v - E)\,\mathrm{Im}(p_{-1}(v))\,dv. \qquad (2.2.3.7)$$

Multiplying (2.2.3.6) by $p_m^* = p_{-m}$ and summing over $m > 0$ we obtain finally

$$\gamma(E) = \sum_{m=1}^\infty \frac{1}{m}\int \frac{p_m^*(v)L_v p_m(v)}{p_0(v)}\,dv. \qquad (2.2.3.8)$$

The same arguments when applied to the DE lead to

$$\gamma(E) = \int [(v_1 + \Delta)\,\mathrm{Im}(p_{-1}(\vec{v})) - v_2\,\mathrm{Re}(p_{-1}(\vec{v}))]\,d\vec{v} \qquad (2.2.3.9)$$

where $\vec{v} = (v_0, v_1, v_2)$ and

$$p_m(\vec{v}) = \int_0^{2\pi} p(\Phi, \vec{v})e^{-im\Phi}\,d\Phi \qquad (2.2.3.10)$$

The corresponding Fokker–Planck equation has the form

$$2\frac{\partial}{\partial \Phi}\{[v_0 - E + (\Delta + v_1)\cos\Phi + v_2\sin\Phi]p(\Phi, \vec{v})\} + L_{\vec{v}}p(\Phi, \vec{v}) = 0 \qquad (2.2.3.11)$$

with $L_{\vec{v}}$ corresponding to the Markov process $\vec{v}$.

Owing to the inequality $|p_m| \leq p_0(\vec{v})$ functions $p_m(\vec{v})$ may be treated as vectors of the Hilbert space $\mathcal{L}_{\vec{v}}$ with the scalar product

$$(f, g) = \int \frac{f(\vec{v})g^*(\vec{v})}{p_0(\vec{v})}\,d\vec{v}. \qquad (2.2.3.12)$$

If the Markov process $\vec{v}(x)$ is statistically isotropic, then the probability density at points $\{\vec{v}(0) = \vec{v}, \vec{v}(x) = \vec{v}'\}$ and $\{\vec{v}(0) = \vec{v}', \vec{v}(x) = \vec{v}\}$ coincide and operator $L_{\vec{v}}$ is positive in the space $\mathcal{L}_{\vec{v}}$, that is

$$(L_{\vec{v}}f, f) \geq 0. \qquad (2.2.3.13)$$

To prove (2.2.3.13) note that the kernel of the operator $\exp\{-tL_{\vec{v}}\}$ is the transition probability density $p(x \mid \vec{v}, \vec{v}')$. The statistical isotropy of the process $\vec{v}(x)$ leads to the symmetry of the operator $L_{\vec{v}}$, and to the inequality $\| \exp\{-tL_{\vec{v}}\}\| \leq 1$ that is equivalent to the positivity of $L_{\vec{v}}$.

In addition note that since we assume the Markov process $\vec{v}(x)$ to be ergodic, zero is a simple eigenvalue of $L_{\vec{v}}$ and the corresponding eigenfunction is the stationary probability density $p_0(\vec{v})$.

Representation (2.2.3.8) and properties of the operator $L_{\vec{v}}$ imply that the Lyapunov exponent $\gamma(E)$ is positive for all E. Indeed if $\gamma(E) = 0$ then all the functions $p_m(\vec{v})$ are proportional to $p_0(\vec{v})$. Then $p(\vec{v}, \Phi)$ has the form

$$p(\vec{v}, \Phi) = F(\Phi)p_0(v)$$

where $F(\Phi)$ satisfies the equation (following from (2.2.3.11))

$$\frac{\partial}{\partial \Phi}\{[v_0 - E + (\Delta - v_1)\cos\Phi + v_2\sin\Phi]F(\Phi)\} = 0$$

and that is impossible.

2.2.4. The Lyapunov Exponent. Simplest Asymptotics

The calculation of $\gamma(E)$ in some limiting cases is an interesting problem. In this section we give some examples of such calculations in the high-energy case.

We start with considering the SE. Introducing a representation that is convenient in this case (cf. (2.2.2.10) and (2.2.2.15))

$$\psi = \begin{pmatrix} \psi_+ \\ \psi_- \end{pmatrix} = \frac{1}{2i}\begin{pmatrix} e^{\xi+i\theta} \\ e^{\xi-i\theta} \end{pmatrix} \tag{2.2.4.1}$$

we obtain the following equations for variables θ and ξ

$$\theta' = \frac{v(x)}{2k}(-1 + \cos 2(\theta + kx))$$

$$\xi' = \frac{v(x)}{2k}(\sin 2(\theta + kx)) \tag{2.2.4.2}$$

with $k^2 = E$. In the high-energy case we can use v/k^2 as a small parameter and obtain in leading order the following expression for the Lyapunov exponent:

$$\gamma(E) = \frac{1}{4E}\int_0^\infty \langle v(x)v(0)\rangle \cos 2kx \, dx. \tag{2.2.4.3}$$

If the correlation radius r_c of the potential satisfies the condition $kr \ll 1_c$ then it follows from (2.2.4.3) that

$$\gamma(E) = \frac{1}{2l(E)} = \frac{2D}{E} \tag{2.2.4.4}$$

where $l(E)$ is the so-called localization length and

$$D = \int_0^\infty \langle v(x)v(0) \rangle \, dx. \tag{2.2.4.5}$$

Formula (2.2.4.4) holds for the δ correlation limit, that is, when

$$\langle v(x)v(0) \rangle = 2D\delta(x) \tag{2.2.4.6}$$

and coincides with that for the white noise potential.

The Helmholtz equation

$$u'' + k^2[1 + \varepsilon(x)]u = 0$$

describing the propagation of electromagnetic wave in the media with the dielectric constant $1 + \varepsilon(x)$ can be reduced to the SE (2.1.2.1) by the change of variables $u \to \psi$, $v(x) \to -k^2\varepsilon(x)$. Therefore if the dielectric constant fluctuations $\varepsilon(x)$ are described by the Gaussian white noise with

$$\langle \varepsilon(x) \rangle = 0, \qquad \langle \varepsilon(x)\varepsilon(x') \rangle = 2d\delta(x - x'),$$

the corresponding coefficient D in (2.2.4.6) equals

$$D = dk^4,$$

and the localization length

$$l(k^2) = \frac{2}{dk^2} \tag{2.2.4.7}$$

being inversely (instead of directly) proportional to $E = k^2$.

In the case of the white noise potential and a high-energy limit one can calculate not only the Lyapunov exponent $\gamma(E)$, but the whole probability density of the quantity

$$\gamma(x) = \xi(x)/x$$

which is equal to $\gamma(E)$ in the limit $x \to \infty$.

The simplest way to find this density is to use the so-called resonance approximation and some version of the method of averaging over the fast variable formalism [105].

Let the potential be the white noise with zero mean and the correlation function be given by (2.2.4.6). It can be considered as small if $D^{2/3} \ll E$. The structure of the equation (2.2.2.3) leads to the conclusion that the dynamics of the system is influenced mainly by the Fourier components with wave numbers close to 0 or to $\pm 2k$. Therefore we can make "the resonance approximation" that is to consider the potential of the form

$$v(x) = v_0(x) + v_2(x)e^{2ikx} + v_2^*(x)e^{-2ikx} \tag{2.2.4.8}$$

with slowly varying $v(x)$, $v_2(x)$, and $v_2^*(x)$. If we substitute (2.2.4.8) into (2.2.4.2), average the fast motion, and use the perturbation theory, we shall obtain the closed equation for $\xi(x)$:

$$\xi' = \frac{1}{4k}U(x) + \frac{1}{2l(E)}. \tag{2.2.4.9}$$

Here $U(x)$ is the white noise potential with the correlation function

$$\langle U(x)U(0)\rangle = 4D\delta(x),$$

and $l(E)$ is defined by (2.2.4.4).

The Fokker–Planck equation for the probability density $p(\xi, x)$ has then the form

$$\frac{\partial p}{\partial x} + \frac{1}{2l} \cdot \frac{\partial p}{\partial \xi} - \frac{1}{4l} \cdot \frac{\partial^2 p}{\partial \xi^2} = 0. \qquad (2.2.4.10)$$

Its solution satisfying the initial condition $p(\xi, 0) = \delta(\xi)$ is the Gaussian distribution with mean $x\gamma(E)$ and variance $x/2l$. Introducing the variable $\gamma = \xi/x$ we obtain

$$p(\gamma, x) = \sqrt{\frac{x}{2\pi\gamma(E)}} \exp\left\{-\frac{x}{2\gamma(E)}(\gamma - \gamma(E))^2\right\}. \qquad (2.2.4.11)$$

This formula clearly shows that $\gamma(x)$ self-averages as $x \to \infty$.

The similar results can be obtained for the DE. In the simplest case with $v_{1,2,3} = 0$ passing to the $\vec{A} = (A_+, A_-)$ variables (see (2.2.2.12)) we obtain the equation

$$-i\sigma_3\vec{A}' + \frac{v_0(x)}{p}\left[E\sigma_0 + \Delta(\cos 2px \cdot \sigma_1 + \sin 2px \cdot \sigma_2)\right]\vec{A} = 0. \qquad (2.2.4.12)$$

Note that the terms mixing the components $A_\pm$ in (2.2.4.12) coincide with those in (2.2.2.3) if we substitute $v_0(x)\Delta/p$ by $v(x)/2k$ (here $p = \sqrt{E^2 - \Delta^2}$ plays the role of k in (2.2.2.3)). The analog of (2.2.4.3) for this equation is also valid:

$$\gamma(E) = \frac{\Delta^2}{p^2} \int_0^\infty \langle v_0(x)v_0(0)\rangle \cos 2px\, dx. \qquad (2.2.4.13)$$

If $pr_c \ll 1$ (r_c being the correlation radius) then (2.2.4.13) yields the analog of (2.2.4.4):

$$\gamma(E) = \frac{d\Delta^2}{p^2} = \frac{d\Delta^2}{E^2 - \Delta^2} \qquad (2.2.4.14)$$

where

$$d = \int_0^\infty \langle v_0(x)v_0(0)\rangle\, dx. \qquad (2.2.4.15)$$

The polar parametrization for $\vec{A}$ has the form (cf. (2.2.4.1))

$$\vec{A} = \begin{pmatrix} \exp[\xi - i(\psi + \varphi + px)] \\ \exp[\xi - i(\psi - \varphi - px)] \end{pmatrix} \qquad (2.2.4.16)$$

with φ and ξ satisfying the equations

$$\varphi' = -p + \frac{v_0(x)}{p}(E + \Delta\cos 2\varphi)$$

$$\xi' = \frac{v_0(x)\Delta}{p}\sin 2\varphi. \qquad (2.2.4.17)$$

The formalism used above to derive (2.2.4.11) when applied to these equations gives the same formula for probability density with $\gamma(E)$ defined by (2.2.4.14).

This result in more general case of MDE was obtained in [19].

2.2.5. The Concentration Expansion of the Lyapunov Exponent

In physical applications, in particular in solid state physics, the usual sources of disorder are randomly positioned impurities of various types. One of the main characteristics of these impurities is their concentration. Often it can be used as a small parameter in various expansions.

The potential generated by the system of such impurities has the form

$$v(x) = \sum_j u(x - x_j) \qquad (2.2.5.1)$$

where $u(x)$ is the potential of a single isolated impurity and random points x_j are the coordinates of impurities.

In the case of a short-range (compared with the wavelength of the particle) electron–impurity interaction it can be approximated by the sum of δ-functions, and (2.2.5.1) takes the form

$$v(x) = \sum_j k_j \delta(x - x_j) \qquad (2.2.5.2)$$

with $k_j = k_0 > 0$. This leads to one of the most popular models of the one-dimensional disordered system known as the Frish–Lloyd model [50]. Random points x_j are supposed to form the Poisson point process, that is the distances $y_j = x_{j+1} - x_j$ are independent identically distributed random variables with the density

$$p(y) = a^{-1} \exp\{-a^{-1}y\}. \qquad (2.2.5.3)$$

We shall construct now the low-concentration expansion for the Lyapunov exponent following [104]. Rewrite the definition of the Lyapunov exponent (2.2.3.1) and (2.2.2.10) in the form

$$\Xi(x) = \int_0^x z(x')dx' + \ln|\psi(0)| + \frac{1}{2}\ln(1 + z^2(x)), \qquad (2.2.5.4)$$

where the logarithmic derivative $z(x)$ of ψ (see (2.2.2.9)) solves the Ricatti equation

$$z'(x) = -(z^2 + E) + v(x). \qquad (2.2.5.5)$$

Since the probability density of z becomes stationary when $x \to \infty$ we can write (2.2.3.2) for $\gamma(x)$ in the limit $x = \infty$ as follows:

$$\gamma(x) = \langle z \rangle_{\text{st}}. \qquad (2.2.5.6)$$

Let $k^2 = E$ and ζ be $z \cdot k^{-1}$. Then the probability density $p(\zeta)$ satisfies the equation

$$(\zeta^2 + 1)p(\zeta) = c \int_{\zeta - \zeta_0}^{\zeta} p(\zeta')\, d\zeta' + \nu \qquad (2.2.5.7)$$

where $\zeta_0 = k_0/k$, $c = (k_0 a)^{-1}$ is a dimensionless concentration, $\nu = k^{-1}N(E)$ and $N(E)$ is the integrated density of states, i.e., the number of states with energy less than E per unit length of the system. We shall construct a solution of (2.2.5.7) as a perturbation series in $(c\zeta_0)$

$$p(\zeta) = \nu \sum_{m=0}^{\infty} (c\zeta_0)^m f_m(\zeta) \tag{2.2.5.8}$$

with

$$f_0(\zeta) = (\zeta^2 + 1)^{-1}.$$

From (2.2.5.7) we obtain the recurrence formulas

$$f_m(\zeta) = \varphi'(\zeta) \int_{\zeta-\zeta_0}^{\zeta} f_{m-1}(\zeta') \, d\zeta', \qquad m > 0 \tag{2.2.5.9}$$

with $\varphi'(\zeta) = f_0(\zeta)$ and (see above $f_0(\zeta)$) $\varphi(\zeta) = \arctan \zeta + \pi/2$. Then for the Lyapunov exponent $\gamma = k\langle\zeta\rangle$ and the integrated density of states $N(E)$ we have

$$\gamma(E) = k\nu \sum_{m=1}^{\infty} B_m (c\zeta_0)^m = k \frac{\sum_{m=1}^{\infty} B_m (c\zeta_0)^m}{\sum_{m=0}^{\infty} A_m (c\zeta_0)^m}, \tag{2.2.5.10}$$

$$kN^{-1}(E) = \nu^{-1} = \sum_{m=0}^{\infty} A_m (c\zeta_0)^m, \tag{2.2.5.11}$$

with

$$A_m = \int_{-\infty}^{\infty} f_m(\zeta) \, d\zeta, \qquad B_m = \int_{-\infty}^{\infty} \zeta f_m(\zeta) \, d\zeta. \tag{2.2.5.12}$$

From (2.2.5.8), (2.2.5.9) it follows that

$$0 \le f_m(z) \le \varphi'(\zeta) \int_{-\infty}^{\zeta} f_{m-1}(\zeta') \, d\zeta', \tag{2.2.5.13}$$

and from this inequality one readily derives that

$$0 \le f_m(z) \le \varphi'(\zeta) \frac{\varphi^m(\zeta)}{m!}. \tag{2.2.5.14}$$

This inequality shows that series (2.2.5.8) converges for any value of parameter $c\zeta_0$, that is for any c and ζ_0. Moreover, using (2.2.5.13) and the recurrence (2.2.5.9) we obtain the bounds for A_m and B_m:

$$0 \le A_m \le \frac{\pi^{m+1}}{(m+1)!}, \qquad |B_m| \le \frac{\zeta_0 \pi^m}{2m!}. \tag{2.2.5.15}$$

These bounds show that the series in (2.2.5.11) converges with the speed of an exponent, so its sum $kN(E)^{-1}$ is the entire analytic function of c for any $E > 0$. Since all the $A_j > 0$, the integrated density of states $N(E)$ is real analytic in $E > 0$. As for $\gamma(E)$, formula (2.2.5.10) shows that it is analytic whenever its denominator differs from zero. This is the case at least for $c\zeta_0 > 0$, so $\gamma(E)$ is real analytic on the semiaxis $c\zeta_0 > 0$.

To obtain some explicit formulas let us write several first terms in the expansions (2.2.5.10) and (2.2.5.11):

$$\gamma(E) = ck_0 \left[\frac{B_1}{A_0} + c\zeta_0 \left(\frac{B_2}{A_0} - \frac{B_1}{A_0} \cdot \frac{A_1}{A_0} \right) + O(c^2) \right] \qquad (2.2.5.16)$$

$$N(E) = N(E_0) \left\{ 1 - c\zeta_0 \frac{A_1}{A_0} - c^2\zeta_0^2 \left[\frac{A_2}{A_0} - \left(\frac{A_1}{A_0} \right)^2 \right] + O(c^3) \right\} \qquad (2.2.5.17)$$

where $N_0(E) = \pi^{-1}E^{1/2}$, and the coefficients A_0, A_1, B_1, and B_2 are defined as follows:

$$A_0 = \pi;$$
$$A_1 = \pi \cdot \arctan(\zeta_0/2);$$
$$B_1 = \frac{\pi}{2} \ln \left(1 + \frac{\zeta_0^2}{4} \right)$$
$$B_2 = \pi \int_0^{\zeta_0} \left\{ \zeta \arctan \frac{\zeta_0 + \zeta}{2} + \ln \left[4 + (\zeta + \zeta_0)^2 \right] \right\} \frac{d\zeta}{\zeta^2 + 4}$$
$$- \frac{\pi}{2} \arctan \left(\frac{\zeta_0}{2} \right) \ln(4 + \zeta_0^2) \qquad (2.2.5.18)$$

The expression for the A_2 is more cumbersome.

For $E \gg k_0^2/c$ (that is $c \gg \zeta_0^2$) we have then explicit formulas:

$$\gamma(E) = \frac{ck_0^3}{8E} \left[1 - \frac{k_0^2}{8E}(1 - 8c) \right], \qquad (2.2.5.19)$$

$$N(E) = \frac{\sqrt{E}}{\pi} \left\{ 1 - \frac{ck_0^2}{2E} \left[1 - \frac{k_0^2}{12E}(1 - 3c) \right] \right\}. \qquad (2.2.5.20)$$

Equalities (2.2.5.16)–(2.2.5.18) give the first terms of the expansions in powers of the concentration c and fixed energy E of the Lyapunov exponent γ and the number of states N. Formulas (2.2.5.19) and (2.2.5.20) include the dependence on the inverse energy (with large E) explicitly.

As $E \to 0$ (and $\zeta_0 \to \infty$) the convergence of the series (2.2.5.10) for $\gamma(E)$ deteriorates. This is caused firstly by the fact that the expansion parameter is not c but $c\zeta_0$, and secondly by the dependence of the coefficients B_m on ζ_0 (for instance, $B_1 \propto \ln \zeta_0$ when $\zeta_0 \to \infty$). Thus one may expect some singularities at $E = 0$ in the Lyapunov exponent considered as a function of c and E. We demonstrate this effect in the exactly solvable case when the coefficients k_j in (2.2.5.2) are not constants but a set of identically distributed independent random variables with the same probability density

$$p(k) = k_0^{-1} \exp(-k/k_0)$$

The equation for the probability density $p(z)$ then has the form

$$\frac{d}{dz}[z^2 p(z)] = \frac{1}{a}\left(p(z) - \int_0^\infty (z - k)\exp\left(-\frac{k}{k_0}\right)\frac{dk}{k_0}\right). \quad (2.2.5.21)$$

According to (2.2.5.5) we have in this case $p(z) = 0$ for all $z < 0$. Together with the obvious equality $N(0) = 0$ this fact enables us to rewrite (2.2.5.21) in the form

$$z^2 \exp\left(\frac{z}{k_0}\right) p(z) = \frac{1}{a}\int_0^z \exp\left(\frac{t}{k_0}\right) p(t)\, dt, \quad (2.2.5.22)$$

and solve it explicitly:

$$p(z) = A z^{-2}\exp\left(-\frac{z}{k_0} - \frac{1}{az}\right). \quad (2.2.5.23)$$

Here A is a normalization constant. Substituting (2.2.5.23) in (2.2.5.6) we find that for $c \ll 1$

$$\gamma(0) \propto -c k_0 \ln c. \quad (2.2.5.24)$$

Thus we can see explicitly the logarithmic singularity of γ at zero concentration and zero energy.

2.2.6.   Localization. Brief Survey of Results

Let us consider a sufficiently long interval (a, b), $-\infty < a < b < \infty$, and the boundary value problem for the SE on it:

$$\psi(a)\cos\alpha - \psi'(a)\sin\alpha = 0 \quad (2.2.6.1)$$

$$\psi(b)\cos\beta - \psi'(b)\sin\beta = 0 \quad (2.2.6.2)$$

The spectrum $\{E_j\}$ of this problem is discrete and can be found following the traditional scheme (see, e.g., [28]) by substituting the solution $y_a^{(\alpha)}(x, E)$ of the Cauchy problem

$$y(a, E) = \sin\alpha$$
$$y'(a, E) = \cos\alpha$$

for the SE that evidently satisfies (2.2.6.1) into (2.2.6.2). We shall modify this traditional scheme and consider two Cauchy problems determined by the conditions at each end of the interval (a, b). Denoting their solutions by $y_a^{(\alpha)}(x, E)$ and $y_b^{(\beta)}(x, E)$, we shall find the eigenvalues from the condition that both the solution and their derivatives coincide at zero:

$$\left(y_a^{(\alpha)}\frac{dy_a^{(\beta)}}{dx} - y_b^{(\beta)}\frac{dy_b^{(\alpha)}}{dx}\right)\Bigg|_{x=0} = 0. \quad (2.2.6.3)$$

From the relations (2.2.3.1) and (2.2.2.10) it follows that if $\gamma(E) > 0$, $E \in (E_1, E_2)$, then envelopes $(y^2 + y'^2)^{1/2}$ of both the solutions for these E grow exponentially with probability 1 from the endpoints of the interval (a, b) to its middle. Thus the corresponding eigenfunctions must decay exponentially from the middle of the interval to its endpoints. If such behavior is preserved in the limit $a \to -\infty$, $b \to \infty$, then in the energy interval (E_1, E_2) we should obtain a pure point spectrum with eigenfunction decaying at infinity as $\exp\{-\gamma(e)|x|\}$.

A pure point character of the spectrum of one-dimensional second-order operators with random coefficients was first conjectured in [115] on the base of heuristic arguments similar in form to those outlined above. These arguments, however, are not complete at several points (see, e.g., the books [27] and [125] for their detailed analysis). One of the formal reasons for their incompleteness is that the exponential growth of the solutions of Cauchy problem for each E with probability 1 implies such a growth in almost all realizations not for all but only for almost all energies with respect to the Lebesgue measure. In the same time the (discrete) spectrum of the problem has zero Lebesgue measure so eigenvalues may as well lie in the exceptional set where eigenfunctions do not decay.

In fact the positivity of the Lyapunov exponent guarantees only the absence of the absolutely continuous part of the spectrum, but leaves place for its singular (not only point!) part. However, the arguments given and examples of the SE with quasiperiodic potential having a pure singular continuous spectrum [8] show that the more independent and smoothly distributed the parts of potential are the more unlikely eigenvalues fall in the exceptional set where we cannot guarantee the exponential decay of the SE solutions. These properties of the random potential seem important for the existence of the point spectrum (of solutions decaying as $|x| \to \infty$) also from the viewpoint of quantum mechanics, since it diminishes the probability of the so-called resonance tunneling that can arise when two similar patterns are present on the potential graph. This effect displays itself in the existence of solutions having amplitudes of the same order on similar patterns independently of the distance between them. Thus such a solution does not decay at infinity.

The first rigorous proof of the exponential localization for all the energies in the spectrum of the one-dimensional SE (i.e., pure point character of its spectrum and the exponential decay of respective eigenfunctions) was given in [57] for the case of the ergodic smoothly distributed Markov potential. The authors used the "two-sided" scheme outlined above for the construction of eigenfunctions and proved the positivity of (2.2.1.1) for all $\alpha \in (1, 2]$, which is sufficient for the spectrum to be the pure point one. However, earlier in [15] a perturbation method of calculation of physical quantities for the one-dimensional SE with the "white noise" potential (2.2.4.6) in the high energy or weak potential regimes was developed and, in particular the complete localization in these regimes on the physical level of rigor was proved. An elegant and simple proof of the complete exponential localization for all energies lying in the spectrum of the SE with short correlated and smoothly distributed potentials was proposed in [91].

These proofs are based essentially on the positivity of the Lyapunov exponent of the corresponding equation. Using the well-known expression of the Green function of the SE via its solutions (e.g., [28]) it can be easily shown that the positivity of the Lyapunov exponent implies the exponential decay of the Green function at fixed energy with probability 1. It turns out that this fact and the smoothness of the probability distribution of the potential guarantee the exponential localization [34], [137]. Since the Green function can be defined for any dimension of the space, one of the ways to prove the exponential localization in the multidimensional case is to establish the exponential decay of the Green function. This was proven in an important paper [51] for the large disorder regime. The authors developed powerful perturbation method combining ideas of the KAM-theory and the renormalization group technique. For the modern form of their approach see the book [125] and references therein. In this article we only mention that the adaptation of the ideas, developed in the study of the multidimensional localization, allowed in [26] to prove the exponential localization for all energies in the one-dimensional SE whose potential being short correlated may not be smoothly distributed (its probability distribution may have atoms or even be purely discrete, as, for example, for the independent identically distributed random potential, assuming only two values in the discrete analog of the SE).

2.3. TRANSMISSION COEFFICIENT

2.3.1. Formulas

Consider the SE (2.1.1.1) with the potential $v(x)$ that differs from zero only in the interval $[0, L]$ and introduce two basic scattering states:

$$\psi_1(x) = \begin{cases} e^{ikx} + r_+ e^{-ikx}, & x < 0 \\ t_+ e^{ik(x-L)}, & x > L; \end{cases} \tag{2.3.1.1}$$

$$\psi_2(x) = \begin{cases} t_- e^{-ikx}, & x < 0 \\ e^{-ik(x-L)} + r_- e^{ik(x-L)}, & x > L. \end{cases} \tag{2.3.1.2}$$

Since the Wronskian

$$W(\chi, \eta) = \chi' \eta - \chi \eta' \tag{2.3.1.3}$$

of any two solutions of the SE with the same energy does not depend upon x, we can get some identities for transmission and reflection amplitudes substituting ψ's instead of χ and η—namely,

$$\begin{aligned} t_- &= t_+ \equiv t, \\ |r_\pm|^2 + |t_\pm|^2 &= 1, \\ -t^* r_+ &= t r_-^*. \end{aligned} \tag{2.3.1.4}$$

The second relation here is just the current conservation law since according to (2.2.2.8) the current density for the SE is

$$j = -i(\psi^* \psi' - \psi^{*'} \psi) = -i W(\psi, \psi^*). \tag{2.3.1.5}$$

The scattering problem may be also described in terms of the transfer-matrix $\hat{T}$ defined as follows: for the general solution $\psi(x)$ of the SE we have outside the support of the potential

$$\psi(x) = \begin{cases} Ce^{ikx} + De^{-ikx}, & x < 0 \\ Ae^{ik(x-L)} + Be^{-ik(x-L)}, & x > L; \end{cases} \qquad (2.3.1.6)$$

with some constants A, B, C, and D, depending upon ψ. The linearity of the SE implies

$$\begin{pmatrix} C \\ D \end{pmatrix} = \hat{T} \begin{pmatrix} A \\ B \end{pmatrix} \qquad (2.3.1.7)$$

where a matrix $\hat{T}$ does not depend upon the individual solution ψ. It is called the transfer-matrix (in the basis of running waves) propagating the solution to the left. The propagation to the right is naturally described by the inverse matrix:

$$\begin{pmatrix} A \\ B \end{pmatrix} = \hat{T}^{-1} \begin{pmatrix} C \\ D \end{pmatrix} \qquad (2.3.1.8)$$

Both the matrixes may be expressed in terms of reflection and transmission amplitudes t and r:

$$\hat{T} = \begin{pmatrix} \frac{1}{t_*} & -\frac{r_-}{t} \\ -\frac{r_-}{t^*} & \frac{1}{t^*} \end{pmatrix} \qquad (2.3.1.9)$$

and

$$\hat{T}^{-1} = \begin{pmatrix} \frac{1}{t^*} & \frac{r_-}{t} \\ \frac{r_-}{t^*} & \frac{1}{t} \end{pmatrix} = \begin{pmatrix} \frac{1}{t^*} & -\frac{r_+^*}{t} \\ -\frac{r_+}{t} & \frac{1}{t} \end{pmatrix} \qquad (2.3.1.10)$$

The transmission coefficient (the transmittivity) of the segment $[0, L]$ is defined as

$$T = |t|^2. \qquad (2.3.1.11)$$

It can be written in terms of the transfer-matrix:

$$T = |T_{11}|^{-2}, \qquad (2.3.1.12)$$

or

$$T = \frac{4}{2 + \text{Sp}(\hat{T} \cdot \hat{T}^+)}, \qquad (2.3.1.13)$$

useful both in analytical and numerical calculations [103], [1], [135].

Another useful representation for the transmission coefficient is

$$T = |D|^{-2}, \qquad (2.3.1.14)$$

where D is the Fredholm determinant

$$D = \det(I - \hat{\mathcal{G}}_0 \hat{\mathcal{V}}). \tag{2.3.1.15}$$

Here $\hat{\mathcal{G}}_0$ is the integral operator defined as the restriction of the free Green function $\mathcal{G}_0(x,y)$ to the square $0 \leq x, y \leq L$:

$$(E - \hat{H}_0) \cdot \mathcal{G}_0 = \hat{I}, \tag{2.3.1.16}$$

and $\hat{\mathcal{V}}$ is the same restriction of the potential:

$$(\hat{\mathcal{V}})(x,y) = v(x)\delta(x - y)\theta(x)\theta(L - x). \tag{2.3.1.17}$$

To prove (2.3.1.15) note that if points x and y satisfy

$$y \leq 0 < L \leq x \tag{2.3.1.18}$$

the Green function of the scattering problem (2.3.1.1) is

$$\mathcal{G}(x,y) = t\mathcal{G}_0(x,y), \tag{2.3.1.19}$$

where t is the transmission amplitude. Both sides of (2.3.1.19) depend on the length L of the (disordered) segment. Therefore

$$\frac{\partial \mathcal{G}(x,y)}{\partial L} = \frac{\partial t}{\partial L}\mathcal{G}_0(x,y). \tag{2.3.1.20}$$

Using the identity

$$\frac{\partial \hat{\mathcal{G}}}{\partial L} = \hat{\mathcal{G}}\frac{\partial \hat{\mathcal{V}}}{\partial L}\hat{\mathcal{G}},$$

and (2.3.1.17) we obtain

$$\frac{\partial \ln t}{\partial L} = -\frac{\partial \ln D}{\partial L}$$

This in turn leads to (2.3.1.14) if we note that

$$t|_{L=0} = D|_{L=0} = 1.$$

The representation (2.3.1.4) was obtained and used in [53] and [54] but its derivation was more complicated.

All the formulas and relations derived above are valid for any potential. If it is of some special form we can get more information. Suppose that we are interested in the so-called underbarrier transmission (see [103], [105]), when

$$v(x) = u_0 + w(x) \tag{2.3.1.21}$$

with $u_0 = $ const and $w(x)$, for instance, the Poisson potential (2.2.5.1) or (2.2.5.2). The underbarrier scattering is the case when the energy of the incident wave $E = k^2$ is less than the height of the uniform barrier:

$$u_0 - k^2 = q^2 > 0. \tag{2.3.1.22}$$

Then the natural basis for the solution inside the barrier consists of real exponents e^{-qx} and e^{qx} rather than complex (see Section 2.2):

$$\psi = \psi_1(x)e^{-qx} + \psi_2(x)e^{qx}, \qquad 0 \le x \le L. \tag{2.3.1.23}$$

Assume that $w(x) \ne 0$ only for $0 + \varepsilon < x < L - \varepsilon$. Then inside the barrier a wave function has the form

$$\psi(x) = \begin{pmatrix} C'e^{-q(x-\varepsilon)} + D'e^{q(x-\varepsilon)}, & 0 < x < \varepsilon \\ A'e^{-q(x-L+\varepsilon)} + B'e^{q(x-L+\varepsilon)}, & L - \varepsilon < x < L. \end{pmatrix} \tag{2.3.1.24}$$

Introduce the underbarrier transfer-matrix T_u:

$$\begin{pmatrix} C' \\ D' \end{pmatrix} = \hat{T}_u \begin{pmatrix} A' \\ B' \end{pmatrix} \tag{2.3.1.25}$$

Matching the solutions (2.3.1.6) and (2.3.1.25) and sending ε to zero we get

$$\hat{T} = \hat{U}\hat{T}_u\hat{U}^{-1} \tag{2.3.1.26}$$

with

$$\hat{U} = \frac{1}{2\varkappa} \begin{pmatrix} \varkappa + i & \varkappa - i \\ \varkappa - i & \varkappa + i \end{pmatrix}, \qquad \varkappa = \frac{k}{q}. \tag{2.3.1.27}$$

Finally we obtain for the transmission coefficient (2.3.1.12):

$$T = |(f, \hat{T}_u g)|^{-2}. \tag{2.3.1.28}$$

Here the scalar product $(\cdot, \cdot)$ is

$$(f, g) = f_1^* g_1 + f_2^* g_2$$

and

$$f = \frac{1}{2\varkappa}\begin{pmatrix} \varkappa - i \\ \varkappa + i \end{pmatrix}, \qquad g = \frac{1}{2}\begin{pmatrix} 1 - i\varkappa \\ 1 + i\varkappa \end{pmatrix}. \tag{2.3.1.29}$$

Now let us rewrite the solution (2.3.1.1) for $0 \le x \le L$ in the form (2.2.2.1). Then the current conservation law (2.3.1.5) gives

$$|t|^2 = |\psi_+(x)|^2 - |\psi_-(x)|^2 \tag{2.3.1.30}$$

and suggests the parametrization (cf. (2.2.2.10) and (2.2.4.1))

$$\psi_+(x) = |t| \operatorname{sh}\frac{\theta}{2}e^{-i\frac{x+\varphi}{2}}$$
$$\psi_-(x) = |t| \operatorname{ch}\frac{\theta}{2}e^{-i\frac{x-\varphi}{2}} \tag{2.3.1.31}$$

Matching the solutions (2.3.1.1) and (2.2.2.1) at $x = L$ we obtain (cf. (2.3.1.13))

$$T = \frac{1}{2 + e^\theta + e^{-\theta}} = \operatorname{ch}^{-2}\frac{\theta}{2} \tag{2.3.1.32}$$

where θ and φ satisfy the equations (see [89])

$$\theta' = \frac{v(x)}{k} \sin(\varphi - 2kx),$$

$$\varphi' = \frac{v(x)}{k} + \frac{v(x)}{k} \operatorname{cth} \theta \cdot \cos(\varphi - 2kx) \qquad (2.3.1.33)$$

(cf. (2.2.4.2)). In this problem as a rule χ is an irrelevant variable.

Continuing in the same spirit decompose the solution (2.3.1.1) for $0 \le x \le L$ in so-called sine and cosine solutions satisfying the appropriate initial conditions at $x = 0$:

$$\psi_c(0) = 1, \qquad \psi_c'(0) = 0,$$
$$\psi_s(0) = 0, \qquad \psi_s'(0) = 1. \qquad (2.3.1.34)$$

These solutions are currentless so we can parametrize them using (2.2.4.2). Replacing there the slow phase θ by $\varphi - kx$ we arrive at the expressions

$$\psi_c(x) = e^{\xi_c(x)} \sin \varphi_c(x), \qquad \psi_s(x) = k^{-1} e^{\xi_c(x)} \sin \varphi_s(x),$$
$$\psi_c'(x) = k e^{\xi_c(x)} \cos \varphi_c(x), \qquad \psi_s'(x) = e^{\xi_c(x)} \cos \varphi_s(x), \qquad (2.3.1.35)$$

New variables $\xi(x)$ and $\varphi(x)$ satisfy equations (cf. (2.2.4.2))

$$\xi' = \frac{v(x)}{2k} \sin 2\varphi,$$

$$\varphi' = k - \frac{v(x)}{2k}(1 - \cos 2\varphi) \qquad (2.3.1.36)$$

with initial conditions

$$\xi_c(0) = \xi_s(0) = \varphi_s(0) = 0; \qquad \varphi_c(0) = \pi/2. \qquad (2.3.1.37)$$

Now if we write the solution $\psi_1(x)$ for $0 \le x \le L$ in the form

$$\psi_1(x) = A_c \psi_c(x) + A_s \psi_s(x), \qquad 0 \le x \le L \qquad (2.3.1.38)$$

and match solutions at the point $x = L$ we get [124]

$$T = \frac{4}{2 + e^{2\xi_s(L)} + e^{2\xi_c(L)}} \qquad (2.3.1.39)$$

(cf. (2.3.1.13) and (2.3.2.3)).

2.3.2. Large Length Behavior of the Transmittivity

Following (2.2.4.7) introduce the quantities

$$\gamma_c(x) = \xi_c(x)/x, \qquad \gamma_s(x) = \xi_s(x)/x. \qquad (2.3.2.1)$$

Then the transmittivity may be represented as a function of two self-averaging quantities $\gamma_{s,c}(x)$ and the "volume" L:

$$T = \frac{4}{2 + e^{2L\gamma_s(L)} + e^{2L\gamma_c(L)}}. \qquad (2.3.2.2)$$

As we have already mentioned in Section 2.1.3 such quantities characterize the behavior of transmittivity of typical realizations. To prove this for the case note that due to the self-averaging of $\gamma_{c,s}$ for large L we have with logarithmic accuracy[1]

$$T \propto e^{-2\gamma L} = e^{-L/l}. \qquad (2.3.2.3)$$

Thus the transmittivity for a fixed realization decreases exponentially and its decrement equals the inverse localization length l^{-1} defined in (2.2.4.4). Rewriting (2.3.2.2) as

$$-L^{-1}\ln T_L = -L^{-1}\ln 4 + L^{-1}\ln(2 + e^{2L\gamma_c(L)} + e^{2L\gamma_s(L)}) \qquad (2.3.2.4)$$

and passing to the limit $L \to \infty$ we find that with probability 1 the l.h.s. of this equality tends to a limit that is not a random quantity. This enables us to introduce the decrement of the transmittivity on a realization

$$\bar{\gamma}_L = -L^{-1}\ln T_L. \qquad (2.3.2.5)$$

that tends to a nonrandom limit coinciding with the inverse localization length (2.2.4.4) as $L \to \infty$ [124]:

$$\bar{\gamma}_L \xrightarrow[\text{Prob 1}]{} \bar{\gamma} = 2\gamma(E) = l^{-1}. \qquad (2.3.2.6)$$

The proof given above was based on the existence of the limit (2.2.3.1) defining the Lyapunov exponent. There is a more straightforward way to do it—namely, let $\hat{T}(L_1, L_2)$, $t(L_1, L_2)$, and $r(L_1, L_2)$ be the respective quantities for the case when the barrier is situated not at the interval $[0, L]$ but at $[L_1, L_2]$. Then for $L_1 < L_2 < L_3$ we have the obvious multiplicative equality

$$\hat{T}(L_1, L_3) = \hat{T}(L_1, L_2) \cdot \hat{T}(L_2, L_3) \qquad (2.3.2.7)$$

for the transfer-matrixes, and (2.3.1.9) implies

$$\frac{1}{t(L_1, L_3)} = \frac{1}{t(L_1, L_2)t(L_2, L_3)} + \frac{r(L_1, L_2)r^*(L_2, L_3)}{t(L_1, L_2)t^*(L_2, L_3)}$$

As the result we have

$$T(L_1, L_3) \equiv |t(L_1, L_3)|^2 \leq 4^{-1}T(L_1, L_2)T(L_2, L_3). \qquad (2.3.2.8)$$

This inequality together with two basic properties of a random potential $v(x)$ discussed in Section 2.1.3 (that guarantee its ergodicity) implies that $\ln(T(L_1, L_2)/4)$ is a subadditive random process (see, e.g., [68]). Therefore owing to the subadditive ergodic theorem the limit (2.3.2.5) exists and is nonrandom.

---

[1] We use the symbol $\propto$ to denote the logarithmic equivalence as $L \to \infty$.

Inequality (2.3.2.8) yields one more useful fact—namely, if the scattering potential is random enough, then according to the common wisdom of the probability theory the variance of $\bar{\gamma}_L$ should have the order $O(L^{-1})$ and, moreover, for large L, $\bar{\gamma}$ is to be of the form

$$\bar{\gamma}_L = \bar{\gamma} + L^{-1/2}\xi \tag{2.3.2.9}$$

where ξ is the Gaussian random variable with zero mean and the L-independent nonzero variance (see (2.2.4.11) for the same property of random variable (2.3.2.1) defining the Lyapunov exponent). Under these conditions $L^{-1}\ln\langle T(0,L)\rangle$ and $L^{-1}\langle \ln T(0,L)\rangle$ do not coincide even in the limit $L \to \infty$. This makes it natural to study the large length behavior of the mean transmittivity $\langle T(0,L)\rangle$, in particular its decrement

$$\gamma_T = -\lim_{L\to\infty} L^{-1}\ln\langle T(0,L)\rangle. \tag{2.3.2.10}$$

Let us prove the existence of this limit based on inequality (2.3.2.8) (see [106]). Consider for simplicity the random potential with independent values, i.e., white noise (2.2.4.6) or the Poisson potential (2.2.5.2). Averaging (2.3.2.8) we find that for $l_1 = L_2 - L_1$ and $l_2 = L_3 - L_2$

$$\langle T(0, l_1 + l_2)\rangle \leq 4^{-1}\langle T(0, l_1)\rangle\langle T(0, l_2)\rangle.$$

That means that $\ln(\langle T(0,l)\rangle/4)$ is a subadditive function of the length of the random barrier; thus according to [129] the limit (2.3.2.10) exists.

The same arguments prove (2.3.2.10) for any random potential which correlations vanish rapidly enough.

Decrements (2.3.2.6) and (2.3.2.10) are the simplest characteristics of the transmission coefficient of the one-dimensional scattering problem for long random barriers. By considering the special quasi-one-dimensional scattering problem one can construct the nonrandom quantity that interpolates between $\bar{\gamma}$ and γ_T (see Section 2.6.2).

The composition law (2.3.2.8) may also be used to derive the criterion of the one-dimensionality of the system—namely, calculate the square modulus of this relation. Since the term

$$2\operatorname{Re}\{r(L_1, L_2)r^*(L_2, L_3)/t^2(L_1, L_2)T(L_2, L_3)\}$$

contains the phase varying on the microscopic scale it may be neglected for samples of the macroscopic length. The resulting relation after using the Landauer formula (2.1.1.7) can be written in the form

$$\rho_{12} = \rho_1 + \rho_2 + 2\frac{e^2}{h}\rho_1\rho_2 \tag{2.3.2.11}$$

where ρ_1, ρ_2, and ρ_{12} are the resistances of the segments (L_1, L_2), (L_2, L_3), and (L_1, L_3) respectively. It shows that if the resistance of the sample is much smaller than $he^{-2} \cong 30$ kΩ the third term in (2.3.2.11) can be neglected and the resistance becomes additive, i.e., obeys the Ohm law. Therefore our sample is a conductor and should not be treated as a one-dimensional one for which the complete localization takes place. In the opposite case when the resistance is much larger than he^{-2} we neglect $\rho_1 + \rho_2$ in (2.3.2.11) and the resistance becomes the multiplicative function of the length growing exponentially when $L \to \infty$. This is the localization regime typical for the one-dimensional disordered systems.

2.3.3. Decrement of the Transmittivity (the Formal Expansion Scheme)

In the case of the nonresonant transmission through the random point scatterers the formal expansion of the decrement of the mean transmission coefficient in the powers of the concentration of scatterers was obtained in [104].

We remind the reader that the potential in this case has the form

$$v(x) = u_0 + k_0 \sum_{0 \leq x_j \leq L} \delta(x - x_j), \qquad k_0 > 0, \qquad (2.3.3.1)$$

and the point scatterers at points x_j are uniformly and independently distributed with the density $n = a^{-1}$ (a being the mean distance between the adjacent scatterers). Then the mean transmission coefficient can be written in the following way:

$$\langle T \rangle = e^{-nL} \sum_{m=0}^{\infty} \frac{n^m}{m!} \int_0^L T(x_1, \ldots, x_m)\, dx_1 \cdots dx_m. \qquad (2.3.3.2)$$

The main idea is to interpret (2.3.3.2) as a partition function of the grand canonical ensemble in statistical mechanics with $T(x_1, \ldots, x_m)$ playing the role of the Boltzmann factor and to apply the technique used in statistical mechanics to derive the virial expansion (see, e.g., [134]). Keeping this in mind we write $T(x_1, \ldots, x_m)$ identically as the sum

$$T(x_1, \ldots, x_m) = T^{(0)} + \sum_i \tilde{T}(x_i) + \sum_{i<j} \tilde{T}(x_i, x_j)$$
$$\ldots + \tilde{T}(x_1, \ldots, x_m) \qquad (2.3.3.3)$$

where $T^{(0)}$ is the transmission coefficient of the uniform barrier without point scatterers.

$$\tilde{T}(x_i) = T(x_i) - T^{(0)}, \qquad (2.3.3.4)$$

$$\tilde{T}(x_i, x_j) = T(x_i, x_j) - T(x_i) - T(x_j) + T^{(0)} \qquad (2.3.3.5)$$

and so on are the analogs of the Ursell functions [95]. Rewriting (2.3.3.2) in terms of $\tilde{T}(\cdot)$ we get

$$\langle T \rangle = T^{(0)} + n \int_0^L \tilde{T}(x_1)\, dx_1 + \frac{n^2}{2} \int_0^L \tilde{T}_1(x_1, x_2)\, dx_1\, dx_2 + \ldots, \quad (2.3.3.6)$$

Expanding the logarithms of both sides of (2.3.3.6) in powers of n we obtain the equality

$$-L^{-1}\ln\langle T\rangle = -L^{-1}\ln T^{(0)} - L^{-1}\int_0^L \left(\frac{\tilde{T}(x_1)}{T^{(0)}} - 1\right)dx_1$$
$$-\frac{n^2}{2}L^{-1}\int_0^L \left(\frac{\tilde{T}(x_1,x_2)}{T^{(0)}} - \frac{\tilde{T}(x_1)}{T^{(0)}}\cdot\frac{\tilde{T}(x_2)}{T^{(0)}}\right)dx_1\,dx_2 \quad (2.3.3.7)$$

To calculate the functions $T(\cdot)$ write first the transfer-matrix for the underbarrier case ($E < u_0$) as a product

$$\hat{T}_u = \hat{G}_0(x_1)\hat{V}\hat{G}_0(x_2-x_1)\hat{V}\cdots\hat{G}_0(x_m-x_{m-1})\hat{V}\hat{G}_0(L-x_m), \quad (2.3.3.8)$$

where

$$\hat{G}_0(x) = \begin{pmatrix} e^{qx} & 0 \\ 0 & e^{-qx} \end{pmatrix}, \qquad \hat{V} = \begin{pmatrix} 1+\frac{k_0}{2q} & \frac{k_0}{2q} \\ -\frac{k_0}{2q} & 1-\frac{k_0}{2q} \end{pmatrix}, \qquad (2.3.3.9)$$

$E = k^2$ is the energy of the incident wave and $q^2 = u_0 - E$. After some calculations we come to the equality (in the case $qL \gg 1$)

$$T(x_1,\dots,x_m) = 4\sin^2\frac{2k}{q}e^{-2qL}\frac{1}{F(x_1,\dots,x_m)}. \qquad (2.3.3.10)$$

The functions F_0, $F_1(x_1)$, $F_2(x_1,x_2),\dots$ in this case can be approximated as follows:

$$F_0 = 1,$$
$$F_1 \approx (1+\mu)^2[1+f(x_1)+f(L-x_1)],$$
$$F_2 \approx (1+\mu)^4[1+g(x_2-x_1)+f(L-x_2)],\dots \qquad (2.3.3.11)$$

where $\mu = -k_0(k_0+2q)^{-1}$ is the underbarrier scattering amplitude, and

$$f(x) = -2\mu\cos\frac{2k}{q}e^{-2qx} + \mu^2 e^{-4qx},$$
$$g(x) = -2\mu^2 e^{-2qx} + \mu^4 e^{-4qx}. \qquad (2.3.3.12)$$

Using (2.3.3.7)–(2.3.3.12) we can find the expansion of the decrement γ_T of the mean transmission coefficient in powers of the dimensionless concentration $c = n/q$:

$$\gamma_T = q[2 - c\mu(2+\mu) + c^2 S + \dots],$$
$$S = -\mu^2(1+\mu)^4\int_0^\infty \frac{e^{-2t}(2-\mu^2 e^{-2t})}{(1-\mu^2 e^{-2t})^2}dt$$
$$= \frac{(1+\mu)^4}{2}\ln(1-\mu^2) - \frac{\mu^2(1+\mu)^3}{2(1-\mu)} \qquad (2.3.3.13)$$

In a similar manner we can obtain a concentration expansion of the decrement of the transmission coefficient for a typical realization, $\bar{\gamma} = 2\gamma(E)$ where $\gamma(E)$ is the

Lyapunov exponent (2.2.3.1). To this end we must apply the same formalism to the function $-L^{-1} \ln T$. The result is

$$\bar{\gamma}(E) = q[2 - c\ln(1 + \mu)^2 + c^2 \tilde{S} + \dots],$$
$$\tilde{S} = 2 \int_0^\infty \ln(1 - \mu^2 e^{-2t}) \, dt. \tag{2.3.3.14}$$

Note that the mean transmission coefficient (2.3.3.13) and that of a realization (2.3.3.14) differ in the first-order term already. The reason for this apparently lies in the difference between typical and representative realizations.

In Section 2.5 above one can find another way of obtaining the concentration expansions of $\bar{\gamma}(E)$ as well as the Lyapunov exponent $\gamma(E)$ for the model with repulsive point scatterers. The Lyapunov exponent $\gamma(E)$ coincides for $E > 0$ with the real part of the analytic continuation of $\gamma(E)$ from the underbarrier nonresonance region $E < 0$, while $N(E)$ is its imaginary part. To carry out this continuation with the expansion coefficients we must substitute ik instead of q, then μ will be replaced by the reflection amplitude $r = k_0(2ik - k_0)^{-1}$ of the point scatterer. The resulting formulas for $N(E)$ and $\gamma(E)$ coincide with (2.2.5.18), (2.2.5.19).

There still remains the open question about the validity of the concentration expansions in the overbarrier region. Proceeding in the way described above we naturally obtain the same formulas (cf. (2.3.3.7)) but now the integral in the analog of (2.3.3.13) diverges owing to the presence of oscillating integrands instead of exponentially decreasing ones. The analytic continuation does not help here as well since $\gamma_T(E)$, being determined via the mean square of the wave function modulus has much poorer analytical properties than $\bar{\gamma}(E)$. The reason for these formal difficulties lies in the presence of certain resonance situations mentioned above. They happen when the energy of the incoming particle lies in the spectrum of the infinite system with the same potential and result in nonanalyticity (as a rule) of $\gamma_T(E)$ as a function of the density if $E > 0$.

As we have mentioned earlier, our expansion formalism is similar to the virial expansion in statistical physics [95], [141], [134] and likewise enables us to find the expansion coefficients for any concentration power. After some modification it can also be used in the case when the scatterers are not independent. For instance if they cannot approach one another closer than a fixed nonrandom distance a the nonresonant energy region lies below the spectrum of the periodic system with the scatterers positioned with a period a. The other example of such a system is an ideal lattice gas of impurities (which independently occupy the sites of the periodic lattice).

2.3.4. Decrement of the Mean Transmittivity (Calculations)

The formal expansions shown above in certain cases lead to the convergent series. In this section we shall prove such a convergence for the case of nonresonant scattering by a random Poisson-type potential [106]. We also discuss the mean transmission coefficient in the resonant (overbarrier) case [107].

Consider the SE (2.1.1.1) with the potential $v(x)$ given by (2.3.3.1) with repulsive centers, that is with $k_0 > 0$. The analogy between the expression (2.3.3.2) and partition function of the grand canonical ensemble suggests the introduction of the corresponding "correlation functions":

$$\rho(\Gamma_N) = \langle T \rangle^{-1} e^{-nL} \sum_{N=0}^{\infty} \frac{n^{N+m}}{m!} \int T(\Gamma_N \cup Y_N) d^m Y \qquad (2.3.4.1)$$

for any nonempty set $\Gamma_N = \{x_1, \ldots, x_N\}$ ($N \geq 1$). Using the well-known statistical mechanics technique [134] we get the analog of the system of the Kirkwood–Zaltzbourg equations for these functions:

$$\rho(x \cup X_N) = n \sum_{m=1}^{\infty} \frac{1}{m!} \int K(x; X_N, Y_m) \rho(X_N \cup Y_m) \, d^m Y + c\delta_N \; (2.3.4.2)$$

with $X_N = (x_1, \ldots, x_N)$ when $N > 0$ and $X_0 = \emptyset$ and $Y_m = \{y_1, \ldots, y_m\}$ lying in $(0, L)$. The kernel K is given by

$$K(x; X_N, Y_m) = \sum_{\emptyset \subseteq Z_k \subseteq Y_m} = (-1)^{m-k} S(x, X_N \cup Z_k), \qquad (2.3.4.3)$$

here the sum is over all the nonordered sets Z_k, and

$$S(x, \Gamma) = T(x \cup \Gamma)/T(\Gamma). \qquad (2.3.4.4)$$

Following the known approach [134] we shall consider the set of functions ρ as a vector in a Banach space E_ξ with a norm

$$\|\rho\|_\xi = \sup_N \sup_{X_N} |\rho(X_N)| \xi^{-N} \qquad (2.3.4.5)$$

and interpret the system (2.3.4.2) as an operator equation in it. The parameter ξ will be fixed later.

The main technical fact that enables us to estimate the norm of this operator is the inequality

$$|S(x, X_N \cup y) - S(x, X_N| < \begin{cases} 2 & \text{for } q|x - y| \leq 0.5 \\ \min(2, 96(k_0/2q)^2 e^{-2q|x-y|}) & \\ & \text{for } q|x - y| > 0.5. \end{cases} \qquad (2.3.4.6)$$

It leads to the bound for the norm of the operator $\mathcal{K}$ defined in E_ξ by (2.3.4.2):

$$\|\mathcal{K}\|_{E_\xi} < \xi^{-1} \left(e^{2\xi} + 96 \left(\frac{k_0}{2q} \right)^2 (1 - 2\xi)^{-1} \right). \qquad (2.3.4.7)$$

This bound makes it possible to invert the operator $1 - n\mathcal{K}$ for

$$|n| < R = \sup \xi (e^{2\xi} + 96 \left(\frac{k_0}{2q} \right)^2 (1 - 2\xi)^{-1})^{-1} \qquad (2.3.4.8)$$

by a Neuman series, and get the expansion

$$\vec{\rho} = (1 - n\mathcal{K})^{-1} \vec{\delta}_{N0} = \sum_{j=0}^{\infty} n^j \mathcal{K}^j \vec{\delta}_{N0}.$$

($\vec{\delta} = \delta_{N0}$ being considered as a vector in E_ξ). Now we choose ξ so as to get the maximum of R. The identity above implies that all the functions ρ are analytic with respect to n and bounded uniformly in $L > 0$ and X_N. This holds in particular for $\rho(n, x)$. Note that

$$(nL)^{-1} \int_0^\infty \rho(n, x)\, dx = L^{-1} \frac{\partial}{\partial n} \ln(e^{nL} \langle T(n) \rangle) = 1 + L^{-1} \frac{\partial}{\partial n} \ln \langle T(n) \rangle. \tag{2.3.4.9}$$

Since $\rho(0, x) \equiv 0$ the l.h.s. of this equality is analytic for $|n| < R$, hence the r.h.s. is analytic as well. The last step consists of passing to a limit $L \to \infty$ using the boundedness of ρ.

To stay in the nonresonant case for the attracting potential ($k_0 < 0$) the incident wave energy should be outside of the SE spectrum. One way to ensure it is to have the scatterers on the lattice $x_j = (j + 1/2)a$ with integer j. Then the randomness of the potential is provided by the factors c_j that are independent random variables taking the value 1 with the probability p and 0 with the probability $1 - p$. The potential has the form

$$v(x) = u_0 - |k_0| \sum_{j=0}^M c_j \delta(x - ja - a/2), \qquad 0 \le x \le Ma \tag{2.3.4.10}$$

The scattering is a nonresonance one if the energy of the incident wave is outside the spectrum of the Schrödinger operator for any realization of the potential, which is the case if

$$\text{ch } aq - 1 > \frac{|k_0|}{2q} \text{ sh } aq. \tag{2.3.4.11}$$

The statistical mechanics formalism used above with sums instead of integrals leads to the estimate of norm of the corresponding operator:

$$\|\mathcal{K}\|_{E_\xi} < C \sum_0^\infty \left(\frac{\ln \mu_1}{2\xi} \right)^{1-m}. \tag{2.3.4.12}$$

Here $C = C(kq^{-1}, k_0 q^{-1}, aq)$ is a constant depending on k, k_0, q, and a, and μ_1 is the larger eigenvalue of the matrix $\hat{G}_0(aq/2)\hat{V}\hat{G}_0(aq/2)$ (see (2.3.3.9) for the definition of $\hat{G}_0$ and $\hat{V}$).

The final result for this case is the analyticity of γ_T with respect to the variable $c = p/(1 - p)$ for

$$|c| < R = \frac{1 - \ln \mu_1}{C \ln^2 \mu_1}. \tag{2.3.4.13}$$

In this model the concentration equals p/a and $\bar{\gamma}$ is analytic in this variable for $|p/a| < R/(1 + R)$.

The results exposed above remain valid if we suppose the scatterers not to be independent but to form (for example) a Gibbs gas with a short-range potential bounded from below with a hard core.

Now we pass to the resonant case when the energy of the incident wave may be in the spectrum of the corresponding Schrödinger operator. The main result here is the equality

$$\gamma_T = -\mu(-1), \tag{2.3.4.14}$$

relating the decrement γ_T defined in (2.3.2.10) to the decrement of the moments of the solution envelope $r(x) = (\psi^2 + k^{-2}\psi'^2)^{1/2}$:

$$\mu(s) = \lim_{L \to \infty} L^{-1} \ln\langle r^s(L) \rangle. \tag{2.3.4.15}$$

Using (2.3.4.14) one can derive a number of asymptotic formulae. All of them show that in the leading order

$$\gamma_T = \bar{\gamma}/2 \tag{2.3.4.16}$$

if $\bar{\gamma}$ is small (see (2.3.2.6) for its definition).

The model that we shall study now is that with the potential

$$v(x) = Q(\xi(x)) \tag{2.3.4.17}$$

where Q is a "nonflat" (see [57]) function of a diffusion process $\xi(x)$, satisfying the conditions:

1) $\xi(x)$ is a stationary process;

2) $\xi(x)$ is symmetrical with respect to "time" inversion $x \to -x$.

3) $\xi(x)$ possesses the exponential mixing property; that means that zero is the isolated eigenvalue of multiplicity one of its infinitesimal operator A, corresponding to the eigenfunction $1 \in L^2(d\xi)$.

Passing to dimensionless variables $t = kx$, $u = k^{-2}v$, and $l = kL$ and introducing the polar coordinates in the $(\psi, \dot{\psi})$ plane (cf. (2.2.2.10))

$$\psi(t) = r(t)\sin\theta(t), \qquad \dot{\psi}(t) = r(t)\cos\theta(t) \tag{2.3.4.18}$$

we come to the system

$$\frac{d\theta}{dt} = 1 - u(t)\sin^2\theta(t)$$
$$\frac{dr}{dt} = \frac{1}{2}r(t)u(t)\sin 2\theta(t) \tag{2.3.4.19}$$

equivalent to the SE.

Let $\theta_\varphi, r_\varphi$ be the solution of (2.3.4.19) satisfying the initial conditions $\theta_\varphi(0) = \varphi, r_\varphi(0) = 1$.

Integrating (2.3.4.19) we get

$$r_\varphi(l) = \exp\left\{\frac{1}{2}\int_0^l Q(\xi(t))\sin 2\theta_\varphi(t)\, dt\right\}. \tag{2.3.4.20}$$

Then the Feynman–Kac formula gives

$$\langle r_\varphi^s(l) \rangle = \int v(l, \xi, \varphi)\, d\xi \tag{2.3.4.21}$$

where the integration is over all the values of ξ and $v(l, \xi, \varphi)$ solves the equation

$$\frac{\partial v}{\partial t} = A_s \equiv \left(k^{-1}A + (1 - Q(\xi)) \sin^2 \theta \frac{\partial}{\partial \theta} + \frac{s}{2} Q(\xi) \sin 2\theta \right) \quad (2.3.4.22)$$

with the initial conditions $v(0, \xi, \theta) \equiv \pi^{-1}$.

Under our conditions the function v with exponential accuracy equal to

$$\exp\{\mu(s)\} \cdot \Phi(\xi, \varphi) \quad (2.3.4.23)$$

as $T \to \infty$. Here $\Phi(\xi, \varphi)$ is a normalized eigenfunction of the operator A_s corresponding to its largest eigenvalue coinciding obviously with $\mu(s)$.

Consider the random variable

$$\rho(l) = (2l)^{-1} \ln(r_\varphi^2(l) + r_{\varphi + \pi/2}^2(l))^{s/2} \quad (2.3.4.24)$$

with a probability density $F_\rho'(x) \propto \exp\{-lf(x)\}$ (see the footnote on page 89). Then the transmission coefficient is

$$T = 4(\exp 2l\rho + 2)^{-1} \quad (2.3.4.25)$$

and

$$\lim_{l \to \infty} l^{-1} \ln\langle \exp\{s \cdot l \cdot \rho(l)\}\rangle = \mu(s). \quad (2.3.4.26)$$

At the same time $f(x)$ is related to $\mu(s)$ via the Legendre transform:

$$f(x) = \sup_s (xs - \mu(s)), \qquad \mu(s) = \sup_x (sx - f(x)). \quad (2.3.4.27)$$

The function $\mu(s)$ is well studied [113]. We will also need the following properties:
1) $\mu(s)$ is a smooth convex function;
2) $\mu(s) \equiv \mu(2 - s)$;
3) $\mu'(0) = -\mu'(-2) = \bar{\gamma}$.
Then

$$f'(0) = -1, \qquad f(0) = \sup(-\mu(s)) = -\mu(-1). \quad (2.3.4.28)$$

Using (2.3.4.34) we see that

$$\langle T \rangle = 4 \int_\infty^\infty \frac{dF_\rho(x)}{\exp\{2lx\} + 2} \propto \int_{-\infty}^\infty \frac{\exp\{-lf(x)\}}{\exp\{2lx\} + 1} dx \quad (2.3.4.29)$$

and making some easy estimates with the use of (2.3.4.27), (2.3.4.28), come to

$$\langle T \rangle \propto \exp\{l\mu(-1)\}. \quad (2.3.4.30)$$

This equality means that (in dimensionless units)

$$\gamma_T = -\lim_{l \to \infty} l^{-1} \ln\langle T \rangle = -\mu(-1),$$

that is (2.3.4.14).

Since $\mu(-1)$ is the largest eigenvalue of the operator

$$\mathcal{A}_{-1} = k^{-1}A + \frac{\partial}{\partial\theta} - Q(\xi)\sin\theta\frac{\partial}{\partial\theta}\sin\theta \qquad (2.3.4.31)$$

(see (2.3.4.20), (2.3.4.21)) our problem now is reduced to the study of this eigenvalue. Our considerations show that for any $s \geq 1/2$

$$\lim_{l\to\infty} l^{-1}\ln\langle T^s\rangle = \mu(-1).$$

Some modification of this formalism leads to analogous results for other classes of random potentials, e.g., Kronig–Penney, or a singular one, defined by (2.3.3.1) with $u_0 = 0$.

Using (2.3.4.14) we can obtain asymptotic expansions of γ_T with respect to various small parameters.

We can introduce in (2.3.4.17) two natural parameters associated with the potential. Namely, let

$$v(x) = \alpha^2 Q(\xi(\beta x)). \qquad (2.3.4.32)$$

Then the SE will contain three parameters α, β, and k, but only two of them are independent. Eliminating k as was done above we get two dimensionless independent parameters $\varepsilon = \alpha k^{-1}$ and $\delta = \beta k^{-1}$.

Then the system (2.3.4.19) is transformed into

$$\frac{d\theta}{dt} = 1 - \varepsilon^2 u(\delta t)\sin^2\theta(t),$$
$$\frac{d\varphi}{dt} = \frac{\varepsilon^2}{2}r(t)u(\delta t)\sin 2\theta(t), \qquad (2.3.4.33)$$

and $\mathcal{A}_{-1}$ takes the form

$$\mathcal{A}_{-1} \equiv \delta A + \frac{\partial}{\partial\theta} - \varepsilon^2 Q(\xi)\sin\theta\frac{\partial}{\partial\theta}\sin\theta. \qquad (2.3.4.34)$$

Our problem now is reduced to investigation of the largest eigenvalue of $\mathcal{A}_{-1}$ in various limits with respect to ε and δ.

The simplest case is the limit $\varepsilon \to 0$, $\delta = \text{const}$ (small potential). Regarding the last term in (2.3.4.34) as a perturbation of the operator $\delta A + \partial/\partial\theta$ we can use the general theory [79] and find the perturbed eigenvalue as

$$0 + \varepsilon^2\pi^{-1}\cdot 0 + \varepsilon^4/8\int Q(\xi)[\delta A(\delta^2 A^2 + 4)^{-1}Q(\xi)]d\xi + o(\varepsilon^4)$$

Simple calculations enable one to express it in terms of the correlation function $B(t) = \langle Q(\xi(0))Q(\xi(t))\rangle$ of the potential:

$$\gamma_T = \frac{\varepsilon^4}{8\delta^2}\int_0^\infty B(t)\cos 2\delta^{-1}t\,dt + o(\varepsilon^4). \qquad (2.3.4.35)$$

Comparing (2.3.4.35) with the expression (2.2.4.3) for $\bar{\gamma}$ one can see that

$$\gamma_T = \bar{\gamma}/2 \cdot (1 + o(1)). \tag{2.3.4.36}$$

A somewhat more complicated case is the high-energy limit $k \to \infty$ (i.e., $\varepsilon \approx \delta \to 0$). We shall consider a slightly more general case when ε, $\delta \to 0$ and $\zeta = \varepsilon^2 \delta^{-1} \to 0$. The result in this case is

$$\mu(-1) = -\frac{\zeta^2 \delta^2}{8} \int_0^\infty B(t) \cos 2\delta^{-1} t\, dt \cdot (1 + o(1))$$

and (using initial parameters)

$$\gamma_T = \frac{\alpha^4}{8k\beta^4} \int_0^\infty B(t) \cos 2k\beta^{-1} t\, dt \cdot (1 + o(1))$$

In this case as before the equality (2.3.4.36) is again true. Thus all the known asymptotic results yield (2.3.4.36). We have no rigorous proof of (2.3.4.36) in more or less general case. All we can do is to give some bounds and heuristic arguments for its validity.

Equality (2.3.4.14) together with properties 1)–3) of the function $\mu(s)$ (see above) imply that $0 < \gamma_T < \bar{\gamma}$ and $\gamma_T \to 0$ when $\bar{\gamma} \to 0$. Because of the same properties it seems natural to approximate $\mu(s)$ in the interval $[-2, 0]$ by a parabola that intersects the abscissa at the points -2 and 0 and has apex at the point $(-1, \mu(-1))$. Its equation is

$$y = -\mu(-1)s(s + 2) = \gamma_T s(s + 2).$$

Since $\bar{\gamma} = \mu'(0)$ (see the property 3) of $\mu(s)$ above) we have

$$\bar{\gamma} = \mu'(0) \approx y'(0) = 2\gamma_T.$$

To finish this subsection let us discuss the singular potential of the form (2.3.3.1) with $u_0 = 0$.

There are two natural physical formulations of the problem.

1. All the x_j's are distributed according to the Poisson law with the concentration c of scatterers (see (2.3.3.1) and the text below it). This formulation describes a random medium obtained by cutting out a piece of length L from an infinitely long sample, and it is analogous to the formulation of the problem for the grand canonical ensemble in statistical physics.

2. A known number N of scattering centers are positioned randomly in the sample of length L. Then both L and N tend to infinity while the concentration $NL^{-1} \to c > 0$. This problem is an analog of the treatment with the canonical ensemble.

In both cases we are interested in the limit $L \to \infty$, and technically the problem is reduced to the investigation of the leading eigenvalue of the same integral operator, although the final expressions for γ_T differ.

Consider the first case; the second one is analogous though somewhat easier. As before we change the variables to dimensionless ones $t = kx$, $l = kL$, $\varkappa = k_0 k^{-1}$, and $n = ck^{-1}$. The function $\mu(s)$ now has the form

$$\mu(s) = \lim_{l \to \infty} l^{-1} \ln \langle \pi^{-1} \int_0^\pi r_\varphi^s(l)\, d\varphi \rangle, \tag{2.3.4.37}$$

We expand into a series with respect to n:

$$\langle \pi^{-1} \int_0^\pi r_\varphi^s(l)\, d\varphi \rangle = e^{-nl} \int_{j=0}^\infty n^j M_j(l). \tag{2.3.4.38}$$

The function $\mu(s)$ defined by (2.3.4.37) possesses all the properties 1)–3) described above and thus we can use the same arguments to prove the main equality (2.3.4.14).

It is useful to make the Laplace transformation and set

$$\mathcal{L}(p) = \int^\infty e^{-pl} \langle \pi^{-1} \int_0^\pi r_\varphi^s(l)\, d\varphi \rangle = \sum_{j=0}^\infty n^j m_j(n+p). \tag{2.3.4.39}$$

Direct calculations show that

$$m_j(q) = q^{-j-1}(1, B_q^j 1), \tag{2.3.4.40}$$

where $(\cdot, \cdot)$ is the scalar product in $L^2(0, \pi)$ and 1 is the function identically equal to 1. B_q is the integral operator in $L^2(0, \pi)$ with a positive kernel that may be written explicitly (see [107]). Now (2.3.4.39) may be rewritten in the form

$$\mathcal{L}(p) = q^{-1}(1, (E - nq^{-1}B_q)^{-1}1) \tag{2.3.4.41}$$

where E is the unit operator and $q = p+n$. The series (2.3.4.39) converges for Re q large enough.

The existence of the limit (2.3.4.37) with $s = -1$ is equivalent to the analyticity of the function $\mathcal{L}(p)$ for Re $p > \mu(-1)$. The operator B_q is compact and analytic with respect to q for $q \neq 2ki$ ($k = \pm 1,\ \pm 2,\ \dots$). Then $\mathcal{L}(p)$ is analytic if $q \neq 2ki$ and the operator $(E - nq^{-1}B_q)$ is invertible. That is the case if $n^{-1}q$ does not belong to the spectrum of B_q. Since its kernel is positive, the eigenvalue Λ of B_q with the largest real part is positive as well and the operator $(E - nq^{-1}B_q)$ is invertible for Re $q > n\Lambda$. So $\mathcal{L}(p)$ is analytic for Re $p > \alpha - n$ where α is the root of the equation

$$\alpha = n\Lambda(\varkappa, \alpha) \tag{2.3.4.42}$$

and $\mathcal{L}(p)$ has a pole at a point $\alpha - n$.

Returning to γ_T we obtain that $\gamma_T = n - \alpha$, or in dimensional variables

$$\gamma_T = c - k\alpha. \tag{2.3.4.43}$$

When studying the second problem (with a fixed number of scatterers) we do not need the Laplace transformation, and the role of the parameter q is played by the concentration n itself. As a result we get the equality

$$\gamma_T = -\mu(-1) = -n \ln \Lambda(\varkappa, n) \tag{2.3.4.44}$$

where Λ is the leading eigenvalue of the operator B_n (with n instead of q!). In dimensional variables

$$\gamma_T = -c \ln \Lambda(k_0 k^{-1}, ck^{-1}). \tag{2.3.4.45}$$

Note that (2.3.4.43) and (2.3.4.45) coincide in the leading order if $\Lambda \approx 1 + o(1)$, but in the case when Λ substantially differs from 1 they do not.

Using these formulas we can get some asymptotic expansions. The simplest one is when $n \to 0$. Then

$$\Lambda(\varkappa, q) = \pi^{-1} \int_0^\pi (1 - \varkappa \sin 2\theta + \varkappa^2 \sin^2 \theta)^{-1/2} \, d\theta + O(q).$$

Then (2.3.4.43) yields

$$\gamma_T = c \left(1 - \pi^{-1} \int_0^\pi (1 - k_0 k^{-1} \sin 2\theta + k_0^2 k^{-2} \sin^2 \theta)^{-1/2} \, d\theta \right) + O(c^2) \tag{2.3.4.46}$$

and (2.3.4.45) yields

$$\gamma_T = -c \ln \left(\pi^{-1} \int_0^\pi (1 - k_0 k^{-1} \sin 2\theta + k_0^2 k^{-2} \sin^2 \theta)^{-1/2} \, d\theta \right) + O(c^2).$$

We see that these formulas are different.

We can pass to a limit $\varkappa \to \infty$ or $\varkappa \to 0$ in (2.3.4.46). In the first case we get $\gamma_T \approx c$ and in the second we get

$$\gamma_T \approx \frac{1}{16} c k_0^2 k^{-2} (1 + o(1)),$$

in accordance with (2.3.4.35) since the correlation function for our potential equals $c k_0^2 \delta(x)/2$.

Now let $\varkappa \to 0$ while $0 < \varkappa n < b < 1$. Then the result is

$$\gamma_T = \frac{c k_0^2}{16(k^2 - c k_0)}(1 = o(1)).$$

Note that in all the cases described (2.3.4.16) holds whenever $\gamma_T \to 0$.

2.3.5. Underbarrier Tunnelling in Semiconductors

Many important problems of radiophysics, optics, and solid state physics, especially physics of semiconductors, being initially three-dimensional ones can be reduced to one-dimensional problems. All of them possess one characteristic feature—the underbarrier tunneling.

Consider one such problem when a potential barrier describes a thin layer of the p-type semiconductor between two thick layers of the n-type semiconductor [67]. At low temperatures the conductance of the layer is defined by the tunnelling probability. The barrier transparency, being exponentially small, depends strongly on the fluctuations of the barrier parameters. As a result the conductance is defined by the rare regions with the large transparency. These regions make main contributions to the transparency in spite of their exponentially small probability.

From a rigorous point of view we have to study the three-dimensional problem on the beam of electrons transmitting through the infinite semiconducting layer. Owing to the underbarrier character of scattering it can be reduced to the ensemble of an infinite number of one-dimensional problems. Transparency per unit area coincides in this case with the mean quasi-one-dimensional transmission coefficient.

The mean conductance is determined by the competition of two factors depending exponentially on the length of the sample: the small probability and large (compared with the nonfluctuating case) transparency. Therefore we can write both terms with only logarithmic accuracy and look for the maximum of the sum of their logarithms.

Now we proceed to calculations; note that in this section we use a common dimensional system of units.

Let N_A be the equilibrium density of the acceptor impurities in the p-type layer and $N(\vec{r})$ the density in the fluctuation region. If the impurities are distributed uniformly inside the layer then the probability of fluctuations is proportional to $\exp\{-\Omega(N(\vec{r}))\}$ where [136]

$$\Omega(N(\vec{r})) = \int \left[N(\vec{r}) \ln \frac{N(\vec{r})}{N_A} + (N_A - N(\vec{r}))\right] d\vec{r}. \qquad (2.3.5.1)$$

The tunneling transparency of a typical realization with $N(\vec{r}) \approx N_A$ is proportional (with the logarithmic accuracy) to $\exp(-S)$ where

$$S = -2 \int_{-L/2}^{L/2} \sqrt{\frac{2m}{\hbar^2} V(x,0,0)}\, dx. \qquad (2.3.5.2)$$

Here m is the electron mass, $\hbar$ the Plank constant, L the thickness of the p-type layer, and $V(\vec{r})$ the electric potential. The coordinate system is chosen in such a way that the x-axis is perpendicular to the layer, while the $X0Z$ plain lies in the middle of the layer.

The electric potential $V(\vec{r})$ is not an independent quantity because it is formed by impurities themselves. In the equilibrium case, when the density of impurities is $N_A = \text{const}$, $V(\vec{r})$ solves the Poisson equation

$$\frac{d^2V}{dx^2} = -\frac{4\pi e^2}{\varkappa} N_A, \qquad (2.3.5.3)$$

where e is the electron charge and $\varkappa$ is the dielectric permeability. If the concentration of the donor impurities in the n-type semiconductor is much higher than N_A, the boundary conditions to (2.3.5.3) are

$$V|_{x=L/2} = V|_{x=-L/2} = 0. \qquad (2.3.5.4)$$

The solution of the problem (2.3.5.3)–(2.3.5.4) is

$$V(x) = V_0 \left[1 - \left(\frac{2x}{d}\right)^2\right], \qquad V_0 = \frac{\pi e^2 N_A d^2}{2\varkappa} \qquad (2.3.5.5)$$

Plugging it into (2.3.5.2) we find the transparency corresponding to the mean concentration N_A:

$$T_0 \propto \exp(-S_0), \qquad (2.3.5.6)$$

with

$$S_0 = \frac{\pi L}{2} \sqrt{\frac{2mV_0}{\hbar^2}}. \tag{2.3.5.7}$$

Consider now the fluctuation region. Here the potential $V(\vec{r})$ and the density $N(\vec{r})$ are connected by the natural formula

$$V(\vec{r}) = \frac{4\pi e^2}{\varkappa} \int G(\vec{r}, \vec{r}') N(\vec{r}') \, d\vec{r}, \tag{2.3.5.8}$$

where $G(\vec{r}, \vec{r}')$ is the Green function of the Poisson equation

$$\Delta G(\vec{r}, \vec{r}') = -\delta(\vec{r} - \vec{r}'), \tag{2.3.5.9}$$

satisfying the boundary conditions

$$G(\vec{r}, \vec{r}')|_{x=\pm L/2} = 0. \tag{2.3.5.10}$$

Thus for the mean conductance with the logarithmic accuracy we obtain

$$\ln\langle G\rangle \sim \max_{N(\vec{r})}(\Omega(N(\vec{r})) + S[N(\cdot)]) = S_{\max}, \tag{2.3.5.11}$$

where the functional $S[N(\cdot)]$ is determined by the formulas (2.3.5.2), (2.3.5.9), and (2.3.5.10).

Some straightforward calculations lead to the following result. The exponent $S_{\max}$ is the function of the ratio L_t/L, where

$$L_t = 8\sqrt{\frac{\pi}{N_A a_B}} \tag{2.3.5.12}$$

and

$$a_B = \frac{\hbar^2 \varkappa}{\pi^2 m e^2} \tag{2.3.5.13}$$

is the Bohr radius. Thus we have

$$S_{\max} = S_0 g(L_t/L), \tag{2.3.5.14}$$

where the function $g(t)$ has asymptotics

$$g(t) \approx \begin{cases} 1 - \left(\frac{1}{8} - \frac{1}{\pi^2}\right) t, & t \ll 1 \\ \frac{64 \ln^2 t}{\pi^2 t}, & t \gg 1 \end{cases} \tag{2.3.5.15}$$

This result can be simply interpreted from the physical point of view. In the case of a very thin layer, $L \ll L_t$ (or $t \gg 1$) the mean conductance is determined by the fluctuation regions where acceptor impurities are completely absent. They have a volume of the order L^3. It should be contrasted with the thick layers $L \gg L_t$ (or

$t \ll 1$) where the density of acceptor impurities on the optimal fluctuation is slightly less than the mean density:

$$\frac{\delta N_A}{N_A} \approx \frac{L_t}{L} \ll 1 \qquad (2.3.5.16)$$

and the optimal fluctuations are the Gaussian ones. In both limiting cases the mean conductance is exponentially larger than that for a nonfluctuating system with $N(\vec{r}) = N_A = $ const.

The second problem concerns the conductance of a p–n transition under the inverse electric potential. The main idea in this case is exactly the same as in the case considered above. The only difference is the explicit form of the tunneling exponent $S[N(\cdot)]$ (2.3.5.2). The final result is similar to that obtained above: tunnel currents calculated accounting for the fluctuations of the impurities density are exponentially larger than those calculated for the mean density. This result [130] is very important because of convincing experiments implying the existence of unusually large tunnel currents [142,143].

2.3.6. Transmission of the Wave Packets

The problem of the plane wave transmission through a disordered segment discussed above can be regarded as an auxiliary one from the point of view of numerous applications, where as a rule one deals with a wave packet transmission. Its transitivity may be naturally defined as

$$T_{\text{WP}} = \frac{\int_{-\infty}^{\infty} j_t(t)\, dt}{\int_{-\infty}^{\infty} j_{in}(t)\, dt} \qquad (2.3.6.1)$$

where $j_i(t)$ and $j_t(t)$ are the current densities of the initial and transmitted wave packets, respectively.

In the case of the SE (2.1.1.1) we have the current density in the form (2.2.2.8). Let us write down the initial wave packet in the form

$$\psi_{\text{in}} = \int a(k)e^{ikx - i\omega(k)t}\, dk \qquad (2.3.6.2a)$$

and the transmitted one in the form

$$\psi_t = \int a(k)t(k)e^{ikx - i\omega(k)t}\, dk, \qquad (2.3.6.2b)$$

where in the case of the SE

$$\omega(k) = k^2. \qquad (2.3.6.3)$$

Then

$$T_{\text{WP}} = \frac{\int_{-\infty}^{\infty} |a(k)t(k)|^2\, dk}{\int_{-\infty}^{\infty} |a(k)|^2\, dk} \qquad (2.3.6.4)$$

In the case of the Helmholtz equation we should take into account its initial form—the Maxwell equations (with $E \equiv u$):

$$-\frac{1}{c} \cdot \frac{\partial H}{\partial t} = \frac{\partial E}{\partial x}; \qquad \frac{1 + \varepsilon(x)}{c} \cdot \frac{\partial E}{\partial t} = -\frac{\partial H}{\partial x}. \qquad (2.3.6.5)$$

Here the current density is proportional to the product of the electric and magnetic fields:

$$j \propto EH \qquad (2.3.6.6)$$

and for the initial and transmitted fields written down in the forms

$$E_{\text{in}} = \int a(k) e^{ikx - i\omega(k)t} \, dk \qquad (2.3.6.7a)$$

and

$$E_t = \int a(k) t(k) e^{ikx - i\omega(k)t} \, dk \qquad (2.3.6.7b)$$

with

$$\omega(k) = ck, \qquad a(k) = a^*(-k), \qquad t(k) = t^*(-k), \qquad (2.3.6.8)$$

we come again to the expression (2.3.6.4).

In the case of the isotopically disordered chain with the Hamiltonian

$$H = \frac{1}{2} \sum_n [M_n \dot{y}_n^2 + K(y_n - y_{n-1})^2] \qquad (2.3.6.9)$$

the current density has the form

$$j_n = -\frac{K}{2} (\dot{y}_n - \dot{y}_{n-1})(y_n - y_{n-1}). \qquad (2.3.6.10)$$

Let the lattice spacing constant equal 1. Writing down the initial and transmitted waves in the form

$$y_{\text{in}}(n, t) = \int a(k) e^{ikn - i\omega(k)t} \, dk \qquad (2.3.6.11a)$$

and

$$y_t(n, t) = \int a(k) t(k) e^{ikn - i\omega(k)t} \, dk \qquad (2.3.6.11b)$$

respectively with

$$\omega(k) = \left(\frac{4K}{M}\right)^{1/2} \sin \frac{k}{2}, \qquad (2.3.6.12)$$

being the dispersion law in the ordered chain for $M_n \equiv M$ and $a(k)$, $t(k)$ satisfying the conditions

$$a(k) = a^*(-k), \qquad t(k) = t^*(-k), \qquad (2.3.6.13)$$

we obtain for (2.3.6.1)

$$T_{\mathrm{WP}} = \frac{\int |a(k)t(k)|^2 \sin^2 \frac{k}{2} dk}{\int |a(k)|^2 \sin^2 \frac{k}{2} dk}. \qquad (2.3.6.14)$$

Both the formulas (2.3.6.4) and (2.3.6.14) show that the wave packet transmittivity is in a sense the mean transmittivity of a plane wave averaged with the weight proportional to $|a(k)|^2$ (for (2.3.6.4)) or $|a(k)|^2 \sin^2(k/2)$ (for (2.3.6.14)). However, as we have seen above the exponential decay of the transmittivity for almost each realization of random potential does not coincide with that for the mean transmittivity (see [105]). Therefore the result of the averaging is sensitive to the form of the wave packet density $|a(k)|^2$. Here we consider three important examples.

The Schrödinger Equation. The transmittivity of the wave packet has the form (2.3.6.4) with

$$|a(k)|^2 \approx \mathrm{ch}^{-2} \left[\frac{\pi}{2\eta} (k - k_0) \right]. \qquad (2.3.6.15)$$

This form of the wave packet spectral density corresponds in the limiting case with $\eta \to 0$ to the NLSE soliton

$$\psi(x,t) = i\eta \frac{\exp\{ik_0 x - i(k_0^2 - \eta^2)t\}}{\mathrm{ch}[\eta(x - 2k_0 t)]}, \qquad (2.3.6.16)$$

It satisfies the free time-dependent version of the NLSE (2.1.2.4) with $\alpha = 2$:

$$i\dot{\psi} = -\psi'' - 2|\psi|^2\psi. \qquad (2.3.6.17)$$

The parameter k_0 in (2.3.6.16) describes the soliton velocity while η is its amplitude (note that the width of the packet (2.3.6.16) decreases as η grows). These parameters are independent.

Consider the transmittivity of the wave packet, averaged over the realizations of the potential, $\langle T_{\mathrm{WP}} \rangle$. It is defined by (2.3.6.14) with the mean transmittivity $\langle T \rangle$ in the place of that on a realization $|t(k)|^2$. We assume that the potential is Gaussian white noise (2.2.4.6). In the limiting case

$$L \gg \eta^{-1}, \ \frac{\eta^2}{(4\pi)^2 D}, \ \frac{4\pi k_0^3}{D\eta} \qquad (2.3.6.18)$$

where D is the "variance" of the white noise, we get [85]

$$\langle T_{\mathrm{WP}} \rangle = \frac{\pi^2 \eta}{2\sqrt{3}\,\bar{k}} \exp \left[-\frac{3\pi \bar{k}}{2\eta} \left(1 - \frac{2}{3} \xi^{-1/3} \right) \right]. \qquad (2.3.6.19)$$

Here

$$\bar{k} = \left(\frac{DL\eta}{4\pi} \right)^{1/3}, \qquad \xi = (\bar{k}/k_0)^3 \gg 1. \qquad (2.3.6.20)$$

Equality (2.3.6.19) shows that the mean transmittivity of the wave packet (2.3.6.15) decreases as $\exp\{-L^{1/3}\}$ instead of $\exp\{-L\}$ for the mean transmittivity of plane wave. Therefore the decay of $\langle T_{WP} \rangle$ is essentially slower.

The Helmholtz Equation. In this case the above difference is even more pronounced. Consider following ([84], see also [73]) the averaged heat flux through the disordered segment. This flux is also defined by the expression (2.3.6.4) where $|t(k)|^2$ is replaced by $\langle T \rangle$ but the localization length l is described by the formula (2.2.4.7) and the spectral density $|a(k)|^2$ is finite when k tends to zero. It is easy to see that after simple changing of variable $z = kL^{1/2}$ we obtain the result

$$\langle T_{\text{WP}} \rangle \approx L^{-1/2} \tag{2.3.6.21}$$

So in this case the decay of the wave packet transmittivity is not even an exponential one.

The Isotopically Disordered Chain. Owing to the presence of the extra factor $\sin^2(k/2)$ in the formula (2.3.6.14) the same calculations lead to the result

$$\langle T_{\text{WP}} \rangle \approx L^{-3/2}. \tag{2.3.6.22}$$

2.4. REFLECTION AND RESONANCES

2.4.1. Reflection Coefficient

The current conservation law enables us to express the transmission coefficient via the reflection amplitude:

$$T = 1 - |r|^2. \tag{2.4.1.1}$$

One can use this formula in combination with the exact equation describing the behavior of the reflection amplitude as a function of x for a given realization.

In order to obtain this equation consider the SE and its solution $\psi_2(x)$ defined in (2.3.1.2). Its logarithmic derivative

$$z(x) = \frac{\psi_2'(x)}{\psi_2(x)}, \qquad 0 \le x \le L \tag{2.4.1.2}$$

satisfies the Ricatti equation

$$z' = -(z^2 + k^2) + v(x) \tag{2.4.1.3}$$

with boundary conditions

$$z(0) = -ik, \qquad z(L) = -ik\frac{1 - r_-(L)}{1 + r_-(L)}. \tag{2.4.1.4}$$

Define function $r_-(x)$ by the equality

$$r_-(x) = \frac{ik + z(x)}{ik - z(k)}. \tag{2.4.1.5}$$

If $z(x)$ satisfies the first of the conditions (2.4.1.4) $r_-(x)$ may be considered as a reflection amplitude for a segment $[0, x]$ when the incident wave comes from the right. Simple calculations lead to the equation

$$r'_- = 2ikr_- - \frac{iv(x)}{2k}(1 + r_-)^2 \qquad (2.4.1.6)$$

with the initial condition

$$r_-(0) = 0. \qquad (2.4.1.7)$$

Let $-\Delta$ and Φ be respectively the real and imaginary parts of the logarithm of r_- (cf. (2.2.4.1)):

$$\ln r_- = -\Delta + i\Phi. \qquad (2.4.1.8)$$

They satisfy the following system of equations:

$$\begin{cases} \Delta' = \frac{v(x)}{k} \operatorname{sh} \Delta \cdot \sin \Phi \\ \Phi' = 2k - \frac{v(x)}{k}(1 + \operatorname{ch} \Delta \cdot \cos \Phi). \end{cases} \qquad (2.4.1.9)$$

and the initial condition

$$\Delta(0) = +\infty.$$

Let us recall now that for a typical realization of the potential $v(x)$ we have

$$1 - |r_-(x)|^2 \simeq \Delta(x) \simeq \exp(-x/l). \qquad (2.4.1.10)$$

Therefore we can approximate the second equation in (2.4.1.9) by the equation:

$$\Phi' = 2k - \frac{v(k)}{k}(1 + \cos \Phi), \qquad (2.4.1.11)$$

which after the simple substitution $\Phi = \pi + 2\varphi$ coincides with the second equation (2.3.1.36) for the wave function phase φ.

One very often used trick is to treat a single segment $[0, L]$ as a union of two consecutive subsegments $[0, x_0]$ and $[x_0, L]$ and study the relations between the scattering characteristics of the whole segment with those of its parts. Let us denote the reflection amplitudes from the left and the right of the subsegments by $r_{1\pm}$ and $r_{2\pm}$ and those of the whole segment by $r_\pm$. Then

$$r_+ = \frac{r_{1+}}{r_{1-}^*} \cdot \frac{r_{1-}^* - r_{2+}}{1 - r_{1-} \cdot r_{2+}} \qquad (2.4.1.12)$$

The amplitude r_- may be derived from the same formula with the obvious substitution of indexes: $1 \leftrightarrow 2$ and $+ \leftrightarrow -$. The transmission amplitude can be represented in the form

$$t_+ = \frac{t_{1+} \cdot t_{2-}}{1 - t_{1-} \cdot r_{2+}} \qquad (2.4.1.13)$$

with appropriate changes for t_-.

The denominator in (2.4.1.12), (2.4.1.13) is quite common in quantum theory and wave mechanics. It appears if one attempts to match the solutions of the SE on semiaxis. Consider for example the bound states problem on the axis for the decreasing potential. When we try to match at zero the solutions of the SE on the right and on the left semiaxis the continuity condition appears in the form

$$1 - r_-(E)r_+(E) = 0 \qquad (2.4.1.14)$$

where $r_\pm(E)$ are the reflection coefficients for the "halves" of initial potential that is for potentials, coinciding with the initial one-on-one semiaxis and vanishing on the other. The negative roots of this equation comprise the bound states (the discrete spectrum) of the problem [21].

2.4.2. Resonant Transmission

According to (2.4.1.10) we have for the reflection coefficient

$$|r(L)|^2 \simeq 1 - \exp(-L/l) \qquad (2.4.2.1)$$

when $L \to \infty$. However for any finite x there exists some set of realizations of the potential and corresponding energies such that

$$r(L) \simeq 0. \qquad (2.4.2.2)$$

They are called transparency or resonant energies. In the case $r(L) = 0$ we shall speak about the ideal transparency.

In order to study the resonant transmission it is convenient to introduce the moduli $\rho_\pm$ and the phases $\Phi_\pm$ of the reflection amplitudes (cf. (2.4.1.8))

$$\begin{cases} r_{1-} = \rho_- e^{i\Phi_-} = e^{\Delta_- + i\Phi_-} \\ r_{2+} = \rho_+ e^{i\Phi_+} = e^{\Delta_+ + i\Phi_+} \end{cases} \qquad (2.4.2.3)$$

Then (2.4.1.12) leads to

$$|r_+| = \left| \frac{\rho_{1-} - \rho_{2+} e^{i(\Phi_+ + \Phi_-)}}{1 - \rho_{1-} \cdot \rho_{2+} e^{i(\Phi_+ + \Phi_-)}} \right| \qquad (2.4.2.4)$$

and a similar expression for r_-.

For the typical realizations of the potential in both the semisegments we have $\rho_\pm \simeq 1$. Unless $\Phi_+ + \Phi_- \neq 2\pi n$, with integer n, the numerator and the denominator of (2.4.2.4) coincide with the exponential accuracy, hence we have $|r_\pm| \simeq 1$. In the case $\Phi_+ + \Phi_- = 2\pi n$ we have

$$|r_\pm| \simeq \left| \frac{\Delta_+ - \Delta_-}{\Delta_+ + \Delta_-} \right|. \qquad (2.4.2.5)$$

With a probability close to unity Δ_+ and Δ_- are small and one of them is exponentially larger than the other, hence $|r_\pm| \simeq 1$. Note also that if for one of the semisegments

$|\rho| \simeq 0$ and for the other $|\rho| \simeq 1$ (which also happens exponentially rarely) then (2.4.2.4) leads to $|r_\pm| \simeq 1$ as well.

In all these cases we obtain almost total reflection.

There are however two exceptions in the case $\Phi_+ + \Phi_- = 2\pi n$. The first one happens when $\rho_\pm \approx 1$ (that is typical) but the difference between Δ_+ and Δ_- is small compared with $\Delta_\pm$. The second one happens when both $\rho_\pm \approx 0$. In both cases we have $1 - |r_\pm| \simeq 1$. Although these two mechanisms of transparency seem to be different in fact they coincide: it suffices to shift slightly the border point x_0 to destroy the second mechanism preserving the mere fact of transparency of the realization (that does not depend upon x_0 of course).

In the case of the ideal transparency $r_\pm = 0$ the scattering states within the segment show distinct features of localization (while it is senseless to talk about it in the whole axis because of the plane-wave form of these states outside the segment). Indeed, the scattering state obeys fixed boundary conditions (matching it with the superposition of the incident wave and the reflected one from one side and the transmitted wave from the other), so almost surely this solution grows exponentially inside the segment.

These mechanisms were studied in the pioneering paper [103]. The model considered there has a potential (2.3.3.1) consisting of a succession of randomly distributed (positive) delta-functions on a constant nonrandom repulsive (positive) pedestal. The resonant states were found to be of a localized character, and the resonant realizations were classified. They included those corresponding to both mechanisms of transparency: a segment with a single δ-function and a series of such segments in a row. Later in the paper [127] was established that the transparency probability densities in the cases of a single or double well have integrable singularities at a point $|t|^2 = 1$ corresponding to ideal resonance.

The concept of resonant-transparent energies for a fixed realization was used by in [9], where the effects of resonance states on the one-dimensional conductivity were studied. A rather simple case was considered in [31]. It was a disordered system composed of parallel identical layers situated at random distances from one another. If an individual layer is transparent then the whole system is transparent as well.

Recently the interest was increased in the study of the properties of finite systems in the infinite volume limit. The system may be random or translation invariant (in the infinite volume limit). In [92] the resonant transmission through a segment representing a finite part of a periodic system was considered. They demonstrated how the set of resonant energies becomes increasingly dense and in the infinite volume limit forms an allowed band in the energy spectrum. The similar problem for electromagnetic waves in an almost periodic (an intermediate case between a random and a translation invariant) system was considered in [145]. In this case the set of resonant energies in the infinite volume limit form a self-similar (fractal) structure.

Concluding the discussion we note that there are energy regions where the resonant transmission is impossible. For instance this is the case when there is a (lower) boundary of the spectrum that is uniform in realizations (see the underbarrier transmission discussed above in Section 2.3.1). The energy region below this boundary is obviously free of spectrum. Such a region is called the nonresonant transmission region.

2.5. PROBABILITY DISTRIBUTIONS OF TRANSMITTIVITY DECREMENTS FOR HIGH ENERGIES

2.5.1. White Noise Potential

One of the most popular and effective models of the field is the Gaussian white noise model with the potential $v(x)$ being the Gaussian random process with zero mean and the correlation function

$$B(x - x') = \langle v(x)v(x') \rangle = 2D\delta(x - x'). \tag{2.5.1.1}$$

We shall start by showing that in an appropriate energy region a wide class of potentials can be approximated by the white noise. It is quite evident that for such an approximation one must assume that the correlation radius of the potential is small and the energy lies in the vicinity of the mean potential. This means practically that we have to consider some kind of an intermediate asymptotics.

To find the proper parameters consider the SE (2.1.1.1) and assume that the potential has zero mean. Delta-correlated potential (2.5.1.1) can provide an adequate approximation only for such an original one that changes rapidly enough, i.e., has a small correlation radius r_c. The natural length unit to compare with is the de Broglie wavelength λ of the state, which is of the order $E^{-1/2}$. So the condition takes the form

$$\lambda \approx E^{-1/2} \gg r_c. \tag{2.5.1.2}$$

If we choose r_c as a natural scale for x, and k_c^2 for $v(x)$ (k_c has the dimensionality of the inverse length) then its correlation function $B(x)$ has the magnitude k_c^4. Introducing these parameters explicitly we can write

$$v(x) = k_c^2 u\left(\frac{x}{r_c}\right),$$

$$B(x) = k_c^4 b\left(\frac{x}{r_c}\right) \tag{2.5.1.3}$$

where both $u(x)$ and $b(x)$ have magnitudes and characteristic ranges of order of unity, while $k_c \to \infty$ and $r_c \to 0$. Without loss of generality we shall assume that $u(x)$ is normalized by the condition

$$\int_{-\infty}^{\infty} b(t)\, dt = 1. \tag{2.5.1.4}$$

The zero mean white noise has a single parameter

$$D = \frac{1}{2} \int_{-\infty}^{\infty} B(t)\, dt \tag{2.5.1.5}$$

of dimensionality L^{-3}. Therefore to obtain the proper intermediate asymptotics we should relate k_c and r_c as follows:

$$\int_{-\infty}^{\infty} B(t)\, dt = k_c^4 r_c \int_{-\infty}^{\infty} b(t)\, dt = 2D \tag{2.5.1.6}$$

Then the natural dimensionless parameter characterizing the original potential is $D^{1/3}r_c = (k_c r_c)^{4/3}$ and in order to guarantee the white-noise approximation we must assume that it is small:

$$D^{1/3}r_c = (k_c r_c)^{4/3} \ll 1. \tag{2.5.1.7}$$

Therefore the white noise approximation is valid if parameters of the problem with a potential, given by (2.5.1.3) satisfy the following conditions:

$$E \ll r_c^{-2}$$
$$k_c r_c \ll 1$$
$$k_c^4 r_c \simeq D \tag{2.5.1.8}$$

and the obvious intermediate asymptotic that leads to it corresponds to $E, D = O(1)$, $r_c = o(1)$, and $k_c = (2Dr_c^{-1})^{1/4} = O(r_c^{-1/4})$.

This means in particular that the energy region where the potential can be replaced by the white noise is narrow compared with the range of its fluctuations.

Now we shall demonstrate that if the conditions (2.5.1.7) are fulfilled the potential can be considered as a Gaussian one. Owing to (2.5.1.2) we can average the SE over a distance Δx that is much less than the de Brogle wavelength but large compared with r_c. Then the SE will include not a local but an averaged potential:

$$v(x) \rightarrow \frac{1}{\Delta x} \int_{-\Delta x/2}^{\Delta x/2} v(x + x')\, dx' = \frac{D^{1/3}}{\Delta x} \int_{-\Delta x D^{1/3}/(2\xi)}^{-\Delta x D^{1/3}/(2\xi)} \xi^{1/2} u(t + t')\, dt'$$

with $\xi = D^{1/3}r_c \ll 1$. It is a sum of a large number $N \simeq \xi^{-1}$ of independent (since $\Delta x \gg r_c$) small ($\xi^{1/2} \simeq N^{-1/2}$) random variables, hence has a Gaussian distribution.

In the case of the Helmholtz equation (see Section 2.2.4) when $v = -k^2 \varepsilon(x)$ we get

$$k_c^2 \approx k^2, \qquad D \approx k^4.$$

Therefore when $k \rightarrow 0$ inequalities (2.5.1.2) and (2.5.1.6) hold automatically so the white noise approximation (2.5.1.1) remains valid.

Consider now the DE (2.1.2.1) and corresponding phase-amplitude dynamic equations (2.2.2.16). They have two natural length scales:

$$l_E = E^{-1}, \qquad l_\Delta = \Delta^{-1} \tag{2.5.1.9}$$

characterizing the free system and four scales l_i ($i = 0, \dots, 3$) characterizing the random potentials that we shall write in the form:

$$v_i(x) = \tilde{l}_i^{-1} \cdot u_i(x/r_c), \tag{2.5.1.10}$$

where each $u_i(\cdot)$ varies on a distance of the order of unity. To replace these potentials by white noises $w_i(x)$ we must be sure that the smallest length scale is the correlation radius r_c. Every white noise $w_i(x)$ has a single parameter l_i with the dimensionality of length:

$$\langle w_i(x) w_i(x') \rangle = 2l_i^{-1} \delta(x - x')$$

so we must have (cf. (2.5.1.5))

$$l_i^{-1} \simeq \int \langle v_i(x) v_i(x') \rangle dx' \simeq \tilde{l}_i^{-2} r_c. \tag{2.5.1.11}$$

And we obtain the analog of conditions (2.5.1.7) for the DE:

$$r_c \ll l_\Delta; \ l_i \quad (i = 0, \dots, 3)$$
$$r_c \ll l_E \simeq E^{-1} \tag{2.5.1.12}$$

The first one determines the range of parameters where the replacement is valid while the second one gives the bounds for E.

The replacement of an arbitrary potential by the white noise that is correct under the conditions (2.5.1.7) considerably simplifies the problem. Nevertheless it still remains too hard to obtain a complete enough description. Therefore to study the transmission coefficient we restrict ourselves to the high energy limiting case when $E \gg D^{2/3}$ (for SE).

We start from the system (2.4.1.9) for the amplitude and phase of the transmission coefficient and our aim is to find the probability density $p(\Delta, L)$ of the quantity $\Delta(L) = -\ln |r_-(L)|$. Then

$$\langle T \rangle = \int_0^\infty (1 - e^{-2\Delta}) p(\Delta, L) \, d\Delta. \tag{2.5.1.13}$$

In order to find $p(\Delta, L)$ we use the resonant approximation formalism (see Section 2.2). We take the potential in the form (2.2.4.8) neglecting its nonresonant harmonics. Then we obtain the dynamic equation

$$\Delta' = -\frac{1}{l} \operatorname{sh} \Delta \operatorname{ch} \Delta + \frac{U(x)}{2k} \operatorname{sh} \Delta \tag{2.5.1.14}$$

where $U(x)$ is the white noise potential with the correlation function (2.5.1.1) and l is the localization length (cf.(2.2.4.4)).

The probability density $p(\Delta, L)$ satisfies the Fokker–Planck equation (see [118], [105]):

$$l \frac{\partial p}{\partial x} = \frac{\partial^2}{\partial \Delta^2} (\operatorname{sh}^2 \Delta \cdot p(\Delta, x)) \tag{2.5.1.15}$$

with the obvious initial condition

$$\forall \Delta_0 > 0 \quad \lim_{x \to +0} \int_{\Delta_0}^{+\infty} p(\Delta, x) \, d\Delta = 1. \tag{2.5.1.16}$$

The last equation means that for $x = 0$ the whole probability is concentrated at infinity.

The system (2.5.1.14)-(2.5.1.15) can be solved explicitly [56], [78], [23]. The result is

$$\langle T \rangle = \frac{2}{\sqrt{\pi}} \left(\frac{1}{L} \right)^{3/2} e^{-L/l} \int_{-\infty}^\infty \frac{\exp(-y^2 l / L)}{\operatorname{ch} y} dy, \tag{2.5.1.17}$$

and for $L \gg l$ we obtain the asymptotic formula

$$\langle T \rangle \simeq \frac{\pi^{5/2}}{2} \left(\frac{l}{L} \right)^{3/2} e^{-L/4l}. \tag{2.5.1.18}$$

This formula shows that in the high-energy limit the mean transmission coefficient decrement γ_T (see (2.3.2.10)) equals

$$\gamma_T = \frac{1}{4l} = \frac{\gamma(E)}{2} = \frac{\bar{\gamma}}{4}. \tag{2.5.1.19}$$

The last result of course contains much less information on the reflection coefficient

$$R = |r_-|^2 = \exp(-2\Delta) \tag{2.5.1.20}$$

than the probability distribution $p(\Delta, L)$. So it is no wonder that it can be obtained in a much simpler way.

We start from the expression (2.3.2.2) for the transmission coefficient. As follows from the result of Section 2.2.4, both the variables $\gamma_{c,s}(L)$ are Gaussian with the variance proportional to L^{-1}. If the segment is long enough ($L\gamma(E) \gg 1$) the probability $\Pr\{\gamma(L) < 0\}$ with logarithmic accuracy equals $\exp(-L/4l)$. We use (2.3.2.2) and omit the integration over the negative semiaxis:

$$\langle T \rangle \leq 4 \int_{-\infty}^{\infty} \frac{p_L(\gamma)\, d\gamma}{2 + e^{2L\gamma}} \cong \int_{0}^{\infty} e^{-2L\gamma} p_L(\gamma)\, d\gamma. \tag{2.5.1.21}$$

The right hand side with logarithmic accuracy equals $\exp(-L\gamma_T)$.

The most interesting fact that follows from these arguments is that the main contribution to the integral (2.5.1.20) is made not by the most probable typical realizations of the potential that have large $\gamma(L)$, but by exponentially rare realizations that lead to $\gamma(L) \simeq 0$, so-called representative realizations.

As was shown in the end of Section 2.2.4 all the results obtained by means of the resonant approximation and averaging over the fast variables technique are valid not only for the SE, but for the DE as well. The only difference is connected with the explicit form of the localization length $l(E) = 1/2\gamma(E)$. In the simplest case $v_{1,2,3} = 0$ it is given by the formula (2.2.4.14).

2.5.2. Point Scatterers

The high-energy asymptotics for the point scatterers was obtained in [127].

Consider the interval $[0, x_0]$ containing random point scatterers with the Poisson distribution. Outside of $[0, x_0]$ we shall write the wave function in the form (2.2.2.1):

$$\psi(x) = \begin{cases} \psi_+(0)e^{ikx} + \psi_-(0)e^{-ikx}, & x < 0 \\ \psi_+(x_0)e^{ik(x-x_0)} + \psi_-(0)e^{-i(x-x_0)}, & x_0 < x. \end{cases} \tag{2.5.2.1}$$

Then

$$\begin{pmatrix} \psi_+(x_0) \\ \psi_-(x_0) \end{pmatrix} = \hat{T}^{-1} \begin{pmatrix} \psi_+(0) \\ \psi_-(0) \end{pmatrix}, \tag{2.5.2.2}$$

where $\hat{T}^{-1}$ is defined by the formula (2.3.1.10). It is convenient to parametrize the scattering amplitudes t and r_- in the following way:

$$t = \sqrt{\frac{2}{\eta+1}}\, e^{i\varphi_\alpha}, \qquad r_- = \sqrt{\frac{\eta-1}{\eta+1}}\, e^{i(\varphi_\beta+\varphi_\alpha)}. \tag{2.5.2.3}$$

The parameter η is directly connected with the transmission coefficient:

$$\eta = \frac{2}{T} - 1. \tag{2.5.2.4}$$

Since T can be expressed as a function of η, φ_α, φ_β via (2.5.2.3), we can obtain the probability density of η by integrating the joint probability density of η, φ_α, φ_β over φ_α, φ_β.

$$p(\eta \mid x_0) = \int p(\eta, \varphi_\alpha, \varphi_\beta \mid x_0)\, d\varphi_\alpha\, d\varphi_\beta. \tag{2.5.2.5}$$

Using the current conservation law

$$|\psi_+|^2 - |\psi_-|^2 = \text{const}, \tag{2.5.2.6}$$

we can introduce the new parametrization for the vector $\binom{\psi_+}{\psi_-}$:

$$\psi_+ = \sqrt{\frac{I+J}{2}}\, e^{i(\varkappa+\varphi)}, \qquad \psi_- = \sqrt{\frac{I-J}{2}}\, e^{i(\varkappa-\varphi)}. \tag{2.5.2.7}$$

Here $I(z) = |\psi_+(z)|^2 + |\psi_-(z)|^2$ is the wave intensity and $I(z) = |\psi_+(z)|^2 - |\psi_-(z)|^2 = \text{const}$ is the conserved current. The dynamics of variables I and φ can be separated from J ($J \equiv \text{const}$) and is described by explicit formulas

$$I(x_0) = \eta I_0 + \sqrt{\eta^2-1}\sqrt{I_0^2-J^2}\cos\psi, \quad \psi = 2\varphi_0 + \varphi_\alpha - \varphi_\beta \tag{2.5.2.8}$$

and

$$e^{2i\varphi(x_0)}\sqrt{I(x_0)^2 - J^2} = e^{i(\varphi_\alpha+\varphi_\beta)}$$
$$\times \left\{ \sqrt{\eta^2-1}\cdot I_0 + \sqrt{I_0^2-J^2}\cdot(\eta\cos\psi + i\sin\psi) \right\} \tag{2.5.2.9}$$

where $I_0 = I(0)$ and $\varphi_0 = \varphi(0)$. It is easy to check that the Jacobian $D(I,\varphi)/D(I_0,\varphi_0)$ of the transformation $(I_0,\varphi_0) \to (I,\varphi)$, defined by (2.5.2.8)–(2.5.2.9) is identically equal to unity, so this transformation preserves the phase volume $dI\, d\varphi$ of the system.

Now let $p_J(\Gamma \mid x)$ be the probability density of the pair $\Gamma = \{I, \varphi\}$ for the fixed current J, and $\tilde{\eta} = \{\eta, \varphi_\alpha, \varphi_\beta\}$. Then p_J solves the integral equation

$$p_J(\Gamma \mid x) = \int \delta(\Gamma - \hat{U}\Gamma_0) p(\tilde{\eta} \mid x) p_J(\Gamma_0 \mid 0)\, d\tilde{\eta}\, d\Gamma_0. \tag{2.5.2.10}$$

Here $\hat{U}$ is the operator defined by the transformation (2.5.2.8)–(2.5.2.9) with $x_0 = x$ and with the parameters $\tilde{\eta}$ corresponding to the $\hat{T}^{-1}$ matrix of the segment $[0, x]$.

It is natural to consider the point scatterer as the limit of the finite radius one when the radius tends to zero. The resulting limit matrix $\hat{T}_1^{-1}$ describing the passage of the wave across the point scatterer preserves the current and therefore can be parameterized by the same parameters as $\hat{T}^{-1}$. We shall mark these parameters with a subindex 1: $\tilde{\eta}_1 = \{\eta_1, \varphi_{\alpha 1}, \varphi_{\beta 1}\}$.

Let n be the density of the scatterers in the interval $[0, x_0]$, then the interval $[x, x + dx]$ contains one scatterer with the probability $n\,dx$ and it is free of scatterers with the probability $1 - n\,dx$ (both with accuracy up to $o(dx)$). This gives the equation for the probability density $p_J(\Gamma \mid x)$ (see [127] or [105]):

$$\frac{\partial p_J}{\partial x} + k \frac{\partial p_J}{\partial \varphi} + n(p_J - \tilde{p}_J) = 0, \qquad (2.5.2.11)$$

where

$$\tilde{p}_J = p_J(I_0(I, \varphi) \mid x) \qquad (2.5.2.12)$$

and the initial intensity $I_0(I, \varphi)$ is defined by (2.5.2.8) with parameters $\tilde{\eta}_1$ corresponding to a single point scatterer.

In the high-energy limit we can find the solution of this equation in the form of the series in the powers of k^{-1}. The zero-order approximation is independent of φ, i.e.,

$$p_J^{(0)} = p_J^{(0)}(I \mid x)$$

and the nontrivial equation for it can be obtained from the existence of 2π periodic solutions of the first-order approximation equation:

$$\frac{1}{n} \frac{\partial p_J^{(0)}(I \mid x)}{\partial x} = \frac{1}{2\pi} \int_0^{2\pi} p_J^{(0)}(I_0(I, \varphi) \mid x)\, d\varphi - p_J^{(0)}(I \mid x). \quad (2.5.2.13)$$

Using (2.5.2.6)–(2.5.2.8) and an obvious condition

$$p_1^{(0)}(\eta \mid 0) = \delta(\eta - 1) \qquad (2.5.2.14)$$

it is simple to see that $p_1^{(0)}(\eta \mid x)$ satisfies the condition

$$p_1^{(0)}(\eta \mid x) = p^{(0)}(\eta \mid x). \qquad (2.5.2.15)$$

In the case $J = 1$ equation (2.5.2.13) becomes

$$\frac{1}{n} \frac{\partial p^{(0)}(\eta \mid x)}{\partial x} = \frac{1}{2\pi} \int_0^{2\pi} p^{(0)}\left(\eta \eta_1 - \sqrt{\eta^2 - 1}\sqrt{\eta_1^2 - 1}\cos 2\varphi \mid x\right) d\varphi$$
$$- p^{(0)}(\eta \mid x). \qquad (2.5.2.16)$$

All the arguments of $p^{(0)}$ are greater than 1 therefore it is natural to represent the solution in the form of the integral over the cone functions [12]:

$$p^{(0)}(\eta \mid x) = \int_0^\infty \tilde{p}^{(0)}(t \mid x) P_{-1/2 + it}(\eta)\, dt. \qquad (2.5.2.17)$$

Using the addition theorem for Legendre functions we obtain from (2.5.2.16) the equation for $\tilde{p}^{(0)}(t \mid x)$:

$$\frac{1}{n}\frac{\partial \tilde{p}^{(0)}(t \mid x)}{\partial x} = [P_{-1/2+it}(\eta_1) - 1]\tilde{p}^{(0)}(\eta \mid x) \qquad (2.5.2.18)$$

and finally

$$p^{(0)}(\eta \mid x) = \int_0^\infty P_{-1/2+it}(\eta) \cdot t \cdot \operatorname{th} \pi t \cdot e^{nx[P_{-1/2+it}(\eta_1)-1]}dt. \quad (2.5.2.19)$$

Now the mean transmission coefficient of the disordered segment of the length L can be found using (2.5.2.4) and (2.5.2.19):

$$\langle T \rangle = 2\pi \int_0^\infty t\frac{\operatorname{th}(\pi t)}{\operatorname{ch}(\pi t)} \exp\{[P_{-1/2+it}(\gamma_1) - 1]nL\}dt. \qquad (2.5.2.20)$$

In the weak scattering limit the reflection coefficient of any single scattered tends to zero and

$$R_1 = 1 - T_1 = \frac{\eta_1 - 1}{\eta_1 + 1} \ll 1$$

so $\eta_1 \simeq 1$. In this case we have

$$P_{-1/2+it}(\eta_1) \simeq 1 - (t^2 + 1/4)R_1 \qquad (2.5.2.21)$$

and

$$\langle T \rangle = 2\pi e^{-nR_1 L/4} \int_0^\infty t\frac{\operatorname{th}(\pi t)}{\operatorname{ch}(\pi t)}e^{-t^2 nR_1 L}dt. \qquad (2.5.2.22)$$

For large systems (that is when $nR_1 L \gg 1$) we obtain the well-known result (2.5.2.6) with the localization length that equals

$$l = (nR_1)^{-1}. \qquad (2.5.2.23)$$

and has the simple physical meaning. In the general case ($\eta_1 \geq 1$) we have

$$P_{-1/2+it}(\eta_1) = P_{-1/2}(\eta_1) - t^2 f(\eta_1), \qquad (2.5.2.24)$$

instead of (2.5.2.21) and (cf. (2.5.2.6)!)

$$\langle T \rangle \simeq \frac{\pi^{5/2}}{2}\left(\frac{1}{f(\eta_1)nL}\right)^{3/2} e^{-nL[1-P_{-1/2}(\eta_1)]}. \qquad (2.5.2.25)$$

This leads to the following formula for the decrement of the mean transmission coefficient:

$$\gamma_T = n \cdot [1 - P_{-1/2}(\eta_1)]. \qquad (2.5.2.26)$$

All the results of this section are valid for the DE as well [62], modulo the fact that the form of η_1 dependence on the parameters of the DE differs from that of the SE parameters.

2.6.   OBSERVABLE TRANSPARENCY

2.6.1.   Averages and Observables

As was mentioned in the Introduction (see Section 2.1.3) the relationship between
the calculated averaged quantity and that experimentally observed deserves special
discussion. They coincide only if the quantity under discussion is the self-averaging
one and only in the infinite volume limit. If this is not the case one needs some special
experiment or/and some special result processing to interpret them.

One such trick uses the time-ergodicity of the system. In this case instead of
measuring the quantity on an ensemble of systems one observes the single system but
for a long period of time. If the microscopic state of the system changes rapidly enough
and the system is ergodic with respect to these changes one gets a fairly rich ensemble
of realizations during the experiment. For example in radiophysical measurements that
is the usual situation. The procedure proves to be quite natural and easy to perform
since the variation required is provided by the time evolution of the parameters of
natural media, e.g., the atmosphere refractive index [94], or the sea surface [20].
An example of utilization of such ergodicity in the laboratory is the backscattering
enhancement observed on two model systems [43]. The first system was a static
"fluff" of submicron-sized balls of SiO_2 in the air; to obtain the intensity peak in the
backward direction it was necessary to measure it on ten different realizations and
average the result. The other system consisted of polystyrene balls suspended in a
liquid. The thermal motion of liquid induced the Brownian vibrations of balls and a
distinct peak was formed even after a single measurement on this sample. The situation
here is essentially the same as one often use to explain why the observed value of a
thermodynamic quantity is "nonrandom." The spatial analog of the picture will be
used below in Section 2.6.2 when investigating the mechanism of the self-averaging
of the transition coefficient of a layer on an infinitely large area.

If there is a certain parameter dependence of the measurement one can use it to
provide some extra averaging. For example consider the wave propagation through a
random medium. The scattering characteristics depend not only on the spatial and time
variables but also on the frequency, some parameters of a transmitter and a receiver,
etc. This allows for the replacement of the ensemble averaging by the one over some
parameters. The averaging interval should be chosen so as to satisfy the following
conditions: it must be large enough as compared with the corresponding correlation
radius, but the ensemble averaging in every given value of a parameter(s) should give
almost the same result in this interval of parameter(s) values [138], [140]. For example
when investigating the light scattering on the rough surfaces of solids the parameter is
the light beam aperture that is the illuminated area of the surface. If it is large enough
the averaged indicatrix of the light scattering may be obtained on a single sample [33],
[37], [75], [76].

The problem of the underbarrier tunnelling outlined briefly in Section 2.3.5 provides a good example demonstrating the difference between the mean value of the
transparency and the value measured on a single realization. The result of the measurement depends on the balance of the thickness $L \gg 1$ of the layer and the area S
of its cross-section. Roughly speaking the mechanism of formation of the observable

value is the following: the transparency of the layer is of the order $\exp\{-L/L_0\}$ for the majority of fluctuations of the medium, but there are some that have transparency close to 1. The number of such fluctuations is of the order $\exp\{-L/L_1\} \cdot S/S_1$ (here L_0 and L_1 are some characteristic constants with the dimensionality of the length and S_1 a characteristic area). Therefore the resulting transparency is composed of two terms:

$$\exp\{-L/L_0\} + \exp\{-L/L_1\} \cdot S/S_1 \qquad (2.6.1.1)$$

and (for fixed L_0, L_1, and S_1) depending on the balance between L and S the first or the second term dominates in the sum.

In the next two sections we shall study this mechanism in more details for the models of the underbarrier tunnelling and that of independent filaments or wires.

2.6.2.   The Layer with the Finite Cross-Section

Consider the layer with a finite cross-section area S consisting of a large number $M = Sb^{-2}$ of separate parallel filaments with a cross-section b^2 and a length $L \gg b$. Let the flow of quantum particles be crossing it along the filaments. We suppose that the system is strongly anisotropic so the free path of the particle between two jumps from one filament to another is large compared with L so every particle follows one and the same filament throughout the layer. This situation is called a quasi-one-dimensional one. Since the transmission of particles through different filaments is independent we can write the layer transparency σ as a sum

$$\sigma = M^{-1} \sum_{i=1}^{M} T_i, \qquad (2.6.2.1)$$

where T_i is the transmission coefficient of the ith filament. Owing to the presence of the arithmetic mean over all the filaments the sum clearly is self-averaging when $M \to \infty$ while L remains constant.

We are interested in the behavior of σ as $L \to \infty$. The discussion in Sections 3.2 and 6.1 makes natural the consideration of the quantity

$$\gamma_\sigma = -L^{-1} \cdot \ln \sigma. \qquad (2.6.2.2)$$

Formula (2.6.1.1) shows that the proper scale for the number M of individual filaments in the layer must also be exponential in L:

$$M = \exp\{qL\}. \qquad (2.6.2.3)$$

The parameter q here is the natural measure of the "ensemble size." Using it we can make considerations of the previous section more precise. When $q \cong 0$ all the terms in (2.6.2.1) for large L with logarithmic accuracy are equal to $\exp\{-\bar{\gamma}L\}$. Therefore we have for this case

$$\gamma_\sigma|_{q \to 0} \sim \bar{\gamma}. \qquad (2.6.2.4)$$

where $\bar{\gamma}$ is specified by (2.3.2.5). In the opposite limiting case when $q \to \infty$ the arithmetic mean in (2.6.2.1) is close to the ensemble average, so

$$\gamma_\sigma|_{q\to\infty} \sim \gamma_T. \tag{2.6.2.5}$$

where γ_T is specified by (2.3.2.10).

The main goal of this section is to demonstrate that for large L and M defined by (2.6.2.3) the decrement (2.6.2.2) of the observed transparency is a self-averaging, i.e., asymptotically nonrandom quantity:

$$\gamma_\sigma|_{L\to\infty} = \langle\gamma_\sigma\rangle|_{L\to\infty} \equiv \gamma(q). \tag{2.6.2.6}$$

The limiting nonrandom quantity $\gamma(q)$ has asymptotics (2.6.2.4) and (2.6.2.5) in the limits $q \to 0$ and $q \to \infty$ respectively. In the general case of an arbitrary q the decrement $\gamma(q)$ is a monotonous decreasing function of q with

$$\gamma_T \leq \gamma(q) \leq \bar{\gamma}, \tag{2.6.2.7}$$

that can be obtained in the explicit form [104].

We outline the proof of the above-formulated statements. Let

$$\ln T_i = -L\gamma_i, \qquad \gamma_i = \bar{\gamma} - \xi_i, \qquad \langle\xi_i\rangle = 0, \qquad \xi_i \leq \bar{\gamma} \tag{2.6.2.8}$$

and rewrite (2.6.2.1) in the form

$$\sigma = e^{-\bar{\gamma}L} M^{-1} \sum_1^M e^{-L\xi_i}. \tag{2.6.2.9}$$

Since the potential configurations on different filaments are statistically independent, the random variables ξ_i are also independent. The probability density $p(\xi)$ of ξ_i (for any i) can be written in the form

$$p(\xi) \cong Ae^{-L\varphi(\xi)}, \qquad \varphi(0) =), \qquad \varphi(\xi) > 0 \text{ for } \xi \neq 0, \tag{2.6.2.10}$$

where $A \cong [L\varphi''(0)/2\pi]^{1/2}$ is the normalization constant. This representation remains valid when $\xi < \bar{\gamma}$ for any random potential with the finite correlation radius r_c. In this case all the events occurring on intervals separated by distances larger then r_c are independent and respective probabilities are simply multiplied.

The function $\varphi(\xi)$ is the analog of the entropy in statistical mechanics; and like entropy it is a convex function of ξ, that is, $\varphi''(\xi) \geq 0$. In the quasiclassical region for the white noise potential (see Section 2.5.1) φ has the form

$$\varphi(\xi) = \frac{\xi^2}{4\bar{\gamma}}, \qquad \xi \leq \bar{\gamma}. \tag{2.6.2.11}$$

One can find an explicit form of $\varphi(\xi)$ in another simple model—that of independent scatterers. In this model we neglect the interference between the incident and reflected waves so that transmission coefficient for a given filament is equal to the product of one-scatterer coefficients T_1:

$$T = T_1^{N_1}, \tag{2.6.2.12}$$

where the number of scatterers N_1 on a filament is a Poisson random variable, that is

$$P\{N_1 = N\} = e^{-nL}\frac{(nL)^{N_1}}{N!}. \tag{2.6.2.13}$$

Here n is a one-dimensional density of scatterers. For this model

$$\varphi(\xi) = n \cdot \tilde{\varphi}\left(\frac{\xi}{\bar{\gamma}}\right),$$

$$\tilde{\varphi}(t) = (1-t)\ln(1-t) + t \qquad (t \leq 1). \tag{2.6.2.14}$$

Now we can express the decrement γ_T of the mean transmission coefficient in terms of $\varphi(\xi)$. According to (2.6.2.8) and (2.6.2.10)

$$\langle T \rangle = A e^{-\bar{\gamma}L} \int_{-\infty}^{\bar{\gamma}} e^{L[\xi - \varphi(\xi)]}\, d\xi. \tag{2.6.2.15}$$

Thus

$$\gamma_T = \bar{\gamma} - \max_{\xi \leq \bar{\gamma}}[\xi - \varphi(\xi)]. \tag{2.6.2.16}$$

These formulas clearly demonstrate the simple mathematical mechanism that produces the differences between the two decrements of multiplicative quantities. The essence of this mechanism lies in the fact that the value of γ_T is controlled by the point where the maximum in the exponent is achieved. As we can show, the observable transparency σ (2.6.2.9) can be represented by an integral similar to (2.6.2.15), but with integration region depending on q. Therefore the corresponding maximum position also depends on the parameter q (see (2.6.2.3)). The final formula (2.6.2.6) now takes the form

$$\gamma(q) = \bar{\gamma} - \max_{\varphi(\xi) \leq q}[\xi - \varphi(\xi)], \tag{2.6.2.17}$$

that demonstrates the explicit dependence on q.

To prove (2.6.2.17) we start with formulas (2.6.2.1) and (2.6.2.8) and write γ_σ in the form

$$\gamma_\sigma = \bar{\gamma} - \frac{1}{L}\ln \int_{-\infty}^{\bar{\gamma}} e^{L\xi} \cdot M^{-1} n(\xi)\, d\xi, \tag{2.6.2.18}$$

where

$$n(\xi) = \sum_{i=1}^{M} \delta(\xi - \xi_i). \tag{2.6.2.19}$$

For large L we can replace $n(\xi)$ by $n_a(\xi)$ smoothed over any fixed interval a. We denote it by $n_a(\xi)$. Then

$$\begin{cases} \langle n_a(x)\rangle = M\int \delta_a(\xi - \xi')p(\xi')\, d\xi' \cong e^{L[q-\varphi(\xi)]}, \\ \frac{\langle n_a^2\rangle}{\langle n_a\rangle^2} - 1 \cong (\langle n_a\rangle)^{-1} \cong e^{-L[q-\varphi(\xi)]}. \end{cases} \tag{2.6.2.20}$$

This means that if $\varphi(\xi) \leq q$, $\langle n_a \rangle$ is exponentially large, while the relative fluctuations of $n_a(\xi)$ are small. On the contrary if $\varphi(\xi) \geq q$ then $\langle n_a \rangle$ is exponentially small. Therefore the integral in (2.6.2.18) asymptotically equals

$$\int_{\varphi(\xi) \leq q} e^{L[\xi - \varphi(\xi)]} d\xi, \qquad (2.6.2.21)$$

which is equivalent to (2.6.2.17) and (2.6.2.6) (for a rigorous version of the above arguments see [123].

Now we can get some corollaries. Using the above-mentioned properties of $\varphi(\xi)$ we can rewrite (2.6.2.17) in the form

$$\gamma(q) = \begin{cases} \bar{\gamma} + q - \xi_0(q) & \text{for } q \leq q_1, \\ \gamma_T & \text{for } q \geq q_1. \end{cases} \qquad (2.6.2.22)$$

Here $\xi_0(q) = \varphi^{-1}(q)$ (that is $\varphi(\xi_0(q)) \equiv q$), and $q_1 = \varphi(\xi_1)$ where ξ_1 is the minimum point of the function $\varphi(\xi) - \xi$ (that is $\varphi'(\xi_1) = 1$). In the case when it has no solutions $(\varphi'(\xi) < 1$ for all $\xi \leq \bar{\gamma}) \xi_1 = \bar{\gamma}$.

Formulas (2.6.2.17) and (2.6.2.22) show that if q is the smallest parameter with the inverse length dimensionality, then $\gamma(q) \approx \bar{\gamma}$ and (2.6.2.4) holds. If on the contrary $q \gg q_1$, then the maximum in (2.6.2.17) coincides with the maximum over all $\xi \leq \bar{\gamma}$, that is with γ_T and (2.6.2.5) holds.

It is possible to make some estimates of the fluctuations $\Delta\gamma_\sigma$ of $\gamma(\sigma)$ (see [104]). There appear to be three regimes of $\Delta\gamma_\sigma$ dependence on L:

—if $q > q_1$ the fluctuation $\Delta\gamma_\sigma$ is exponentially small and in the simplest case when $q = \varphi(\gamma)$ is of the order $\exp\{-L(q - q_1)\}$;
—if $L^{-1} < q < q_1$ we have $\Delta\gamma_\sigma \approx [L\varphi'(\xi_0)]^{-1}$;
—if at last $q < L^{-1}$ then $\Delta\gamma_\sigma \approx [L\varphi'(\xi_0)]^{-1/2}$.

This shows that the decrement of the mean transition coefficient as well as its fluctuations change continuously between the infinitely thick fiber and the single filament chosen from the ensemble.

We have obtained the expected behavior of transparency. Summing up we can state that if the number M of filaments (that is, of terms in the sum (2.6.2.1)) grows more slowly than $\exp\{L\}$ then up to the logarithmic accuracy the sum behaves like a single term (the transparency of an individual filament). If M grows exponentially with respect to L then in spite of the exponential smallness of almost all terms in the sum (2.6.2.1) the averaged transparency differs from an individual one. Finally if the growth is faster than $\exp\{q_1 L\}$ the value of transparency stabilizes at γ_T. In this respect it resembles the behavior of the thermodynamic potential in the presence of a phase transition.

In the independent scatterer approximation (see (2.6.2.12)–(2.6.2.14)) $\bar{\gamma}$ and γ_T may be calculated explicitly:

$$\bar{\gamma} = -n \cdot \ln T_1, \qquad \gamma_T = n(1 - T_1).$$

Moreover, according to (2.6.2.14) we have $q_1 = n(1 - T_1(1 - \ln T_1))$, and $\gamma(q) = \bar{\gamma}$ for $q \ll n$.

In the case described by (2.6.2.11) the maximum in (2.6.2.17) is always reached at the edge of the interval, that is at the point $\xi_0(q)$ and has the form

$$\gamma(q) = \begin{cases} \bar{\gamma}\left(1 - \sqrt{q/\bar{\gamma}}\right) & \text{for } q \leq \bar{\gamma}/4, \\ \bar{\gamma}/4 & \text{for } q > \bar{\gamma}/4. \end{cases} \qquad (2.6.2.23)$$

Here $\gamma(q)$ approaches the stable value $\gamma_T = \bar{\gamma}/4$ at $q = q_1 = \bar{\gamma}/4$ while for $q \ll \bar{\gamma}$ we have $\gamma(q) \approx \bar{\gamma}$.

2.6.3. Mesoscopic Effects

An important role in the previous discussion was played by the passage to the limit $L \to \infty$. It enabled us to replace the exact density n by the smoothed one n_a (see (2.6.2.19)), as well as to calculate the integral (2.6.2.18). In the case of finite L and M the whole picture admits the so-called mesoscopic structure.

There are two approaches to the study of these effects: the dynamical approach and the statistical approach.

We start with the dynamical approach. To get the qualitative understanding of the situation let us fix some L and some realization of the entire layer with infinitely many filaments. Then we choose at random M filaments and measure the transparency of this fiber. Formula (2.6.2.17) shows that the entire transparency is due to the presence in the fiber of (rather few) transparent filaments. If we now increase M the growth of the transmitted energy is due to the growth of the number of such filaments, and this number increases by 1 from time to time. Thus for not very large M the transparency as a function of M (or q) is not smooth, but is rather a sequence of jumps. Their positions on the q-scale and amplitudes are individual for any realization of the layer, and therefore the observed transparency manifests a typical mesoscopic behavior.

The second approach—the statistical one—supposes averaging over the ensemble of realizations of the whole layer and deals with various statistical characteristics of the transparency as a random variable.

Often an external parameter exists, that influences the transparency. In some interesting cases the most transparent filaments change from one value of parameter to another; this leads to mesoscopical oscillations of transparency. On the other hand these changes allow us to use the averaging over the external parameter instead of that over the layer realizations (see discussion on the results of measurements processing in Section 2.6.1).

In this section we shall reproduce the results on the model describing a p–n transition in the external electric field [131] (see also survey [132]).

Consider the underbarrier tunnelling of electrons through the sample (layer) of finite cross-section area. We suppose that the transparency of the layer is due to separate fluctuation regions where the density of charged impurities deviates from its mean value. The transparency of the layer can be written in the form

$$\sigma_F = \frac{1}{S} \sum_i e^{-v_i(F)} \qquad (2.6.3.1)$$

where the index i runs over all fluctuations, S is the dimensionless layer area measured in typical areas of fluctuation cross-sections, and $\exp\{-v_i(F)\}$ is the transparency of a fluctuation, depending on an external parameter F. We suppose that fluctuations are separated by the distances large compared to the fluctuation region size.

The probability density

$$f_F(Q) = \langle \delta(Q - \ln \sigma_F) \rangle \qquad (2.6.3.2)$$

of the transparency logarithm can be represented in the form

$$f_F(Q) = \frac{e^{-Q}}{2\pi} \int_{-\infty}^{\infty} \exp\left\{ it \cdot e^{-Q} + S \int_0^\infty \left(e^{-\frac{it}{S}e^{-u}} - 1 \right) \rho_F(u)\, du \right\} dt. \qquad (2.6.3.3)$$

Here $\rho(u)$ is the probability density of fluctuation with fixed value of the transparency logarithm u:

$$\rho_F(u) = \int \delta(u - v\{N(\vec{r}), F\}) e^{-\Omega\{N(\vec{r})\}} \mathcal{D}N(\vec{r}). \qquad (2.6.3.4)$$

In this formula we denote by $\exp\{-\Omega\{N(\vec{r})\}\}$ the probability density of the fluctuation $N(\vec{r})$ while its transparency equals $\exp(-v\{N(\vec{r}), F\})$.

The further calculations use the explicit form of both the exponents as functionals of $N(\vec{r})$.

We consider the p–n transition in the case of the strong compensation of the n-type semiconductor. This means that the mean densities of the donor $\bar{N}_D$ and the acceptor $\bar{N}_A$ impurities are approximately equal, i.e., $\bar{N}_D - \bar{N}_A \ll \bar{N}_D, \bar{N}_A$. Let $N_A(\vec{r})$ and $N_D(\vec{r})$ be the corresponding density fluctuations. The fluctuation $N(\vec{r})$ that increases the transparency equals

$$N(\vec{r}) = N_D(\vec{r}) - N_A(\vec{r}) - \bar{N}_D + \bar{N}_A. \qquad (2.6.3.5)$$

As was shown in [131] (see also Section 2.3.5) the main contribution to the transparency is due to Gaussian fluctuations, whence

$$\Omega\{N(\vec{r})\} = \frac{1}{2(\bar{N}_A + \bar{N}_D)} \int N^2(\vec{r})\, d\vec{r}. \qquad (2.6.3.6)$$

If we denote by $V(\vec{r})$ the potential inducted by the fluctuation $N(\vec{r})$ that is the solution of the equation

$$\Delta V(\vec{r}) = -4\pi N(\vec{r}) \qquad (2.6.3.7)$$

normalized by the condition $V(0) = 0$, the transparency logarithm in this model can be written as

$$v\{N(\vec{r}), F\} = \left(\frac{8m}{\hbar^2 E_g} \right)^{1/2} \int_0^{x_t} dx \sqrt{(Fx - V(x))(V(x) - Fx + E_g)} \qquad (2.6.3.8)$$

where m is the effective mass of the electron in the semiconductor, $\hbar$ is the Planck constant, E_g is the semiconducting energy gap, and x_t is the root of the equation $V(x_t) = Fx_t - E_g$.

To obtain an explicit form of the probability density $\rho(u)$ we must find the fluctuation $N(\vec{r})$ where the function Ω has the conditional minimum under the restriction $v\{N(\vec{r}), F\} = u$ (see formula (2.6.3.4)). The result of calculations is

$$\rho_F(u) = \exp\left\{-\frac{Q_0^2}{4u} + \frac{Q_0 F}{F_0}\right\}, \qquad (2.6.3.9)$$

where Q_0 is the large parameter of the theory, and F_0 is the scale electric field. They can be expressed via the Bohr radius $a_B = \hbar^2 \varkappa / me^2$ and the energy $E_B = me^4/2\hbar^2 \varkappa^2$ of the electron (here $\varkappa$ is the dielectric permeability of the medium):

$$F_0 = \frac{2\Gamma^2(1/4)}{3} E_g^{1/4} \cdot E_B^{3/4}(\bar{N}_D + \bar{N}_A)^{1/2} a_B^{1/2} \qquad (2.6.3.10)$$

and

$$Q_0 = \frac{\Gamma^2(1/4)}{24\pi}\left(\frac{E_g}{E_B}\right)^{5/4}[(\bar{N}_D + \bar{N}_A)a_B^3]^{-1/2}. \qquad (2.6.3.11)$$

The result (2.6.3.9) is valid for small ($F \ll F_0$) fields only.

The formula (2.6.3.9) for the probability density being established we can find and study the mean transparency as well as the probability density of the logarithm of the observable transparency (2.6.3.4). Moreover the approach used above can also be used to find the correlation coefficients of such logarithms:

$$\varphi_{F_1 F_2}(Q_1, Q_2) = \langle \delta(Q_1 - \ln\sigma_{F_1})\delta(Q_2 - \ln\sigma_{F_2})\rangle \qquad (2.6.3.12)$$

as well as the correlation function of the single fluctuation transparencies:

$$\rho_{F_1 F_2}(u_1, u_2) = \int \delta(u_1 - v\{N(\vec{r}), F_1\})\delta(u_2 - v\{N(\vec{r}), F_2\})e^{-\Omega\{N(\vec{r})\}}\mathcal{D}N(\vec{r}). \qquad (2.6.3.13)$$

For example

$$\rho_{F_1 F_2}(u_1, u_2) = \rho_{\bar{F}}(\bar{u})\left(\frac{Q_0 H}{\pi}\right)^{1/2}\frac{F_0}{\delta F \cdot \bar{u}}$$

$$\cdot \exp\left\{-\frac{Q_0 H}{\bar{u}^2}\left(\frac{F_0}{\delta F}\right)^2\left(\delta u + 4\frac{\bar{u}^2 \delta F}{Q_0 F_0}\right)^2\right\}. \qquad (2.6.3.14)$$

Here $\bar{F} = (F_1 + F_2)/2$, $\bar{u} = (u_1 + u_2)/2$, $\delta F = F_1 - F_2$, $\delta u = u_1 - u_2$, and H is some dimensionless function of the ratio F/F_0. This result is valid for $\delta u \ll \bar{u}$ and $\delta F \bar{F}$.

As we can see, the additional part of the transparency logarithm δu has the Gaussian distribution with both the mean value and the variance proportional to the deviation δF of the external field.

Consider now two fluctuations with corresponding values $u_1 > u_2$ of the transparency logarithm. For δF large enough approximately in half of all cases the inequality can change sign. This leads to mesoscopic changes of transparency due to deviations of external field. The same kind of behavior is known for the oscillations in magnetic field of a hopping and metal conductivity of small samples that arise from the multiple scattering of electrons [4], [5].

2.6.4. Random Matrices

We discuss briefly one more important mesoscopic effect that can also be described in the framework of the scattering problem. This is the so-called universal conductance fluctuations that were already mentioned in Section 2.1.3. Consider, as in Section 2.6.2, the sample of length L and the cross section S as a collection of N one-dimensional wires (N is of the order of Sk_F^2, where k_F is the Fermi momentum of the electron). But now, unlike in Section 2.6.2, we treat a sample as a N-channel one-dimensional system, taking into account the interactions between the channels.

   The corresponding scattering formalism is the natural extension of the one-channel formalism outlined in Section 2.3.1. Namely, each channel can carry two waves, propagating in opposite directions. The wave function outside the scattering region (disordered sample) is specified by $2N$ parameters. Considering these parameters as a $2N$-component vector we can define a $2N \times 2N$ transfer-matrix $\hat{T}$ by a formula analogous to (2.3.1.7), and express its entries via entries of the scattering matrix

$$S = \begin{pmatrix} r_+ & t_- \\ t_+ & r_- \end{pmatrix}, \qquad (2.6.4.1)$$

where $r_\pm$ and $t_\pm$ are the $N \times N$ matrixes of the reflection and transmission amplitudes of waves travelling to the right and to the left respectively. We obtain the N-channel analog of (2.3.1.8). By using these formulas and assuming that our sample is long enough one can show that the multichannel extension of the Landauer formula (2.1.1.6) is [24], [139]

$$g = he^{-2}G = \operatorname{tr} \frac{2}{2+X} \qquad (2.6.4.2)$$

with

$$X = \hat{T}\hat{T}^* + (\hat{T}\hat{T}^*)^{-1}. \qquad (2.6.4.3)$$

   The naive probabilistic intuition and the classical approach to the nonhomogeneous conduction [94] predict the reverse square-root-of-the-size dependence of the relative fluctuations of the dimensionless conductance (2.6.4.2). However, according to the perturbative calculations ([2], [98]) the quantum (wave) interference yields completely different relation:

$$\operatorname{Var}(g) \equiv \langle g^2 \rangle - \langle g \rangle^2 \approx 1 \qquad (2.6.4.4)$$

in the mesoscopic size range, i.e., $\operatorname{Var}(g)$ is independent of the sample size and the mean conductance (disorder). In other words, a measurement at low enough temperatures for a specific sample will show sample-specific reproducible variations with a universal amplitude. Thus the quantum (wave) nature of the low-temperature transport manifests itself in much larger than classically predicted fluctuations of the conductance. The physical reason for the quantum fluctuations is that for low enough temperatures the interference effects producing strong interchannel correlations are important and that interference or speckle patterns [126], [104] resulting from the strengthening of particular channels are sensitive to the specific arrangement of the disorder in a given sample.

According to [72] (see also [116], [112]), the explanation of this remarkable result can be obtained by using formula (2.6.4.2) and the property of spectra of random matrixes known as the spectral rigidity (see [40], [22]). This is a rather strong regularity of long sequences of eigenvalues of such matrixes, a sort of long-range order that is valid in spectral ranges where the density of eigenvalues does not change much. As a result for a wide class of linear statistics of eigenvalues the variance is independent of both the order of matrixes and the mean level distance (inverse density of eigenvalues).

According to (2.6.4.2)

$$g = \sum_{1}^{N} (1 + \lambda_i)^{-1}$$

where λ_i are N nondegenerate eigenvalues of the matrix (2.6.4.3), i.e., g is the linear statistics. Therefore $\mathrm{Var}(g)$ is not of the order of $\langle g \rangle \approx N$ as one would expect for independent λ's when the fluctuations are not universal and typically much larger than (2.6.4.4). Thus we can speak of the considerable reduction of active channels, that can be defined by the relation $\langle g \rangle^{-1} \mathrm{Var}(g) \approx N_{\mathrm{eff}}^{-1}$. For short (microscopic) samples $N_{\mathrm{eff}} \approx N$ while for long (macroscopic) ones $N_{\mathrm{eff}} \approx 1$, where the systems becomes a purely one-dimensional and single-channel. With the self-averaging $\ln g$, exponentially decaying $\langle g \rangle$ and exponentially large fluctuations (see Section 2.3.2) the assertion (2.6.4.4) now means that in the mesoscopic size we have $1 \ll N_{\mathrm{eff}} \ll N$.

It should be noted that it is far from being obvious that the random matrixes (2.6.4.3) of our problem, i.e., coming from the multicomponent SE with the random potential, satisfy the condition under which the level repulsion and the spectra rigidity were proven. It is believed, however, that these properties are valid for a much wider class of random matrices (see, e.g., [46], [77]) for the supports of this belief).

2.7. WAVES IN THE RANDOMLY STRATIFIED MEDIUM

2.7.1. The Field of the Point Source

Localization and transmission properties of one-dimensional disordered systems manifest themselves in the problem of the propagation of waves in the randomly stratified system. We demonstrate this by studying the point source field in such a layer [47]. The field $G(\vec{R}, \vec{R}_0)$ at a point $\vec{R}$ produced by the monochromatic source positioned at a point $\vec{R}_0$ satisfies the equation

$$\Delta G(\vec{R}, \vec{R}_0) + \frac{\omega^2}{c^2}\varepsilon(x)G(\vec{R}, \vec{R}_0) = \delta(\vec{R} - \vec{R}_0), \qquad (2.7.1.1)$$

with

$$\vec{R} = (\vec{\rho}, x).$$

Here ω is the source frequency, c is the light velocity, x and $\vec{\rho}$ are transversal and tangent (longitudinal with respect to the layer) coordinates, $\varepsilon(x)$ is the dielectric constant, which we suppose to depend only upon the x-coordinate. Suppose for simplicity that the source is at a point $\vec{R}_0 = (0, x_0)$ above the ideally reflecting plane $x = 0$. Then the field $G(\vec{R}, \vec{R}_0)$ satisfies the radiation condition at infinity and a self-adjoint (impedance type) boundary condition

$$\left(G + a\frac{\partial G}{\partial x}\right)\bigg|_{x=0} = 0 \qquad (\text{Im } a = 0) \tag{2.7.1.2}$$

at $x = 0$. This condition means that the reflection coefficient of the plane $x = 0$ for the plane wave with the wave number $k = \omega/c$ equals

$$r_- = \frac{ika - 1}{ika + 1}, \qquad (|r_-| = 1). \tag{2.7.1.3}$$

Let the dielectric constant $\varepsilon(x)$ in (2.7.1.1) have the form

$$\varepsilon(x) = \varepsilon_0 + \delta\varepsilon(x), \tag{2.7.1.4}$$

where $\delta\varepsilon(x)$ is a random function of x with the mean 0, correlation radius r_c and variance $\langle\delta\varepsilon^2(x)\rangle = \sigma_\varepsilon^2$.

The Fourier transform

$$\tilde{G}(\vec{k}, x) = \int G(\vec{\rho}, x)e^{-i\vec{k}\cdot\vec{\rho}}\, d\vec{\rho} \tag{2.7.1.5}$$

of the field $G(\vec{R}, \vec{R}_0)$ with respect to $\vec{\rho}$ as a function of x coincides with the Green function for the SE (2.1.1.1)

$$-\psi'' + v(x)\psi = E\psi \tag{2.7.1.6}$$

with the boundary conditions at $x = 0$ induced by (2.7.1.2) (in fact the same). In (2.7.1.6)

$$v(x) = -E_0\frac{\delta\varepsilon(x)}{\varepsilon_0}, \qquad E = E_0 - k^2,$$

$$E_0 = \varepsilon_0\frac{\omega^2}{c^2}, \qquad \psi(x) = \tilde{G}(\vec{k}, x). \tag{2.7.1.7}$$

We shall restrict our considerations to the case when fluctuations of the dielectric constant take place in a layer of the finite thickness L, i.e., in (2.7.1.4)

$$\varepsilon(x) = \begin{cases} \varepsilon_0 + \delta\varepsilon(x) & \text{for } 0 < x < L \\ \varepsilon_0 & \text{for } x \geq L \end{cases} \tag{2.7.1.8}$$

where $\delta\varepsilon(x)$ is a homogeneous random process on the whole axis with correlations that vanish at infinity (see Section 2.1.3).

Since the complete localization takes place in the transverse direction of the layer when $L \to \infty$ one can expect some manifestations of the waveguide phenomenon in such a medium. The characteristic feature of such a propagation is the cylindrical divergence of the energy flux density

$$\vec{S}(\vec{R}) = 2 \,\text{Im } G^*(\vec{R})\nabla G(\vec{R})] \tag{2.7.1.9}$$

at a large distance from the source:

$$|\vec{S}(\vec{\rho}, x)|\begin{array}{c} x = \text{const} \\ \rho \to \infty \end{array} \simeq \rho^{-1} \qquad (2.7.1.10)$$

This divergence is unambiguously associated with the existence of the nonzero flux $\Phi_d(x) > 0$ through the side surface of a cylinder of infinite radius that is bounded by the reflecting plane $x' = 0$ and a plane $x' = x$:

$$\Phi_d(x) = \lim_{\rho \to \infty} \rho \int_0^x dx' \int_0^{2\pi} d\varphi S_\rho(\rho, x'). \qquad (2.7.1.11)$$

The flux $\Phi_c(x)$ through the surface $x = x_0$

$$\Phi_c(x) = \int S_x(\vec{\rho}, x) \, d\vec{\rho} \qquad (2.7.1.12)$$

is obviously related to $\Phi_d(x)$ by the formula

$$\Phi_c(x) + \Phi_d(x) = -2 \operatorname{Im} G(\vec{R}_0, \vec{R}_0) \equiv \Phi_0 \qquad (2.7.1.13)$$

for $x > x_0$ and

$$\Phi_c(x) + \Phi_d(x) = 0$$

for $x < x_0$.

Sometimes it is useful to expand the Green function G and the fluxes in eigenfunctions of the one-dimensional boundary value problem:

$$4G(\vec{R}, \vec{R}_0) = \sum_j \psi_j^*(x)\psi_j(x_0) H_0^{(1)}(\rho\sqrt{E_0 - E_j})$$

$$+ \int_0^{E_0} +\psi_E^*(x)\psi_E(x_0) H_0^{(1)}(\rho\sqrt{E_0 - E}) dE$$

$$+ \int_{E_0}^\infty \psi_E^*(x)\psi_E(x) H_0^{(1)}(i\rho\sqrt{E - E_0}) dE \qquad (2.7.1.14)$$

and

$$\Phi_d(x) = \frac{1}{2} \sum_j \psi_j^2(x_0) \int_0^x \psi_j^2(\xi) \, d\xi, \qquad (2.7.1.15)$$

$$\Phi_c(x) = \Phi_c + \frac{1}{2} \sum_j \psi_j^2(x_0) \int_x^\infty \psi_j^2(\xi) \, d\xi \qquad (x_0 < x), \qquad (2.7.1.16)$$

where

$$\Phi_c = \Phi_c(\infty) = \frac{1}{2} \int_0^{E_0} \psi_E^2(x_0) \, dE, \qquad (2.7.1.17)$$

$$\Phi_d = \Phi_d(\infty) = \frac{1}{2} \sum_j \psi_j^2(x_0). \qquad (2.7.1.18)$$

Here $\psi_j(x)$ and $\psi_E(x)$ are the eigenfunctions of discrete $(E_j < 0)$ and continuous $(E > 0)$ spectrum normalized to the Kronecker-symbol or to the δ-function, respectively.

As we mentioned before, the criterion of the wave-guide regime is ρ^{-1}-behavior of $S(R)$ as $\rho \to \infty$. From (2.7.1.15) we see that it takes place only in the case of a nonempty discrete spectrum. But for large L we have a massive enough discrete spectrum with probability exponentially close to 1. On the other hand according to (2.7.1.16)

$$\Phi_c'(x) = -\frac{1}{2} \sum_j \psi_j^2(x_0)\psi_j^2(x).$$

Thus the flux through the plane $x = \text{const}$ decreases as $x \to \infty$ as a result of energy channelling in the transverse directions. On the other hand in view of (2.7.1.17) the discrete spectrum does not contribute to the "upward" flux Φ_c. Therefore the energy of transverse modes is "locked" within the layer and we have a "fluctuation" waveguide.

Another interesting property of this problem is the existence in its one-dimensional version of a large number of scattering states of a quasistationary nature. Namely, for a finite L the positive spectrum of the problem is absolutely continuous, i.e., consists of scattering states only. But if L is large enough those states for energies below some $E(L)$ $(E(L) \to \infty$ as $L \to \infty)$ behave as exponentially localized ones for $0 \le x \le L \to \infty$. It should be fairly interesting to construct the explicit picture of the "evolution" of these states into localized ones that only survive when $L = \infty$.

2.7.2. Mesoscopic Properties of the Energy Fluxes

From (2.7.1.18) one can see that the waveguide propagation takes place when the SE (2.7.1.6) has a discrete spectrum. In this case $\Phi_d \ne 0$ and

$$\Phi_c'(x) = -\Phi_d'(x) = -\frac{1}{2} \sum_j \psi_j^2(x_0)\psi_j^2(x). \qquad (2.7.2.1)$$

This relation means that the flux through the plane $x' = x$ decreases as $x \to +\infty$. It is the result of the energy canalization along the layer by waveguide modes associated with the discrete spectrum of SE (2.7.1.6). Each mode corresponds to an eigenvalue $E_j < 0$ and represent normal waves running along the layer. These waveguide modes depend upon ρ for $\rho\sqrt{E_0 - E_j} \gg 1$ as $H_0^{(1)}(\rho\sqrt{E_0 - E_j}) \approx \exp(i\rho\sqrt{E_0 - E_j})$. As is seen from (2.7.1.17) they do not contribute to the upward flux Φ_c when $x \to \infty$, i.e., their energy is locked within a layer of finite thickness.

It is known that in the one-dimensional case each potential well contains at least one discrete level. Therefore the realizations that have no waveguide propagation are those represented by functions $\delta\varepsilon(x) \le 0$ $(v(x) \ge 0)$ for all $x < L$. Their probability is exponentially small in the parameter L/r_c. In this respect the system under consideration is equivalent for $L \gg r_c$ to a dielectric layer that is optically denser then the environment (a potential well of a finite depth); note however that in this case the energy canalization is the purely fluctuation effect that vanishes when $\delta \varepsilon = 0$.

Now we shall consider in some more detail the distribution of the wave fields and energy flux $\Phi_d(x)$.

As (2.7.1.14) shows the height distribution of normal wave fields may be described in terms of $\psi_j(x)$. So we must study eigenfunctions $\psi(x)$.

For $E < 0$ and $x > L$ (outside the layer) $\psi(x) \approx \exp(-\sqrt{-E}(x - L))$, so the effective boundary condition for ψ at the point L can be written in the self-adjoint form

$$\left.\frac{\psi'}{\psi}\right|_{x=L} = -\sqrt{-E}. \tag{2.7.2.2}$$

Hence ψ has the same properties as eigenfunctions of closed disordered system, in particular they are localized. This means that $|\psi|$ essentially differs from 0 in the regions of a size $\approx l(E_j)$ around localization centers ($l(E_j)$ being the respective localization length). The characteristic distance Δx between these centers is of the order of $N^{-1}(E)$, where $N(E)$ is the number of states with energy less than E per unit thickness of the layer (cf. Section 2.5.2).

If the dielectric constant fluctuations σ_ε are sufficiently small and the smallest length of the problem is the correlation radius r_c, that is, if

$$r_c\sqrt{E}, \quad \frac{r_c}{\lambda_0}\sqrt{\frac{\sigma_\varepsilon}{\varepsilon_0}} \ll 1 \tag{2.7.2.3}$$

with $\lambda_0 = c \cdot \omega_0^{-1}\varepsilon_0^{-1/2}$, the random function $v(x)$ may be considered as the Gaussian white noise (see (2.2.4.6)). Such a process has a single parameter with the dimension of a length: $D^{-1/3}$ where

$$D \approx \lambda_0^{-3}\frac{r_c}{\lambda_0}\left(\frac{\sigma_\varepsilon}{\varepsilon_0}\right)^2. \tag{2.7.2.4}$$

By dimensional arguments $N(0) \approx D^{1/3}$ and $\Delta x \approx D^{-1/3}$. So the essential difference between a randomly stratified layer and a regular dielectric waveguide is the highly inhomogeneous dependence of the wave mode of fields on the transverse coordinate.

The difference in behavior shows also the energy flux Φ_d carried by the discrete spectrum waves along layers. From (2.7.1.15) it follows that the contribution to Φ_d of the jth state on the level x equals $\frac{1}{2}\psi_j^2(x_0)\psi_j^2(x)$ and is exponentially small when $|x - x_j| > l_j$ ($\psi_j^2(x)$ is small) or $|x_0 - x_j| > l_j$ ($\psi_j^2(x_0)$ is small). Therefore the total flux along the thick ($L \gg l$) layer is generated by a comparatively few waves for which $|x_0 - x_j| \le l_j$ and essentially differs from zero only for $|x - x_0| < l$. The integral (total) flux is of the order of

$$\Phi_d \approx (\Delta x)^{-1} \approx D^{1/3}. \tag{2.7.2.5}$$

The flux $\Phi_d(x)$ is not a self-averaging quantity and therefore has a mesoscopic fine structure depending on the individual properties of the realization. This structure may be observed by studying the differential flux at the height x:

$$\Phi_d'(x) = \frac{1}{2} \sum_j \psi_j^2(x_0)\psi_j^2(x). \tag{2.7.2.6}$$

It is combined of a number of small peaks near all localization centers that grow near x_0 and produce there a large peak that also has some fine structure.

The distances between these peaks are of the same order as the distances between the localization centers, that is Δx. One can pose a kind of an inverse problem: by the analysis of the flux for different positions of the source x_0 and its frequency to restore the localization centers x_j of the modes (i.e., the most transparent in the longitudinal direction parts of the layer that concentrate the canalization energy) and the wavefunction values $\psi_j(x_j)$ at these centers. The set of these values is a characteristic of the realization of our random layer such as the so-called "magnetofingerprints" in mesoscopic conductors (that is the dependence of conductivity on the magnetic field).

So far we discussed the propagation due to the discrete part of the spectrum $E_j < 0$ that resulted in waveguide propagation observed in almost (modulo exponentially rare) all realizations of the random medium. However the system is open. This fact manifests itself by the presence of the continuum spectrum (above zero). The part of the energy belonging to it propagates outside the system. At the same time it greatly enhances (as compared with the regular structures) the waveguide effect (see below Section 2.7.3).

Let us consider the flux Φ_c (2.7.1.17) emerging from the layer and rewrite it in the form

$$\Phi_c = \int_0^{E_0} \rho(E)\, dE \tag{2.7.2.7}$$

where $\rho(E)$ is the radiation energy current density per unit interval of the spectrum parameter E. It can be interpreted also as the density of the angular (in $\theta = \arccos\sqrt{E/E_0}$, $dE = E_0 \sin 2\theta\, d\theta$) distribution of the upward (outgoing) flux. In the particular case $x_0 \to +0$ and $r_- \to 1$ $(a \to \infty)$ this density is

$$\rho(E) = \frac{1 - |r_+(E)|^2}{2\pi\sqrt{E}\,|1 - r_+(E)|^2}, \tag{2.7.2.8}$$

where $r_+(E)$ is the reflection amplitude of the plane wave with energy E incident from the left to the disordered segment $[0, L]$.

It follows from (2.4.1.9) that for $E \gg |v|$ the phase $\Phi_+(E) = \mathrm{Arg}\, r_+(E)$ of the reflection amplitude behaves as $2L\sqrt{E}$. Since at a typical realization $|r_+(E)| \approx 1 - O(\exp\{-L/l\})$ we have for fixed E as a rule

$$\rho(E) \approx e^{-L/l}. \tag{2.7.2.9}$$

There are however exceptions—the values of the parameter E for which

$$\Phi_+(E_n) = 2\pi n. \tag{2.7.2.10}$$

Then

$$E_n = \frac{n^2\pi^2}{L^2}, \tag{2.7.2.11}$$

the denominator in (2.7.2.8) becomes small:

$$|1 - r(E)|^2 \approx e^{-2L/l} \tag{2.7.2.12}$$

and $\rho(E)$—large:

$$\rho(E_n) \approx \exp\left(\frac{L}{l(E_n)}\right). \tag{2.7.2.13}$$

This forms the peaks of $\rho(E)$. The distance $\Delta E_n = E_{n+1} - E_n$ between the peaks is according (2.7.2.11) of the order of $\Delta E_n \approx 2\pi^2 n/L^2$ and the half-width of peaks δE_n as related to the difference of $|r_+|$ from unity is

$$\delta E_n \approx \frac{1}{2\pi}\Delta E_n e^{-L/l(E_n)}. \tag{2.7.2.14}$$

All the arguments presented above show that the flux emerging from the layer has a strongly inhomogeneous angular distribution and together with Φ_d possesses a fine mesoscopic structure. The radiation undergoes a sort of focusing near the values $\theta_n = \arccos\sqrt{E_n/E_0}$ that corresponds to the values of E where $\rho(E)$ has maxima [49].

2.7.3. *Quasihomogeneous Waves. Fluctuation Waveguide*

The meaning of the values E_n in (2.7.2.11) becomes clear if we rewrite the solution of (2.7.1.6) for $x > L$ in the form

$$\begin{aligned}\psi(E, x) = {}&[1 - r_+(E)]t^*(E)\exp\{-i\sqrt{E}\,(x - L)\}\\ &+ [1 - r_+^*(E)]t(E)\exp\{i\sqrt{E}\,(x - L)\},\end{aligned} \tag{2.7.3.1}$$

where $t(E)$ is the transmission amplitude. For real E we can find both the incident (from the right) and the reflected wave in (2.7.3.1). If $\mathcal{E}_n = E_n - \delta_{1n} - i\delta_{2n}$ is complex some important changes take place. In this case $|t(\mathcal{E}_n)|^2 \neq 1 - |r_+(\mathcal{E}_n)|^2$ and $r_+^*(\mathcal{E}_n) \neq (r_+(\mathcal{E}_n))^*$, so if we choose $\mathcal{E}_n$ to satisfy the condition

$$r_+(\mathcal{E}_n) = 1, \tag{2.7.3.2}$$

$1 - r_+(\mathcal{E}_n)$ becomes zero and there remains only the outgoing wave in (2.7.3.1):

$$\psi(\mathcal{E}_n, x) = [1 - r_+^*(\mathcal{E}_n)]t(\mathcal{E}_n)\exp\{i\sqrt{E_n}(x - L)\}. \tag{2.7.3.3}$$

Such a wave-function is known in quantum mechanics as a decay state. Due to the time dependence $\propto \exp\{-i\mathcal{E}_n t\}$ it decreases with a characteristic time $\tau \propto \delta_{2n}^{-1}$. It is quasistationary if $\delta_{2n} \ll E_n - \delta_{1n}$. Such states play an important role in the theory of nuclear reactions and decays [13]. Note that in our case such states appear even for an energy that is high compared to the barrier, and the localization is generated by the constructive interference of the waves multiply scattered on the fluctuations of the potential [48].

In the case $L \gg l(E)$ the reflection amplitude differs from unity by an exponentially small quantity, and we can approximate $\mathcal{E}_n$ by $E_n - i\delta_{2n}$ with

$$\delta_{2n} = \frac{1}{2}\delta E_n = \frac{1}{4\pi}\Delta E_n \exp[-L/l(E_n)]. \tag{2.7.3.4}$$

(the shift δ_{1n} of the real part of E_n with the same accuracy equals zero). Thus the values E_n for which the radiation energy-current density $\rho(E)$ see (2.7.2.7)) has peaks are the real parts of the solutions of (2.7.3.1).

Note that although the wave-functions $\psi(E_n, x)$ are exponentially localized within the layer they do not decay to zero at infinity due to oscillating tails outside the layer. Therefore they are not normalizable and localization is similar to that of resonant states inside the layer (see [103], [9]). To explain this result we note that for $x < L$ and $r_- = 1$ according to (2.2.2.10) the eigenfunction $\psi(E_n, x) \propto \psi_c(E_n, x) = \exp\{\xi_c\}\sin\varphi_c$ grows exponentially from the point $x = 0$ (see Section 2.2.3), and use the identity

$$\psi_c^2(E_n, L) + E_n^{-1}\psi_c'^2(E_n, L) = \frac{|1 - r_+(E_n)|^2}{1 - |r_+(E_n)|^2} \approx e^{-L/l(E)}. \tag{2.7.3.5}$$

Once we have understood the quasistationary states we can study the dependence of the field on the longitudinal coordinate $\vec{\rho}$. The study of the formula (2.1.1.6) shows that integrals there can be calculated as a sum of the residues of the denominator poles and the integrals along the branching-cut sides [21]. Replacing $k^2\sin^2\theta$ by $E_0 - \mathcal{E}$ we can rewrite the residue sum as follows:

$$G = \sum_n G_n(x, x_0, \mathcal{E}_n)e^{ik_n\rho}, \tag{2.7.3.6}$$

where $k_n = \sqrt{E_0 - \mathcal{E}_n}$ and $\mathcal{E}_n$ is the root of the dispersion equation

$$1 - r_+(x_0)r_-(x_0) = 0, \tag{2.7.3.7}$$

(here $r_\pm(x_0)$ are the reflection amplitudes for the regions $[o, x_0]$ and $[x_0, \infty]$ with the right (left) incident wave).

It is not difficult to prove that the set of solutions $\mathcal{E}_n$ of the equation (2.7.3.7) does not depend on the point x_0, so for $r_- = 1$ it coincides with that of equation (2.7.3.2). This set includes solutions corresponding to quasistationary states. In our case the role of time is played by the distance ρ of the source from the observation point in the plane (y, z). Therefore quasistationary states appear in our situation as quasihomogeneous waves attenuating at L/l exponentially large distances D_n from the source:

$$D_n \approx \lambda_0 \exp(L/l). \tag{2.7.3.8}$$

This attenuation is due not to the dissipation but rather to the upward emergence of the field (for $x > L$). However for $\rho < D_n$ quasihomogeneous waves are locked within the layer.

To sum up the discussion we can state that the randomly stratified layer is an effective fluctuation waveguide. It differs from the usual dielectric one by the essential role of the continuum spectrum. The flux from this spectrum is formed by quasistationary states whose energy emerges from the layer but does so only at exponentially large distances from the source. The part of the field that is associated with the continuous spectrum and is described by the integral terms in the expansion (2.7.1.14) contains the sum over quasistationary states and corresponds to weakly attenuating quasihomogeneous waves canalizing energy along the layer for extremely long distances.

2.8. DISORDER AND NONLINEARITY

Two words appearing in the title of this chapter are related to different but very complex and interesting branches of modern theoretical physics. The section contains several problems that deal with more or less the same type of the disorder but various types of nonlinearity.

We begin in Section 2.8.1 with the problem of nonlinear sound amplification in superconductors. This section deals with a linear transmission problem and the nonlinearity manifests itself in the final physical results only. The problem considered in Section 2.8.2 is nonlinear from the very beginning, but all the relevant information can be extracted from the solution of a certain linear transmission problem. In Section 2.8.3 the transmission of the plane wave through a nonlinear disordered segment is studied. This problem contains all the most essential features of nonlinearity. And in the last section we study the "most" nonlinear problem: the soliton transmission in a nonlinear medium through disordered segment.

2.8.1. Nonlinear Sound Absorption in Superconductors

In this section we discuss the kinetic phenomena associated with the transmission of irregular signals of a finite length through translation-invariant media. We mention the sound propagation in metals, superconductors, and superfluid ^{3}He, the propagation of helical waves in metals and superconductors, etc. All these problems possess a common small parameter—the interatomic distance divided by a characteristic wavelength of the signal. It seems to enable one to use a semiclassical description of quasiparticles in terms of the dispersion law. But Keldysh [80] noted that even a regular signal (i.e., a periodic train of ultrasonic waves) affects the quasiparticle spectrum and gives rise to a special band structure with a number of forbidden energy bands near the Landau resonance. In the resonance region the dynamics of quasiparticles in a one-dimensional field of a plane wave depends on their longitudinal velocity. This velocity is of the same order as the wave velocity and is small compared with the characteristic Fermi velocities. This violates the semiclassical condition and leads to essentially quantum behavior of quasiparticles. In the collisionless limit when the relaxation times of the system are large compared with characteristic dynamic times and the wave interaction with the quasiparticles of the environment is coherent, quantum interference effects are predominant. The system density matrix is far from being equilibrium one and for this reason the resonance wave characteristics (such as the signal wave energy dissipation rate) are nonlinear in the amplitude of the signal. These effects should be even more pronounced in the case of the irregular signal.

At the beginning of the more detailed study note that since resonance particles occupy a small volume in the phase space, their effect on the wave parameters should be characterized by times large in the scale of the characteristic inverse frequencies of the wave packet. Therefore we can assume that in the case of dispersionless medium the wave field is stationary and write the wave equation in the coordinate system that moves with a signal in the form [17]

$$[\hat{\varepsilon}(p_\perp, \hat{p}) - s\hat{p} + \hat{V}(p_\perp, x) - E]\vec{\psi} = 0. \qquad (2.8.1.1)$$

Here $\hat{p} = -i\frac{\partial}{\partial x}$, $p_\perp$ is the conserved transverse momentum, $x = X - st$, s is the wave velocity, and $\hat{\varepsilon}$, $\hat{V}$ are the operators of the longitudinal kinetic and potential energies (in the general case they are matrixes). Note that the potential energy depends on the transverse momentum as a parameter and plays the role of a static random potential.

In the general case a complicated form of the kinetic energy operator $\hat{\varepsilon}$ implies the multichannel scattering resulting in a complex structure of wave-functions. However even in the case of the two-channel scattering we have two possibilities. They describe the cases when free particles of the same energy belonging to different channels move in the same or in opposite directions (in the moving coordinates).

In the former case (when waves describing free particles may propagate in opposite directions) backscattering may appear, and thus in a one-dimensional system with a random potential all the states are localized [19]. The possibility of localization follows also from the form of the current-conservation law (see below). Namely, in this case for an arbitrary state that is a linear combination of two free wave functions with the same energy, the difference of the squares of the corresponding coefficients' moduli is fixed. Therefore there may exist states that are zero at infinity and are not zero near the center of the system. This case is realized for example in a metal with an almost spherical Fermi surface. Then for small resonant momenta the dispersion law is quadratic here and in view of the Galilean invariance (2.8.1.1) is reduced to the Schrödinger equation.

In the latter case when waves describing free particles propagate in the same direction there is no backscattering. Therefore all the states in such an infinite one-dimensional system are delocalized even for a random potential [19]. This fact is also evident from the form of the current conservation law (see (2.8.1.15)): in this case the sum but not the difference of the coefficient square moduli is fixed and a state vanishing at infinity equals to zero identically.

This character of dynamics is directly reflected in the amplitude dependence of the nonlinear absorption of a stochastic signal by a medium. The wave packet energy dissipation is determined by the energy transfer to quasiparticles during their inelastic (in the laboratory coordinates) scattering by the wave packet. The rate of the energy transfer from the packet to quasiparticles is [17]

$$Q = (\vec{\psi}, \hat{\varepsilon}\vec{\psi}) = \sum_{p_\alpha, E} \sum_{\alpha, \beta} W_{\alpha, \beta}(\varepsilon_\alpha - \varepsilon_\beta)[n_F(\varepsilon_\alpha) - n_F(\varepsilon_\beta)], \qquad (2.8.1.2)$$

where $\varepsilon_{\alpha, \beta}$ are the branches of the free ($\hat{V} \equiv 0$) particle spectrum in the laboratory coordinate system, $W_{\alpha\beta}$ is the corresponding scattering probability and $n_F(\varepsilon)$ is the

Fermi function. This formula is similar to the well-known Landauer formula (2.1.1.7) because it relates the kinetic quantity (the energy transfer rate) to the characteristics of the scattering problem. This characteristic is for the equation (2.8.1.1) the scattering probability. It coincides with the reflection coefficient of a quasiparticle from the wave packet in the former case (the backscattering is present) and the transformation coefficient in the latter (with no backscattering). For a stochastic signal this probability is of the order of unity in a wide energy range (up to semiclassical). Therefore when estimating Q one may replace the probability of scattering for each realization by its average $\langle W \rangle$ where the main contribution is provided by typical realizations of the signal [105]. As a result its absorption rate is considerably higher than that for a periodic wave train and does not depend upon the wave amplitude.

We shall illustrate the above arguments by the example of the interaction of electrons in a superconductor with a random longitudinal sound wave packet. This system is of an interest for two reasons. First, both cases discussed above take place in this system. Second, in view of the recently developed technique of pulse injection of the strong sound into a sample it can be compared with an experimental situation.

To begin with we shall consider the absorption of a signal with a short correlation radius. The wave equation in this model is [17] the modified Dirac-type equation ((MDE), see (2.1.2.3)):

$$-i(\sigma_3 - \beta)\frac{d\vec{\psi}}{dx} + \Delta\sigma_1\vec{\psi} + v(x)\vec{\psi} = E\vec{\psi}. \qquad (2.8.1.3)$$

Here β is the sound velocity measured in x-projections of the electron (Fermi) velocity (from now on we set this projection equal to unity), $v(x) = \beta\Phi(x)$, where $\Phi(x)$ is the electric potential of the sound wave (see [55]). Other notation coincides with that from Section 2.1.2. In fact (2.8.1.3) is nothing but the Dirac equation written in the coordinate system moving with the velocity of sound. It is convenient to introduce (as was done for the DE in (2.2.2.13)) the two-component vector $\vec{A} = (A_+, A_-)$ of coefficients from the representation of $\vec{\psi}$ as a linear combination of free solutions $\vec{\psi}_\pm^{(0)}(x) \propto e^{ixp_\pm(E)}$ of MDE (2.8.1.3).

1. When $\beta < 1$ quasiparticles with the same energy E but belonging to two different branches of the spectrum $p = p_\pm(E)$ have velocities $v_\pm = (dp_\pm/dE)^{-1}$ with opposite signs. In $\vec{A}$ variables the current conservation law (see Section 2.2) has the form

$$j = (\vec{A}, \sigma_3\vec{A}) = |A_+|^2 - |A_-|^2 = \text{const}. \qquad (2.8.1.4)$$

Therefore in this case all the states in a random potential have to be localized, i.e., the spectrum of equation (2.8.1.3) on the whole axis is pure point. The proof of this statement consists of two steps. The first one is the proof of the positiveness of the Lyapunov exponent, which follows closely the proof for DE (2.1.2.1) in Section 2.2.3. In accordance with [91] the second step for Markov random potential consists of the proof of boundedness of conditional expectations of the density of spectral measure of the problem for fixed values of potential on each of the semiaxes. This fact also holds for a wide class of Markov potentials [19]. For the Gaussian white noise potential the quantity in interest is not just bounded but also a continuous function of the energy.

The current conservation law (2.8.1.4) implies the following parametrization for $\vec{A}$:

$$\vec{A} = j^{1/2} \left(e^{-i\varphi} \, \text{ch} \, \frac{\theta}{2}, \, \text{sh} \, \frac{\theta}{2} \right). \tag{2.8.1.5}$$

Let the "potential" $\Phi(x)$ be a section of a spatially homogeneous random function with

$$\langle \Phi \rangle = 0, \qquad \langle \Phi(x) \Phi(x') \rangle = B(x - x'). \tag{2.8.1.6}$$

We suppose that its correlation radius r_c is the smallest length of the problem and has the order of the characteristic wave length of the sound packet: $r_c \propto s/\omega$. Then for high energy the probability density $p(\theta, x)$ satisfies the Fokker–Planck equation

$$l \frac{\partial p}{\partial x} = \frac{\partial}{\partial \theta} \left(\text{sh} \, \theta \frac{\partial}{\partial \theta} \frac{p}{\text{sh} \, \theta} \right) \tag{2.8.1.7}$$

with the characteristic length

$$l = (2b^2 \tilde{B}(p))^{-1}, \tag{2.8.1.8}$$

where

$$b^{-2} = s^2 \frac{|1 - \beta^2|}{(\beta \Delta)^2} [E^2 - \Delta^2 (1 - \beta^2)], \tag{2.8.1.9}$$

and

$$p = p_+(E) - p_-(E). \tag{2.8.1.10}$$

$\tilde{B}(p)$ is the Fourier transform of the correlation function (2.8.1.6):

$$B(p) = \int_{-\infty}^{\infty} B(x) \cos px \, dx. \tag{2.8.1.11}$$

The equation (2.8.1.7) is close to the equation (2.5.1.15) of the theory of the propagation through a one-dimensional disordered segment modelled by the white noise [105]. Note that (2.5.1.15) can be reduced to (2.8.1.7) by a simple change of variables. Therefore if $\beta < 1$ (that is, in the region of normal scattering) and the length of the signal $L \gg l$ (see (2.8.1.8)), quasiparticle states are localized within the region of size l and the mean reflectivity coinciding with the scattering probability is exponentially (in $L \to \infty$) close to unity.

Note that in the case $\beta = 0$, $s = 1$ the localization length l from (2.8.1.8) coincides with that (see (2.2.4.14)) for the DE (2.1.2.1). Moreover all the known facts of the one-dimensional Schrödinger equation theory with random potential remain valid for the normal scattering case.

Before calculating the acoustic absorption rate we remark that for any signal parameters the contributions to the rate come both from the states with small and with large localization length. This may be seen from (2.8.1.2) and (2.8.1.8). The contribution from the large localization length states may be calculated in the Born approximation. If they play the dominant role, the absorption may be adequately described in the linear approximation; if they do not, the problem is essentially nonlinear.

Now we can proceed to calculations.

Let the Fourier transform of the correlation function have an exponentially decreasing short-wave asymptotic:

$$\tilde{B}(k) \approx \Phi_0^2 r_c \exp(-\alpha |k| r_c), \qquad \alpha \approx 1 \qquad (2.8.1.12)$$

(Φ_0 is the characteristic signal amplitude). Then at high energies ($p(E)r_c \gg 1$) the localization length grows exponentially (see(2.8.1.8)), so we are in the quasiclassical region. Therefore the nonlinear absorption takes place when $p \approx r_c^{-1}$ and the localization length $l(p)$ does not exceed the wave packet length L.

We conclude that in the low-temperature limit, $T \ll s/r_c \ll \Delta$ the main contribution to the absorption rate is given by the phase region with $l^{-1} \approx \Phi_0^2 r_c p/s\Delta \cdot \exp(-\alpha p r_c)$. Now we see that the nonlinearity parameter here is $\Phi_0^2 Lp/s\Delta$. The final formulas for absorption are

$$Q \approx N(O)(s^2/v_F)T\Delta e^{-\Delta/T} \cdot \sqrt{s/Tr_c}\Phi_0^2 Lp/s\Delta,$$
$$\Phi_0^2 Lp/s\Delta \ll 1$$
$$Q \approx N(O)(s^2/v_F)T\Delta e^{-\Delta/T} \cdot \sqrt{a/Tr_c}\ln(\Phi_0^2 Lp/s\Delta),$$
$$\Phi_0^2 Lp/s\Delta \gg 1. \qquad (2.8.1.13)$$

Here $N(O)$ is the density of states at the Fermi energy and v_F is the Fermi velocity.

We see that in the nonlinear case ($\Phi_0^2 Lp/s\Delta \gg 1$) the absorption dependence on the signal parameters Φ_0 and L is weak. The reason is contribution of all the quantum states whose size does not depend on these parameters being entirely determined by the correlation radius r_c. For the same reason the nonlinear absorption for an irregular signal essentially exceeds that for a periodic signal that effectively reflects quasiparticles only in a narrow energy band of the order of Φ_0 [80].

At higher ($s/r_c \ll T \ll \Delta$) temperatures in the strong nonlinearity limit ($\Phi_0^2 Ls \gg \Delta^4 r_c^3 T^3$) the absorption becomes qualitatively similar to (2.8.1.12):

$$Q \approx N(O)(s^2/v_F)T\Delta e^{-\Delta/T} \cdot \ln(\Phi_0^2 Ls^2/\Delta^4 r_c^3 T^3). \qquad (2.8.1.14)$$

The region $\beta < 1$ physically corresponds to scattering of thermal excitations in the superconductor. This is reflected by the presence of temperature exponents in (2.8.1.12) and (2.8.1.13), associated to the freezing of thermal excitations in the superconductor at low temperatures. This absorption mechanism exists also in normal metals where for quadratic electron dispersion law the expression for nonlinear absorption is

$$Q \approx N(O)(s^2/v_F)(m^{*2}r_c^4)^{-1} \cdot \ln(\Phi_0^2 Lm^{*2}r_c^4), \qquad (2.8.1.15)$$

where m^* is the effective mass of electron. In this case the nonlinearity parameter is $\Phi_0^2 Lm^{*2}r^4 \gg 1$.

At zero temperature the nonlinear absorption (2.8.1.12) due to thermal excitations scattering is zero. But there is another mechanism of sound absorption associated with destruction of the Cooper pairs and creation of excitations. In the problem (2.8.1.3) this process is described by the scattering for $\beta > 1$. Then the coordinates system moving with the velocity s "leaves particles behind itself," so velocities corresponding to both branches of the spectrum have the same sign. Therefore the scattering is anomalous: instead of a reflected wave there appears a transformed one moving in the same direction as the transmitted one. Accordingly the current conservation law now looks like

$$j = -|A_+|^2 - |A_-|^2 = \text{const} \qquad (2.8.1.16)$$

instead of (2.8.1.4).

This allows us to regard all the states in this case as delocalized irrespective of the nature of the potential (whether it is random or not).

To prove this statement consider the Green function $\mathcal{G}^{(0)}(x - x', \mathcal{E})$ of the free operator

$$\mathcal{H}_0 = -i(\sigma_3 - \beta)\frac{d}{dx} + \Delta\sigma_1, \qquad (2.8.1.17)$$

at complex values of the energy $\mathcal{E}$. The Green function satisfies the equation

$$(\mathcal{E} - \mathcal{H}_0)\mathcal{G}^{(0)}(x - x', \mathcal{E}) = \hat{I} \cdot \delta(x - x'), \qquad (2.8.1.18)$$

where $\hat{I}$ is the 2×2 unit matrix. The matrix elements of the Green function are

$$(\mathcal{G}^{(0)}(x - x', \mathcal{E}))_{ij} \equiv G_{ij}^{(0)}(x - x', \mathcal{E}) = \langle x_i | (\mathcal{E} - \mathcal{H}_0)^{-1} | x_j' \rangle. \qquad (2.8.1.19)$$

For $x \neq 0$, $\mathcal{G}^{(0)}(x)$ is a linear combination of the free states $\vec{\psi}_\pm^{(0)}$ and the coefficient of the growing exponent must be zero. But for $\delta > 0$ we always have $\text{Im } p_\pm(\mathcal{E}) > 0$ and both the waves grow as $x \to -\infty$. Therefore $\mathcal{G}^{(0)}(x) = 0$ when $x < 0$, i.e., the Green function is triangular for complex energy in the coordinates representation:

$$G_{ij}^{(0)}(x, x'; \mathcal{E}) = 0, \qquad (x < x', \text{ Im } \mathcal{E} < 0). \qquad (2.8.1.20)$$

Now consider the Green function $\mathcal{G}(x, x'; \mathcal{E})$ of the operator

$$\mathcal{H} = -i(\sigma_3 - \beta)\frac{d}{dx} + \Delta\sigma_1 + v(x) \qquad (2.8.1.21)$$

with a potential that satisfies the condition

$$\int_a^b |v(x)|dx < \infty \qquad (2.8.1.22)$$

for any $-\infty < a < b < \infty$ (that is locally summable).

The Green function $\mathcal{G}$ is also triangular. To demonstrate this we use the identity $\mathcal{G} \equiv \mathcal{G}^{(0)} + \mathcal{G}^{(0)}v\mathcal{G}$ to expand $\mathcal{G}$ in a perturbation theory series. Then we use (2.8.1.19) to get the estimate for the nth order term when $x < x'$:

$$\left| \left(\mathcal{G}^{(0)} \underbrace{v\mathcal{G}^{(0)}v \cdots v}_{n} \mathcal{G}^{(0)} \right)_{ik} (x, x') \right| \leq \frac{\left(2G_0 \int_{x'}^x |v(t)| \, dt \right)^n}{n!}, \qquad (2.8.1.23)$$

where
$$G_0 = \max_{y,z \in [x',x]} |G_{ij}^{(0)}(y,z)|.$$

Straightforward calculations show that G_0 is finite, so the series (2.8.1.22) converges for all x, x' and $\mathcal{E}$ with Im $\mathcal{E} < 0$ to an analytic function for Im $\mathcal{E} < 0$ (x, x' being fixed). It is triangular together with $\mathcal{G}^{(0)}(x, x'; \mathcal{E})$ in the coordinate representation:

$$G_{ij}(x, x'; \mathcal{E}) = 0 \qquad \text{if } x < x', \ Im\mathcal{E} < 0. \qquad (2.8.1.24)$$

This property is valid for any potential that is locally summable (see (2.8.1.21)). A random potential for which the mean value of the l.h.s. of (2.8.1.21) is finite is locally summable with probability 1. By virtue of the spatial homogeneity of $v(x)$ the last condition takes the form

$$\langle |v(0)| \rangle < \infty. \qquad (2.8.1.25)$$

It is well known (see [105]) that the condition for the absence of the point spectrum in the problem is the vanishing of the Anderson function

$$p(x, E) = \lim_{\delta \to +0} \frac{\delta}{\pi} \langle \sum_{ij} |G_{ij}(x, 0; \mathcal{E})|^2 \rangle. \qquad (2.8.1.26)$$

Since the Green function is triangular (see (2.8.1.23)) $p = 0$ for $x < 0$, and for $x = 0$ it can be derived from the explicit form of $\mathcal{G}^{(0)}(x + 0, x; \mathcal{E})$ and the identity

$$\mathcal{G}^{(0)}(x + 0, x; \mathcal{E}) = \mathcal{G}(x + 0, x; \mathcal{E}). \qquad (2.8.1.27)$$

The representation

$$p(x; E) = \langle \sum_n \delta(E - E_n) \sum_{ik} |\psi_n^{(i)}(x)\psi^{(k)}(0)|^2 \rangle \qquad (2.8.1.28)$$

together with the spatial homogeneity in the mean of potential $v(x)$ shows that $p(x; E)$ is an even function of x. Therefore $p = 0$ for any x and the problem (2.8.1.3) on the whole axis has no point spectrum if $\beta > 1$.

In the case under consideration we can say even more: the spectrum of the Dirac operator (2.8.1.20) with an arbitrary spatially homogeneous random potential satisfying (2.8.1.24) is absolutely continuous with probability 1. This means that with probability 1 its spectral measure has an integrable density with respect to the Lebesgue measure. As is well known from the general spectrum theory [79] this type of spectrum has differential and difference operators with constant and periodic coefficients. What discriminates the random potential in this case is a much more complicated structure of eigenfunctions (compared to simple plane or Bloch waves for constant or periodic coefficients).

In accordance with the principles of spectral theory the absolute continuity of the spectrum means that the quantity Im$\langle \mathcal{G}(\mathcal{E})\varphi, \varphi \rangle$ for any bounded function φ with a compact support has a finite limit when $\mathcal{E} \to E \pm i0$. This property as well as the absolute continuity of the spectrum follows directly from the properties of the Green function of the Dirac operator (2.8.1.20) for $\beta > 1$.

Returning to the calculations for the case $\beta > 1$, note that the current conservation law (2.8.1.15) dictates a new (compared with (2.8.1.5)) representation for the vector $\vec{A} = (A_+, A_-)$:

$$\vec{A} = (-j)^{1/2} \left(e^{-i\varphi} \cos \frac{\theta}{2}, \sin \frac{\theta}{2} \right). \qquad (2.8.1.29)$$

For high energy the probability density $P(\theta, x)$ satisfies the Fokker–Planck equation (c.f. (2.8.1.7))

$$l \frac{\partial p}{\partial x} = \frac{\partial}{\partial \theta} \left(\sin \theta \frac{\partial}{\partial \theta} \frac{p}{\sin \theta} \right) \qquad (2.8.1.30)$$

with the same (cf. (2.8.1.8)) characteristic length l. This equation has a stationary solution $P(\theta) = \sin \theta$ corresponding to the constant probability density with zero mean of the difference $|A_+|^2 - |A_-|^2$. In the anomalous region $\beta > 1$ the amplitudes $A_\pm$ are randomly modulated. However their mean values within the length l (2.8.1.8) (that now plays the role of the mixing length) become equal. The mean transformation coefficient (that now plays the role of reflectivity and coincides with the probability of the excitation creation) for long pulses ($L \gg l$) has the form

$$\langle W \rangle = \frac{1}{2}(1 - \exp(-2L/l)). \qquad (2.8.1.31)$$

The last result has an interpretation in terms of the stochastic paramagnetic resonance. Replace the two-dimensional complex vector $\vec{A}$ by a three-dimensional real vector $\vec{M}$:

$$\vec{M} = (\mathrm{Re}\ A_+^* A_-,\ \mathrm{Im}\ A_+^* A_-,\ \frac{1}{2}(|A_+|^2 - |A_-|^2)) \qquad (2.8.1.32)$$

(with an asterisk denoting the complex conjugation) and introduce the "magnetic field"

$$\vec{H} = (-b\Phi, 0, p - b\Phi \cdot (v_+^{-1} - v_-^{-1})). \qquad (2.8.1.33)$$

Then $\vec{M}$ satisfies the equation

$$\frac{d\vec{M}}{dx} = [\vec{M}, \vec{H}]. \qquad (2.8.1.34)$$

The "field" $\vec{H}$ in addition to the large constant component $\approx p(E)$ contains also a small stochastic term of the order of $\approx \Phi(x)$ that results in a random walk of $\vec{M}$ on the sphere $|\vec{M}| = \mathrm{const}$. Therefore after a sufficiently long "time" $L \gg l$ the moment will be distributed uniformly over the whole sphere, $\langle M \rangle \to 0$ that corresponds to the exponential decrease of the quantity $\langle |A_+|^2 - |A_-|^2 \rangle = 2\langle W \rangle - 1$.

As was mentioned before, at zero temperature the only contribution to absorption is given by the anomalous scattering. Since the momentum transfer satisfies in this case the inequality $p \geq 2\Delta/s$ the mixing length even for small correlation radius r_c contains the exponentially large factor $\exp\{\Delta r_c/s\}$. Nonlinear behavior of the absorption rate is most pronounced when $r_c \approx s/\Delta$ and the nonlinearity parameter is $\Phi_0^2 L/s\Delta$ as in (2.8.1.12). At $T = 0$ the nonlinear absorption rate is

$$Q \approx N(O)(s^2/v_F)\Delta^2 \cdot \ln^2(\Phi_0^2 L/s\Delta). \qquad (2.8.1.35)$$

Now we shall consider the case when the signal consists of a sequence of point-like pulses. The physical results of this section are qualitatively similar to that obtained for the signal with small but finite correlation radius r_c. Therefore we shall discuss now only the scattering problem.

If $\beta < 1$ the current conservation law for the MDE leads to the same parametrization of the transfer-matrix $\hat{T}^{-1}$ (see (2.5.2.3), (2.3.1.9)) as in the SE case. Therefore all the results obtained in Section 2.5.2 for the SE are valid for the MDE as well, i.e., formulas (2.5.2.9) for the probability density of the parameter η that defines the transmittivity, (2.5.2.23) for the localization length, (2.5.2.22) and (2.5.2.25) for the mean transmittivity, and (2.5.2.26) for its decrement.

The only difference is the nature of the potential for the MDE. All the possible self-adjoint extensions of the Hamiltonian (2.8.1.20) corresponding to the point-like potentials are described in the paper [62]. We shall discuss only the simplest and the most natural one.

Let us consider a square well potential

$$v(x) = \begin{cases} v_0, & |x| < a \\ 0, & |x| > a. \end{cases} \qquad (2.8.1.36)$$

In the limit $v_0 \to \infty$, $a \to 0$ with $2av_0 = k_0 = \mathrm{const}$, the transfer matrix $\hat{T}^{-1}$ for this potential is equal to

$$\hat{T}^{-1} = e^{-2i\alpha_-} \left\{ \hat{I} + \left(1 - e^{\frac{2ik_0}{1-\beta^2}}\right) \right.$$
$$\left. \times \begin{pmatrix} \mathrm{sh}^2\,\theta_0 & -\,\mathrm{sh}\,\theta_0\,\mathrm{ch}\,\theta_0 \\ -\,\mathrm{sh}\,\theta_0\,\mathrm{ch}\,\theta_0 & -\,\mathrm{ch}^2\,\theta_0 \end{pmatrix} \right\}. \qquad (2.8.1.37)$$

Here $\hat{I}$ is the unit 2×2 matrix,

$$\mathrm{cth}\,2\theta_0 = \frac{E}{E_0}, \qquad E_0 = \Delta\sqrt{1-\beta^2}, \qquad \alpha_{\pm} = \frac{k}{2(1\pm\beta)}. \qquad (2.8.1.38)$$

The single scatterer reflectivity $R_1 = 1 - |T_{11}^{-1}|^{-2}$ is equal to

$$R = \left(1 + \left[\mathrm{sh}^2\,2\theta_0 \sin^2 \frac{k_0}{1-\beta^2}\right]^{-1}\right)^{-1}. \qquad (2.8.1.39)$$

The parameter η_1 then is

$$\eta_1 = 1 + 2\,\mathrm{sh}^2\,2\theta_0 \sin^2 \frac{k_0}{1-\beta^2}, \qquad (2.8.1.40)$$

and the inverse localization length is

$$l^{-1} = n \ln \left(1 + \mathrm{sh}^2\,2\theta_0 \sin^2 \frac{k_0}{1-\beta^2}\right), \qquad (2.8.1.41)$$

where n is the inverse mean distance between pulses of the signal.

Note that if we use the natural definition of the integral

$$\int_{-\varepsilon}^{\varepsilon} \delta(x)\vec{\psi}(x)\,dx = \frac{1}{2}(\vec{\psi}(+0) + \vec{\psi}(-0)), \qquad (2.8.1.42)$$

the potential producing the matrix (2.8.1.37) corresponds to an additional term in the Hamiltonian of the form

$$k_0\delta(x)\hat{p}, \qquad (2.8.1.43)$$

where

$$\hat{p} = \begin{pmatrix} \frac{\text{tg }\alpha_-}{\alpha_-} & 0 \\ 0 & \frac{\text{tg }\alpha_+}{\alpha_+} \end{pmatrix} \qquad (2.8.1.44)$$

If the scattering is weak ($\alpha_\pm \ll 1$) $\hat{p} \to \hat{I}$ and

$$R = \frac{k_0}{(1 - \beta^2)^2} \text{ sh}^2\, 2\theta_0. \qquad (2.8.1.45)$$

In the limit when $n \to \infty$, $k_0 \to 0$ and $nk_0 \to 2d$, (2.8.1.41) transforms into an expression for the localization length corresponding to a white noise potential that coincides with (2.8.1.8) with $B(x) = 2d\delta(x)$ (and $\tilde{B}(p) = 1$).

From (2.8.1.37) and (2.8.1.39) it follows that in the case

$$\frac{k_0}{1 - \beta^2} = n\pi \qquad (2.8.1.46)$$

the point scatterer is reflectionless ($R_1 = 0$). Then the corresponding transfer matrix $\hat{T} = \exp(2i\alpha_-) \cdot \hat{I}$, though the potential (2.8.1.43) remains a nonzero one.

Note that this is a property of the MDE but not the DE. In the case of DE with $\beta = 0$ the transfer matrix of the reflectionless potential equals $\pm\hat{I}$ and the potential itself—to either zero or infinity, respectively.

Returning to MDE note that in this case the inverse localization length (2.8.1.41) also is equal to zero. Consequently we can construct a random potential that is made of reflectionless scatterers and produces only delocalized states even in the typically "localization" case $\beta < 1$. The structure of such states is rather simple: they have a constant amplitude and a phase at any point is a random variable distributed uniformly in the interval $(0, 2\pi)$. We note that, the nature of reflectionless potentials described above is absolutely different from so-called Bargman potentials (see [108]).

Turning to the case $\beta > 1$ note that in this case both free solutions propagate in the same direction. The transfer matrix in the case has the form

$$\hat{T}^{-1} = \begin{pmatrix} \left(\frac{1+\eta}{2}\right)^{1/2} e^{i\varphi_\alpha} & \left(\frac{1-\eta}{2}\right)^{1/2} e^{i\varphi_\beta} \\ -\left(\frac{1-\eta}{2}\right)^{1/2} e^{i(\lambda-\varphi_\beta)} & \left(\frac{1+\eta}{2}\right)^{1/2} e^{i(\lambda-\varphi_\alpha)} \end{pmatrix}. \qquad (2.8.1.47)$$

This matrix transforms the wave of the first type that is incident from the right end of the disordered segment into the wave of the second type. The sum of the squared moduli of free solitons equals unity while their difference equals η. The transformation is absent if $\eta = 1$ and is complete if $\eta = -1$. This gives the reason to call η the disbalance coefficient of the transformation. It is connected with the transformation coefficient W (2.8.1.31) via the formula

$$\eta = 1 - 2W. \tag{2.8.1.48}$$

Using (2.8.1.47) and the current conservation law we can repeat all the considerations of Section 2.5.1 from (2.5.2.3) up to (2.5.2.16). We obtain the integral equation for the probability density $p^{(0)}(\eta \mid x)$ in the form

$$\frac{1}{n}\frac{\partial p^{(0)}(\eta \mid x)}{\partial x} = \frac{1}{2\pi}\int_0^{2\pi} p^{(0)}(\eta\eta_1 - [(1-\eta^2)(1-\eta_1^2)]^{1/2}\cos 2\varphi \mid x)\,d\varphi$$
$$- p^{(0)}(\eta \mid x). \tag{2.8.1.49}$$

Since all the arguments of $p^{(0)}$ are less than 1, it is natural to look for a solution in a form of series in Legendre polynomials $P_k(\eta)$:

$$p^{(0)}(\eta \mid x) = \sum p_k^{(0)}(x)P_k(\eta). \tag{2.8.1.50}$$

Plugging it into (2.8.1.49) and using the addition formula for Legendre polynomials we obtain equations

$$\frac{1}{n}\frac{\partial p_k^{(0)}}{\partial x} = [P_k(\eta_1) - 1]\cdot p_k^{(0)}. \tag{2.8.1.51}$$

Their solutions sum up to the final formula

$$p^{(0)}(\eta \mid x) = \sum_{k=0}^{\infty}\frac{2k+1}{2}P_k(\eta)e^{nx[P_k(\eta_1)-1]}. \tag{2.8.1.52}$$

From this expression it follows that the mean value of the disbalance coefficient is exponentially small for large x:

$$\langle \eta \rangle = \exp(-2x/l), \qquad l = 2/n[1 - P_k(\eta_1)]. \tag{2.8.1.53}$$

Therefore it is natural to call l the mixing length. In the weak scattering case (when $W_1 \ll 1$, $\eta_1 = 1 - 2W_1$) it is equal to

$$l = (nW_1)^{-1} \tag{2.8.1.54}$$

and the probability density (2.8.1.51) transforms into

$$p^{(0)}(\eta \mid x)_0 = \sum_{k=0}^{\infty}\frac{2k+1}{2}P_k(\eta)e^{-k(k+1)\frac{x}{l}}. \tag{2.8.1.55}$$

Approximating the point potential by the square well one (2.8.1.36) with a shrinking support we find an explicit form of the transfer-matrix:

$$\hat{T}^{-1} = e^{-2i\alpha_-}$$
$$\times \left\{\hat{I} + \left(e^{\frac{2ik_0}{\beta^2-1}} - 1\right)\begin{pmatrix} \cos^2\theta_0 & \sin\theta_0\cos\theta_0 \\ \sin\theta_0\cos\theta_0 & \sin^2\theta_0 \end{pmatrix}\right\}. \tag{2.8.1.56}$$

where

$$\alpha_\pm = \frac{k_0}{2(\beta \pm 1)}, \qquad \text{ctg } 2\theta_0 = -\frac{E}{\Delta\sqrt{\beta^2 - 1}}. \qquad (2.8.1.57)$$

For the transformation coefficient we get

$$W_1 = (T_{12}^{-1})^2 = \sin^2 2\theta_0 \cdot \sin^2 \frac{k_0}{\beta^2 - 1} \qquad (2.8.1.58)$$

that yields in the weak transformation limit an expression for the mixing length

$$l^{-1} = \frac{nk_0^2}{2} \cdot \frac{\sin^2 2\theta_0}{2}, \qquad (2.8.1.59)$$

coinciding with that obtained above for the limit case of the white-noise potential. For

$$\frac{k_0}{\beta^2 - 1} = n\pi \qquad (2.8.1.60)$$

the transformationless propagation $W_1 = 0$ through the particular point scatterer occurs.

2.8.2. *Dark Soliton Generation in Nonlinear Optical Fibers by the Random Input Pulse*

In this section we consider a nonlinear problem that can be reduced to a linear one. It is a propagation of short optical pulses in nonlinear single-mode optical fibers. This propagation is described by the well-known NLSE [69], [70], which in appropriate coordinates has the form

$$i\frac{\partial u}{\partial x} = -\sigma\frac{\partial^2 u}{\partial t^2} - 2|u|^2 u. \qquad (2.8.2.1)$$

Here u is the (complex) amplitude envelope of the pulse, x is the distance along the fiber and the time variable t is a retarded time measured in a frame of reference moving along the fiber at the group velocity. The parameter σ in (2.8.2.1) equals $+1$ if the group velocity dispersion is negative and 1 if it is positive. The inverse scattering problem formalism [117] reduces the solution of (2.8.2.1) with the "initial" condition

$$u(0, t) = u_0(t) \qquad (2.8.2.2)$$

to the finding of the scattering data for an associated linear initial value problem for the DE (cf. (2.1.2.1)):

$$-i\sigma_z\frac{d\vec{\psi}}{dt} + (v(t)\sigma_x - w(t)\sigma_y)\vec{\psi} = \lambda\vec{\psi},$$
$$v(t) = \text{Re } u_0(t), \qquad w(t) = \text{Im } u_0(t). \qquad (2.8.2.3)$$

We shall consider only the case of positive group velocity dispersion (GVD), that is when $\sigma = -1$. It is known [147] that NLSE with positive GVD can be explicitly integrated. The spectrum of the associated linear problem in the case

$$|u_0(t)|\big|_{|t|\to\infty} = u_0 > 0. \qquad (2.8.2.4)$$

contains a number of discrete levels. Each level $\lambda = \tilde\lambda e^{-i\theta_0}$ corresponds to a dark soliton

$$u(x,t) = u_0 \frac{(\tilde\lambda - i\nu)^2 + \exp(Z)}{1 + \exp(Z)} \exp(2iu_0^2 x + i\phi_0), \qquad (2.8.2.5)$$

where

$$Z = 2\nu u_0(t - t_0 - 2\tilde\lambda u_0 x), \qquad \nu = \sqrt{1 - \tilde\lambda^2}, \qquad (2.8.2.6)$$

θ_0 is the phase of the initial condition (2.8.2.2) (we assume that $\lim_{t\to\infty} u(0,t) = u_0 \exp(i\theta_0)$), and t_0, ϕ_0 are arbitrary constants.

In realistic optical experiments the condition (2.8.2.4) is never satisfied and the limit value of u_0 always equals zero. This means that in the real physical world dark solitons do not exist at all. Nevertheless they were recently observed practically simultaneously by three experimental groups: [93], [144], [65]. The input pulse in the first two publications was an extended background with a rather sharp and short time perturbation. The authors of the third paper interpreted their results assuming that a random input pulse can generate dark solitons.

We shall begin by discussing the results of the first two papers. The associated linear problem in these cases has no discrete (stationary) levels but some quasistationary ones [59], [83]. This means that instead of exact solitons the solution of NLSE contains soliton-like excitations with a large life "time." Then the following mechanism of the generation of such excitations by the random input pulse is proposed. All the states of the system, described by (2.8.2.3) on the whole axis with random potentials v and w, are almost surely localized (see Sections 2.3, 2.6). Then the estimate (2.4.1.10) for the modulus of reflection amplitude of the disordered segment described by the random pulse $v(t) + iw(t)$ with the duration T_0 is also valid (the localization length l in (2.4.1.10) being replaced by the localization time τ and the length L—by T_0). Therefore equation (2.7.3.7) must have solutions with real parts like (2.7.2.11) and exponentially (in T_0/τ) small imaginary parts. Such solutions may be treated as soliton-like excitations.

One can calculate approximately the real parts λ_n of complex resonances Λ_n describing the quasistationary states [61]. For simplicity we shall consider the case of real input pulse (that is $w(t) \equiv 0$). Then the imbedding equation for the reflection amplitude r takes the form (cf. (2.4.1.6))

$$\dot r = 2i\lambda r - iv(t)(1 + r^2). \qquad (2.8.2.7)$$

It follows from this equation that for a large enough duration T_0 and the weak potential v we have

$$\lambda_n = \frac{n\pi}{T_0}(1 + o(1)). \qquad (2.8.2.8)$$

The localization time τ can be easily calculated for the case of Gaussian white noise potential $v(t)$ with

$$\langle v(t) \rangle = 0, \qquad \langle v(t)v(t') \rangle = 2\Delta\delta(t - t'). \qquad (2.8.2.9)$$

In the quasiclassical region $\Delta \ll \lambda$ the localization time equals

$$\tau(\lambda) \cong (2\Delta)^{-1} \qquad (2.8.2.10)$$

and is independent (in the leading order) of the spectral parameter. This result is also valid for the random potential with a nonzero but small correlation time τ_c. Precisely, if the correlation function of the potential

$$\langle v(t)v(t') \rangle = \omega_c^2 f\left(\frac{t - t'}{\tau_c}\right) \qquad (2.8.2.11)$$

has the scale ω_c^2 then for

$$\omega_c \tau_c \ll 1 \qquad (2.8.2.12)$$

the result (2.8.2.10) is valid in the spectrum region

$$\Delta \ll |\lambda| \ll \tau_c^{-1}. \qquad (2.8.2.13)$$

The spectral density $|a(\lambda)|^{-2}$ of generated excitations coincides with the transmittivity of the problem (2.8.2.3) with finite random potential $v(t) \neq 0$ only for $t \in [0, T_0]$. Therefore as it was for the SE we can represent the spectral density in terms of reflection and transmission amplitudes $r_\pm$, $t_{1,2}$ (see (2.4.1.13)):

$$|a(\lambda)|^{-2} = \frac{|t_1 t_2|^2}{|1 - r_- r_+|^2}. \qquad (2.8.2.14)$$

Furthermore, using the time-version of an estimate (2.4.1.10) we can write the reflection amplitudes moduli as

$$|r_\pm| = e^{-\Delta_\pm}, \qquad (2.8.2.15)$$

where the values $\Delta_\pm$ are exponentially small in T_0/τ. Then the spectral density (2.8.2.14) can be rewritten in form

$$|a(\lambda)|^{-2} = \frac{4\Delta_1 \Delta_2}{|\Delta_- + \Delta_+|^2}. \qquad (2.8.2.16)$$

The quantities $\Delta_\pm$, being exponentially small, fluctuate also exponentially. Therefore the spectral density (2.8.2.16) as a rule is also exponentially small. Exceptions are those values $\tilde{\lambda}_n$ for which Δ_- and Δ_+ coincide with the exponential accuracy. Then we have

$$|a(\lambda)|^{-2} = T(\tilde{\lambda}_n) \approx 1. \qquad (2.8.2.17)$$

This means that such values $\tilde{\lambda}_n$ of the spectral parameter correspond to a resonant transparency of the realization (see Section 2.4.2). Thus only those solitons are effectively excited by a random pulse $v(t)$ that correspond to *resonant transparent* quasistationary states with $T(\tilde{\lambda}_n) \approx 1$, and *not to all* quasistationary states.

2.8.3. *Transmission of the Plane Wave through the Nonlinear Disordered Segment*

The simplest generalization of the transmission problem to the nonlinear case seems to be one in which we consider a finite nonlinear disordered segment. Such a problem allows us to obtain well-defined transmission characteristics with the same interrelations as in the linear case. This simplifies the problem. Another simplification arises from the opportunity to take into account the single temporal harmonic neglecting all the higher ones. Sometimes (e.g., for the NLSE) this assumption is rather restrictive. We hope, however, that it preserves to some extent the main features of the phenomena.

We begin with a brief discussion on the general properties of transmission through the nonlinear disordered segment.

As we saw above in the linear case the Anderson localization leads to the exponential decrease of transmittivity. The main mechanism generating this phenomenon is the interference that takes place due to the superposition principle valid in this case. In the nonlinear case this principle is not valid and the interference is suppressed. So we may expect that the transmittivity will decrease more slowly than exponentially.

Quite a new problem arises from the fact that in the nonlinear case we must distinguish between some notions that coincide in the linear case. For example consider the problems with a fixed input (incident wave or current intensity) and with output (outgoing wave or current intensity). In the linear case these problems are equivalent while if the nonlinearity is essential the fixed input problem may have several solutions due to bistability phenomena (see, e.g., [45]).

Proceeding to the quantitative investigations we begin with writing the NLSE (cf. (2.8.2.1)) that will be the object of our interest in this section:

$$i\frac{\partial u}{\partial t} - \frac{\partial^2 u}{\partial x^2} - \alpha |u|^2 u + v(x)u = 0, \qquad (2.8.3.1)$$

with the random potential $v(x)$. We shall consider only the single-frequency regime when the solution of (2.8.3.1) can be written in the form

$$u(x,t) = e^{ik^2 t}\psi(x). \qquad (2.8.3.2)$$

Plugging it into (2.8.3.1) we obtain the stationary NLSE:

$$-\psi'' + v(x)\psi - \alpha|\psi|^2\psi = k^2\psi. \qquad (2.8.3.3)$$

We assume that both the nonlinearity and disorder are concentrated within the finite interval $[0, L]$, so if $x < 0$ or $x > L$ we have $\alpha = v(x) = 0$ and (2.8.3.1) is the free linear SE (see (2.1.1.1)). Therefore we can consider a standard transmission problem with a fixed incident (from the right) wave, and a reflected wave for $x > L$ and a transmitted wave for $x < 0$:

$$\psi(x) = \begin{cases} A_0\left(e^{-ik(x-L)} + r_- e^{ik(x-L)}\right), & L < x, \\ A_0 t e^{-ikx}, & x < 0. \end{cases} \qquad (2.8.3.4)$$

The only difference between this representation and the usual one (i.e., in the linear case) is the presence of the amplitude parameter A_0 in the incident wave. It is necessary since nonlinear effects can strongly depend on incident wave amplitude.

The NLSE (2.8.3.3) leads to a conservation law for a current defined as

$$J = -\frac{1}{2i}(\psi^*\psi' - \psi^{*\prime}\psi) = \text{const} \qquad (2.8.3.5)$$

that differs from that defined in (2.3.1.5) by a sign. In the absence of potential (i.e., $v(x) \equiv 0$) there is also an energy conservation law:

$$E = \frac{1}{2}\left(|\psi'|^2 + k^2|\psi|^2 + \frac{\alpha}{2}|\psi|^4\right) = \text{const.} \qquad (2.8.3.6)$$

Conservation laws together with (2.8.3.4) lead to the evident result for this case: $|t|^2 = T = 1$. But in the nonlinear case they yield some nontrivial conclusions. Substituting (2.8.3.4) into (2.8.3.5) and (2.8.3.6) we find that the first motion integral

$$F = \frac{E}{kJ} + 1 \qquad (2.8.3.7)$$

satisfies the equation

$$F = 2 + \frac{\alpha|A_0|^2|t|^2}{4k^2} = \frac{2}{|t|^2} + \frac{\alpha|A_0|^2}{4k^2} \cdot \frac{|1+r|^4}{|t|^2}. \qquad (2.8.3.8)$$

This means that the reflection amplitude r of the nonlinear segment must satisfy some additional equation whereas in the linear case it can be an arbitrary quantity satisfying the condition $|r| \le 1$ only. In the problem with the fixed output when the current

$$J = k|A_0|^2|t|^2 \qquad (2.8.3.9)$$

or, which is the same, the output intensity

$$W_0 = |A_0|^2|t|^2 \qquad (2.8.3.10)$$

is fixed it satisfies the equation

$$\frac{8k^3}{\alpha J} \cdot \frac{|r|^2}{1 - |r|^2} + \frac{|1+r|^4}{(1-|r|^2)^2} = 1. \qquad (2.8.3.11)$$

In the problem with the fixed input when the input intensity

$$W = |A_0|^2 \qquad (2.8.3.12)$$

is fixed the equation is

$$\frac{8k^3}{2W} \cdot \frac{|r|^2}{(1-|r|^2)^2} + \frac{|1+r|^4}{(1-|r|^2)^2} = 1. \qquad (2.8.3.13)$$

Note that the input W and the output W_0 intensities are connected by the equality

$$W = \frac{W_0}{1 - |r|^2}. \qquad (2.8.3.14)$$

Following [88] introduce two real functions q and θ defined by

$$\psi(x) = A_0 q(y)|t|e^{-i\theta(y)}, \tag{2.8.3.15}$$

where $y = kx$ is a new dimensionless coordinate.

Substituting (2.8.3.15) into the current conservation law (2.8.3.5) and (2.8.3.9) and into the NLSE (2.8.3.3) we obtain equations

$$\theta' = q^{-2}, \tag{2.8.3.16}$$

$$q'' - q^{-3} + q(1 + \varepsilon(y) + \beta q^2) = 0 \tag{2.8.3.17}$$

with a new dimensionless nonlinearity parameter

$$\beta = \alpha J k^{-3} \tag{2.8.3.18}$$

and "potential"

$$\varepsilon(y) = -k^{-2}v(x). \tag{2.8.3.19}$$

The matching of solutions (2.8.3.15) and (2.8.3.4) at $x = 0$ and $x = L$ give rise to boundary conditions for these equations. We can choose the boundary condition for θ in the form

$$|t|e^{-i\theta(0)} = t. \tag{2.8.3.20}$$

Then $q(y)$ satisfies the cosine condition at $x = 0$:

$$q(0) = 1, \qquad q'(0) = 0, \tag{2.8.3.21}$$

and

$$[q' - i(q + q^{-1})]\big|_{y=\mathcal{L}} = -\frac{2i}{|t|}, \tag{2.8.3.22}$$

at the point $y = \mathcal{L} = kL$ ($\mathcal{L}$ is the dimensionless length of the segment).

Using (2.8.3.21) we can express the transmittivity $T = |t|^2$ in terms of $q(\mathcal{L})$ and $q'(\mathcal{L})$ (compare to (2.3.1.13) and (2.3.1.39)):

$$T = \frac{4}{2 + [(q'^2 + q^{-2}) + q^2]\big|_{y=\mathcal{L}}}. \tag{2.8.3.23}$$

Direct calculations show that

$$\begin{cases} (q'^2 + q^{-2})\big|_{y=\mathcal{L}} = e^{2\xi_s(L)}, \\ q^2\big|_{y=\mathcal{L}} = e^{2\xi_s(L)}. \end{cases} \tag{2.8.3.24}$$

If disorder is absent ($v(x) = 0$) equation (2.8.3.17) conserves the dimensionless energy.

$$\mathcal{E} = \frac{E}{kJ} \tag{2.8.3.25}$$

or

$$\mathcal{E}(q', q) \equiv \frac{1}{2}(q'^2 + q^{-2} + q^2 + \frac{\beta}{2}q^4) = 1 + \frac{\beta}{4}. \tag{2.8.3.26}$$

Using this formula and (2.8.3.23) we obtain the explicit expression for the transmittivity T and the input intensity W in terms of the solution $q_0(y)$ of the free equation (2.8.3.17):

$$T = \frac{1}{1 + \frac{\beta}{8}(1 - q^4(\mathcal{L}))}, \tag{2.8.3.27}$$

$$W = \frac{J}{k}\left[1 + \frac{\beta}{8}(1 - q^4(\mathcal{L}))\right]. \tag{2.8.3.28}$$

The free equation (2.8.3.17) is equivalent to the Hamiltonian system (with $p \equiv q'$)

$$p' = -\frac{\partial \mathcal{H}_0}{\partial q}, \qquad q' = \frac{\partial \mathcal{H}_0}{\partial p} \tag{2.8.3.29}$$

with the Hamilton function $\mathcal{H}_0(p, q) \equiv \mathcal{E}(p, q)$ (see (2.8.3.23)). In the presence of the random potential equation (2.8.3.17) can be reduced to the system with the "time"-dependent Hamilton function:

$$\mathcal{H}(p, q) \equiv \mathcal{E}(p, q) + \frac{q^2}{2}\varepsilon(y). \tag{2.8.3.30}$$

In this case the energy (2.8.3.24) is not conserved and the expression (2.8.3.27) for the transmittivity is no longer valid. If we now substitute $\mathcal{E}$ from (2.8.3.26) to (2.8.3.26) we shall obtain

$$T = \frac{2}{\mathcal{E}(p, q) + 1 - \frac{\beta q^4}{4}}. \tag{2.8.3.31}$$

The precise asymptotic of $T(\mathcal{L})$ for $\mathcal{L} \to \infty$ is unknown. But in [88] the behavior of $\mathcal{E}$ for large $\mathcal{L}$ was examined both analytically and numerically. Analytical result for the white noise potential is

$$\mathcal{E}(\mathcal{L})|_{\mathcal{L}\to\infty} \propto \mathcal{L}. \tag{2.8.3.32}$$

Numerical simulations show that the ratio $\mathcal{E}/y$ stabilizes and the larger the nonlinearity parameter β is, the quicker it happens. On the other hand for small β we obtain the linear theory result (the mean energy grows exponentially).

If we believe in the leading role of the energy behavior in the denominator of (2.8.3.31) at large $\mathcal{L}$ than the transmittivity must manifest the power (in the length of the system $L \to \infty$) decrease:

$$T \propto L^{-1}. \tag{2.8.3.33}$$

These results agree with those obtained earlier analytically and numerically in [36]; see also [38].

Now we shall discuss briefly the fixed input problem. First of all rewrite (2.8.3.14) in the form

$$W = \frac{W_0}{1 - |r(L, W_0)|^2}.$$ (2.8.3.34)

The reflection amplitude $r(L, W_0)$ is in fact the function of L and W_0 since it satisfies the imbedding equation [10]

$$r' = 2ikr - \frac{i\nu(x)}{2k} \cdot (1 + r)^2 + \frac{i}{2k}\alpha W \frac{|1 + r|^2}{1 - |r|^2}$$ (2.8.3.35)

(cf. (2.4.1.6)). We may consider (2.8.3.34) as an equation for the output intensity W_0 with a fixed input intensity W. It may have many solutions producing as a result the multistability phenomenon.

In a recent paper [89] a new characteristic of the multiplicity of solutions and localization properties of a nonlinear disordered segment was proposed. It is the average multiplicity density $D(T; L, W)$ (of the number of solutions of (2.8.3.23)) considered as a function of their transmittivity T.

Let $N(T_1, T_2; L, W)$ be the mean number of solutions of (2.8.3.23) with transmittivity T lying between T_1 and T_2. The average density of solutions is defined naturally as

$$D(T; L, W)_0 = \lim_{\Delta T \to 0} \frac{N(T, T + \Delta T; L, W)}{\Delta T}.$$ (2.8.3.36)

Then the total number of solutions equals

$$N(0, 1; L, W) = \int_0^1 D(T; L, W)\, dT.$$ (2.8.3.37)

It can be considered as a measure of the likelihood of the appearance of multistability in the nonlinear case. The averaged transmittivity $\langle T \rangle$ can be written as

$$\langle T \rangle = \frac{\int_0^1 D(T; L, W)T\, dT}{\int_0^1 D(T; L, W)\, dT}.$$ (2.8.3.38)

The average multiplicity density D gives a lot of information about localization. If there is localization D has at large L a peak near $T = 0$ that contains the large part of the total mass under the curve. If this is not the case (i.e., if the peak is shifted from zero and the mass is smeared) the localization is not a very likely phenomenon. Note that at large L the behavior of $\langle T \rangle$ in the nonlinear case differs from that in the linear case. It may seem that since the field becomes small far inside the scattering region the linear theory is applicable and produces localization. However, the number of solutions grow together with L and it takes still larger L to localize them. This is reflected in $\langle T \rangle$ that is averaged over all solutions as well as realizations of the potential. The behavior of $\langle T \rangle$ as $L \to \infty$ is a very interesting but unsolved problem.

The numerical results show that nonlinearity suppresses the localization. For example for small L the values of $\langle T \rangle$ in linear and nonlinear cases are almost identical because nonlinear effects are small and no multistability occurs. But as L grows the decay of $\langle T \rangle$ slows down and seems to stop eventually. This indicates the presence of a significant number of delocalized states.

The results of [89] (see also [25]) give a strong evidence of existence of delocalized transmission states for the propagation of waves through a random medium. However, since only a single (temporal) frequency problem was considered some of these states may not be physically observable owing to possible unstability.

2.8.4. Soliton Propagation in the Presence of Weak Random Point Scatterers

The problem considered above was in a sense the simplest nonlinear transmission problem. The next natural step is to study the wave transmission through a disordered segment posed in a medium that is nonlinear at any point (but not inside this segment $[0, L]$ only).

We begin with the study of a nonlinear lattice with the nearest-neighbors interaction. We shall follow the paper [101]. The Hamiltonian of our system is

$$H = \sum_n \left[\frac{1}{2} M_n \dot{y}_n^2 + V(y_n - y_{n-1}) \right]. \qquad (2.8.4.1)$$

The interaction potential $V(y)$ has the form

$$V(y) = \frac{Ky^2}{2} + \frac{Ay^3}{3} + \frac{By^4}{4}. \qquad (2.8.4.2)$$

For the homogeneous system (with $M_n \equiv M$) the dynamical equation

$$M_n \ddot{y}_n = -V'(y_n - y_{n-1}) + V'(y_{n+1} - y_n) \qquad (2.8.4.3)$$

in the continuum limit transforms to the generalized Boussinesq equation [100]

$$u_{tt}(c_0^2 u + pu^2 + qu^3 + hu_{xx})_{xx} = 0, \qquad (2.8.4.4)$$

where

$$u = y_x \equiv \frac{\partial y}{\partial x}, \quad c_0^2 = \frac{Ka^2}{M}, \quad p = \frac{Aa^3}{M}, \quad q = \frac{Ba^4}{M}, h = \frac{Ka^4v^2}{12Mc_0^2}. (2.8.4.5)$$

Here a is the lattice spacing constant and v is the group velocity of the slowly varying solitary wave $y(x - vt)$. The equation (2.8.4.4) has a kink-type solution that for the atomic displacement $y(x, t)$ is

$$y(x, t) = \pm 2(\text{sgn } h)(2h/q)^{1/2} \text{ arctg} \left[W^{-1} \text{ th} \frac{x - vt - x_0}{\lambda} \right] \qquad (2.8.4.6)$$

with

$$W = \left[\frac{[4p^2 + 18(v^2 - c_0^2)q]^{1/2} \pm 2p}{[4p^2 + 18(v^2 - c_0^2)q]^{1/2} \mp 2p} \right]^{1/2} \qquad (2.8.4.7)$$

and

$$\lambda = 2 \left(\frac{h}{v^2 - c_0^2} \right)^{1/2}. \tag{2.8.4.8}$$

The formula (2.8.4.6) can be simplified in two limiting cases.

If $q = 0$ ($B = 0$, cubic nonlinearity only)

$$y(x, t) = A_m \, \text{th}[k_s(x - x_0 - vt)], \tag{2.8.4.9}$$

where

$$A_m = \frac{2 \, \text{sgn}(h)[h(v^2 - c_0^2)]^{1/2}}{p}, \qquad k_s = L^{-1}. \tag{2.8.4.10}$$

The total energy of the kink is equal to

$$E = \frac{4M}{15a} k_s A_m^2 (4v^2 + c_0^2). \tag{2.8.4.11}$$

If $p = 0$ ($A = 0$, quartic nonlinearity only)

$$y(x, t) = A_m \, \text{arctg}\{\exp[k_s(x - x_0 - vt)]\}, \tag{2.8.4.12}$$

with

$$A_m = \pm 2(h/q)^{1/2}, k_s = 2/L, \tag{2.8.4.13}$$

and the total energy equals

$$E_{\text{tot}} = \frac{M}{6a} k_s A_m^2 (2v^2 - c_0^2). \tag{2.8.4.14}$$

The disorder in the segment $[0, L]$ is in the randomness of M_n that independently take the value M with the probability p and value γM with the probability $(1 - p)$. Authors have investigated numerically the transmission of a kink through such a segment. In a homogeneous (free nonlinear) medium the kink is a rather sharp peak in the u-variable (see (2.8.4.5), (2.8.4.6)). Let such a kink represent a wave incident from the left on a disordered segment at $t = 0$. The numerical simulation shows that for a large enough time $t > 0$ on the right semiaxis $[L, \infty[$ there exists a transmitted kink plus some amplitude oscillations and on the left one $] - \infty, 0]$ some reflected wave is present. The transmittivity in this problem may be defined as

$$T = \frac{E_{\text{in}}}{E_t}, \tag{2.8.4.15}$$

where E_t and E_{in} are the total energies of the transmitted and incident kinks, respectively. Numerical results show the power law dependence of transmittivity on the length L of the disordered segment. In the case of purely quartic potential ($p = 0$) we have

$$T \propto L^{-1/2}. \tag{2.8.4.16}$$

In the purely cubic case ($q = 0$) this dependence is valid for long enough segments while for shorter ones

$$T \propto L^{-3/2}. \tag{2.8.4.17}$$

Authors proposed some approach that qualitatively explains these formulas. It is based on the independent scattering approximation (for this approximation in the linear case see [105]).

The independent scattering approximation is valid in the dilute limit when the concentration p of the mass impurities is low and the mean distance between them is larger than the soliton size. Therefore in the dilute limit we have

$$T = \prod_{i=1}^{N} T(E_i), \tag{2.8.4.18}$$

where N is the number of impurities inside the disordered segment, E_i is the incident kink energy for the ith impurity, and $T(E_i)$ is its transmittivity. This becomes obvious if we write

$$T(E_i) = E_{i+1}/E_i \tag{2.8.4.19}$$

and multiply these equations for $i = 1, \dots, N$:

$$T = \frac{E_{N+1}}{E_1} = \frac{E_{N+1}}{E_N} \cdots \frac{E_2}{E_1}. \tag{2.8.4.20}$$

The relation (2.8.4.19) can be rewritten in the form

$$E_{i+1} - E_i = -E_i(1 - T_i(E_i)). \tag{2.8.4.21}$$

If the initial energy (and consequently all the transmitted ones) is sufficiently small then the transmittivity $T_i(E_i)$ is close to unity (see (2.8.4.24) below) and the energy E changes slowly. In this case we can write the macroscopic version of (2.8.4.21):

$$\frac{dE}{dx} = -nE \cdot [1 - T(E)], \qquad n = pa^{-1}. \tag{2.8.4.22}$$

Besides, by averaging (2.8.4.21) we get the following equation:

$$\frac{d\langle E \rangle}{dx} = -n(\langle E \rangle - \langle ET(E) \rangle) \tag{2.8.4.23}$$

that transforms into (2.8.4.22) after decoupling

$$\langle ET(E) \rangle \Rightarrow \langle E \rangle T(\langle E \rangle). \tag{8.4.23^a}$$

The possibility of such a decoupling means that for macroscopic scales the energy E is not random.

In [100] the authors obtained the following expression for the transmittivity $T_i(E)$ of the single scatterer:

$$T_i(E) = 1 - a_0 E^\alpha \cdot |\gamma - 1|^2. \tag{2.8.4.24}$$

The constants a_0 and α for the quartic potential are $a_0 = 0.66$ and $\alpha = 2$, while for the cubic one $a_0 = 8.0$ and $\alpha = 2/3$. To obtain this result the kink was treated as a linear wave packet of the appropriate form defined by (2.8.4.6). The transmittivity $T_i(E)$ was chosen to be equal to the corresponding wave-packet transmittivity (2.3.6.14) with

$$t(k) = \left[1 - i(\gamma - 1)\, \mathrm{tg}\, \frac{ka}{2}\right]^{1/2}. \tag{2.8.4.25}$$

Taking into account the kink structure (2.8.4.6) and formulas (2.8.4.11) and (2.8.4.14) for total kink energy, the authors get (2.8.4.24).

Formula (2.8.4.24) allows us to solve explicitly the equation (2.8.4.22) and to determine the length dependence of the total transmittivity of the disordered segment. Here is the final power law for the asymptotic of the transmittivity when $L \to \infty$

$$T(L) \approx L^{-1/\alpha} \cdot E_{\mathrm{in}}^{-1} \cdot (p|\gamma - 1|^2)^{-1/\alpha},$$
$$L \gg (pa_0\alpha E_{\mathrm{in}}^{-1}|\gamma - 1|^2)^{-1} \tag{2.8.4.26}$$

that for cubic and quartic nonlinearities coincides with (2.8.4.16) and (2.8.4.17) obtained by the numerical experiment.

The methods used above may be applied to a more complicated problem—the transmission of NLSE solitons through the segment with the disordered point scatterers [84]. Consider the nonstationary NLSE

$$iu_t + u_{xx} + 2|u|^2 u = \varepsilon(x)u \tag{2.8.4.27}$$

(cf. (2.1.2.4) and (2.8.3.1)). Here

$$\varepsilon(x) = \sum_n \varepsilon\delta(x - x_n) \tag{2.8.4.28}$$

is the random potential of point scatterers with random shifts x_i. The homogeneous (with $\varepsilon(x) = 0$) equation admits the distortionless propagation of localized excitations in the form of envelope solitons

$$u(x,t) = i\eta\, \frac{\exp\left[\frac{iV}{2}x - i\left(\frac{V^2}{4} - \eta^2\right)t\right]}{\mathrm{ch}[\eta(x - Vt)]}, \tag{2.8.4.29}$$

where η and V are the soliton amplitude and velocity, respectively. Note that after substitution $k_0 \leftrightarrow V/2$ this equation coincides with (2.3.6.16), which was considered earlier. Now we are interested in the propagation of solitons (2.8.4.29) through the disordered segment.

First of all write two known integrals of motion for (2.8.4.27). They are the energy

$$E = \int_{-\infty}^{\infty} \left(|u_x|^2 + \varepsilon(x)|u|^2 - |u|^4\right) dx, \tag{2.8.4.30}$$

and the so-called "number of quasiparticles"

$$N = \int_{-\infty}^{\infty} |u|^2\, dx. \tag{2.8.4.31}$$

Note that these quantities are conserved in time while the quantity (2.8.3.6), which was also called the energy, is the constant in space. These quantities change due to the emission of radiation but they are conserved for the system as a whole. Therefore we can describe the scattering process by the energy transmittivity

$$T^{(E)} = \frac{E_t}{E_i},$$ (2.8.4.32)

and the "number of particles" transmittivity

$$T^{(N)} = \frac{N_t}{N_i}.$$ (2.8.4.33)

The indexes i and t denote the initial and the transmitted quantities.

Since the soliton is described by two quantities E and N we must write for the scattering process on a single impurity two equations instead of one (2.8.4.19):

$$E_{i+1} = E_i \cdot T_i^{(E)}(E_i, N_i),$$
$$N_{i+1} = N_i \cdot T_i^{(N)}(E_i, N_i).$$ (2.8.4.34)

As a result we obtain the system of two differential equations instead of a single one (2.8.4.22):

$$\frac{dE}{dx} = -nE(x)R^{(E)}[E(x), N(x)],$$
$$\frac{dN}{dx} = -nN(x)R^{(N)}[E(x), N(x)],$$ (2.8.4.35)

where $R^{(E,N)} = 1 - T^{(E,N)}$ are the reflectivities for the energy and the "number of particles," respectively and n is the density of impurities.

Soliton reflectivities $R^{(E,N)}$ may be calculated via soliton perturbation theory [86]. Here we briefly describe this procedure.

The reflection coefficients are defined by the reflected wave packets in the form of linear waves emitted by the soliton in the process of scattering. The inverse scattering transformation (IST) gives the expression for the spectral density of emitted waves

$$n_{\mathrm{rad}}(\lambda, t) = \frac{1}{\pi}|b(\lambda, t)|^2, \qquad \text{for } |b|^2 \ll 1,$$ (2.8.4.36)

$b(\lambda, t)$ being the so-called Jost coefficient corresponding to IST for the NLSE. The spectral parameter λ appearing in the IST is connected with the wave number $k(\lambda)$ and the frequency $\omega(\lambda)$ of the generated linear waves by relations $\omega(\lambda) = k^2(\lambda) = 4\lambda^2$. The weak perturbation $P(u)$ leads to a change of the IST spectral coefficients that in turn affects the Jost coefficient $b(\lambda, t)$. This influence may be described in the form

$$\frac{\partial b(\lambda, t)}{\partial t} = 4i\lambda^2 b(\lambda, t) + \int_{-\infty}^{\infty} [P(u)\Phi_1^{(1)}(x, t; \lambda)\Phi_2^{(2)}(x, t; \lambda)$$
$$-P^*(u)\Phi_2^{(1)}(x, t; \lambda)\Phi_1^{(2)}(x, t; \lambda)] \, du,$$ (2.8.4.37)

where $\Phi_{1,2}^{(1,2)}$ are the components of the Jost eigenfunctions, or in another form

$$\frac{\partial b(\lambda, t)}{\partial t} = 4i\lambda^2 b(\lambda, t) + \left[\left(\lambda + \frac{V}{2}\right)^2 + \frac{\eta^2}{4}\right]^{-1}$$

$$\cdot \left\{\frac{\eta^2}{4} \int_{-\infty}^{\infty} P(u) \, \text{ch}^{-2} Z \exp[-2i\lambda x - iVx + 2i\Delta(t)] \, dx \right.$$

$$\left. - \int_{-\infty}^{\infty} P^*(u) e^{-2i\lambda x} \left(\lambda + \frac{V}{4} - \frac{i\eta}{2} \, \text{th} \, Z\right)^2 \right\}, \qquad (2.8.4.38)$$

where

$$Z = \eta(x - Vt) \qquad \text{and} \qquad \Delta = (V^2/4 - \eta^2)t. \qquad (2.8.4.39)$$

The single impurity corresponds to the perturbation $P(u)$ of the form

$$P(u) = \varepsilon\delta(x - x_0) \cdot u, \qquad (2.8.4.40)$$

where x is its position and u comes from (2.8.4.29).

Before the scattering the wave coincides with the soliton (2.8.4.29), so the initial conditions for (2.8.4.38) should be written as

$$b(\lambda, t = -\infty) = 0. \qquad (2.8.4.41)$$

Integrating (2.8.4.38) and using (2.8.4.36) one can find the radiative density after the scattering:

$$n_{\text{rad}}(\lambda, t = +\infty) = \frac{1}{\pi}|b(\lambda, t = +\infty)|^2. \qquad (2.8.4.42)$$

This expression allows us to calculate the "number of particles" and the energy of the reflected waves:

$$N_r = \int_{-\infty}^{0} n_{\text{rad}}(\lambda) \, d\lambda; \qquad E_r = 4 \int_{-\infty}^{0} \lambda^2 n_{\text{rad}}(\lambda) \, d\lambda \qquad (2.8.4.43)$$

and to define the soliton reflectivities

$$R^{(N)} = N_r/N \qquad \text{and} \qquad R^{(E)} = E_r/E. \qquad (2.8.4.44)$$

Here N and E are the "number of particles" and the energy corresponding to the single unperturbed soliton (2.8.4.29):

$$N = 2\eta; \qquad E = \frac{N}{4}(V^2 - N^2/3). \qquad (2.8.4.45)$$

Introduce $\alpha = NV^{-1}$. It is a very convenient parameter for our problem because it is related to the nonlinearity of the incoming soliton: the larger α is, the more "nonlinear" is the soliton. In other words the greater α is the larger becomes the number of quasiparticles contained in the soliton and its shape becomes narrower, see

(2.8.4.29). Besides for small α the wave looks like a narrow (see (2.3.6.15)) linear wave packet.

In the Born approximation, which is valid for $\varepsilon \ll V^2/\eta$, the soliton reflectivities become

$$R^{(N)} = \frac{\pi \varepsilon^2}{64 N V} \int_0^\infty F(y, \alpha) \, dy, \qquad (2.8.4.46)$$

and

$$R^{(E)} = \frac{\pi \varepsilon^2 V}{256 E} \int_0^\infty y^2 F(y, \alpha) \, dy. \qquad (2.8.4.47)$$

Here

$$F(y, \alpha) = \frac{[(y+1)^2 + \alpha^2]^2}{\mathrm{ch}^2 \left[\frac{\pi}{4\alpha} (y^2 + \alpha^2 - 1) \right]}. \qquad (2.8.4.48)$$

As a consequence we obtain the following system of integro-differential equations for slowly varying parameters N and E:

$$\frac{dN}{dz} = -\frac{1}{V} \int_0^\infty F(y, \alpha) \, dy, \qquad (2.8.4.49)$$

$$\frac{dV}{dz} = -\frac{1}{2N} \int_0^\infty (y^2 - 1) F(y, \alpha) \, dy - \frac{N}{2V^2} \int_0^\infty F(y, \alpha) \, dy, \quad (2.8.4.50)$$

where the distance is measured in units of $64/(\pi n \varepsilon^2)$, i.e.,

$$y = \frac{\pi n \varepsilon^2}{64} x. \qquad (2.8.4.51)$$

Note that we used (2.8.4.45) to write down the equation (2.8.4.50) for the soliton velocity V, which is more convenient than that for the energy E. Note also that due to (2.8.4.45) we can interpret (2.8.4.49) as an equation for the soliton amplitude η.

In the linear limit $\alpha \ll 1$ the system (2.8.4.49)–(2.8.4.50) can be solved analytically. Calculating the integrals for small α we see that the derivative of the square velocity is of the order of α^2 and is negligible compared with the derivative of the square of the number of quasiparticles N^2, which is of order of α. Thus we can put $V = V(0) = \mathrm{const}$, and come to a well-known result

$$T^{(N,E)}(x) = \frac{N(x)}{N(0)} = \frac{E(x)}{E(0)} = e^{-x/l} \qquad (2.8.4.52)$$

with the localization length (2.5.2.23)

$$l = (nR_1)^{-1}, \qquad (2.8.4.53)$$

where

$$R_1 = \varepsilon^2 V^{-2}(0) \qquad (2.8.4.54)$$

is nothing else but the reflectivity for a plane wave with a wavenumber $k_0 = V(0)/2$ when scattered by a single impurity. Therefore both the transmittivities $T^{(N)}$ and $T^{(E)}$ coincide with (2.3.6.4) for the linear wave packet (2.3.6.15) in the limiting case $\alpha = \frac{\eta}{2k_0} \ll 1$. Note that in Section 2.3.6 we considered the opposite limiting case—the wide wave packet.

The system (2.8.4.49)–(2.8.4.50) has an unstable fixed point $\alpha_c = 1.28505(4)$ satisfying the transcendent equation

$$\alpha_c^2 + G(\alpha_c) = 2 \qquad (2.8.4.55)$$

where

$$G(\alpha) = \frac{\int_0^\infty (y^2 - 1) F(y, \alpha)\, dy}{\int_0^\infty F(y, \alpha)\, dy} \qquad (2.8.4.56)$$

and therefore the behavior of the velocity V and the number of quasiparticles N depends essentially on the initial value $\alpha(0)$. The system (2.8.4.49)–(2.8.4.50) was investigated numerically for different values of $\alpha(0)$. The results [84] are as follows:

(i) $\alpha(0) < \alpha_c$. The system evolves to a final state where N asymptotically tends to zero while V tends to a nonzero constant. Hence $\alpha(\infty) = 0$. This behavior corresponds to a decay of the transmittivity and as we have seen before for small α this decay is exponential and described quite correctly by the formula (2.8.4.52).

(ii) $\alpha(0) > \alpha_c$. In this case both N and V decay very rapidly up to a point where the two functions become practically constants. The same of course happens to α, which tends to a value ≈ 10 as $z \to \infty$. Initial transients and shapes of the curves $T^{(N)}(z)$ and $T^{(E)}(z)$ are very sensitive to initial conditions $N(0)$ and $V(0)$ but the general property of these curves remains the same: both the transmittivities tend to certain asymptotic values and the rearranged after the scattering wave packet is henceforth transmitted without any reflection.

It is curious that the power law of the decay of transmittivities was not observed. If it is present in the considered model it should take place in the vicinity of the fixed point α_c. This fact and some questions related to the validity of the simple approach described above (in particular the independence of scattering on impurities, the Born approximation, the decoupling (8.4.23^a)) demand further study of the problem. (On the other hand recent results [60] on the sine-Gordon kink transmission also did not show power decay of the transmittivity.)

REFERENCES

[1] E. Abrahams and M. J. Stephen, *J. Phys. C* **13**, L377 (1980).

[2] B. L. Altshuler, *Pis'ma Zh. Eksper. and Teor. Fiz.* **41**, 530 (1985) [*JETP Lett.* **41**, 648 (1985)].

[3] B. L. Altshuler, P. A. Lee, and R. A. Webb (Eds.), *Mesoscopic Phenomena in Solids*, North-Holland, Amsterdam, 1991.

[4] B. L. Altshuler and B. Z. Spivak, *Pis'ma Zh. Eksper. and Teor. Fiz.* **42**, 363 (1985). [*JETP Lett.* **42**, 447 (1985)]

[5] B. L. Altshuler and D. Ye. Khmelnitsky, *Pis'ma Zh. Eksper. and Teor. Fiz.* **42**, 291 (1985) [*JETP Lett.* **42**, 359 (1985)].

[6] P. W. Anderson, *Phys. Rev.* **109**, 1492 (1958).

[7] P. W. Anderson, D. J. Thouless, E. Abrahams, and D. Fisher, *Phys. Rev. B* **22**, 3519 (1980).

[8] J. Avron and B. Simon, *Bull. Am. Math. Soc.* **6**, 81 (1982).

[9] M. Ya. Azbel', *Phys. Rev. B* **28**, 4116 (1983).

[10] G. I. Babkin and V. I. Klyatskin, *Zh. Eksper. and Teor. Fiz.* **79**, 817 (1980) [*Sov. Phys. JETP* **52**, 416 (1980)].

[11] V. Baluni and J. Willemsen, *Phys. Rev. B* **31**, 3358 (1985).

[12] H. Bateman and A. Erdelyi, *Higher Transcendental Functions*, v. **1**, McGraw Hill, New York, 1953.

[13] A. I. Baz', Ya. B. Zel'dovich, and A. M. Perelomov, *Scattering, reactions and decay in nonrelativistic quantum mechanics*, NASA Tecn. Transl., F-510 (1969).

[14] M. Belzons, P. Devillard, F. Dunlop, E. Guazelli, O. Parodi, and B. Souillard, *Europhys. Let.* **4**, 909 (1987).

[15] V. L. Berezinsky, *Zh. Eksper. and Teor. Fiz.* **65**, 1251 (1973) [*Sov. Phys. JETP*, **38**, 620 (1973)].

[16] V. L. Bonch-Bruevich, I. Zvyagin, R. Kaiper, A. Mironov, R. Enderline, and B. Esser, *Electronic Theory of Disordered Semiconductors*, Nauka, Moscow, 1981 (in Russian).

[17] E. N. Bratus' and V. S. Shumeiko, *J. Low Temp. Phys.* **60**, 109 (1985).

[18] E. N. Bratus', S. A. Gredeskul, L. A. Pastur, and V. S. Shumeiko, *Phys. Lett. A* **131**, 449 (1988).

[19] E. N. Bratus', S. A. Gredeskul, L. A. Pastur, and V. S. Shumeiko, *Teor. Matem. Fizika.* **76**, 401 (1988) [*Theor. Math. Phys. (USSR)* **76**, 945 (1988)].

[20] S. Ya. Braude (Ed.), *Radiooceanographic Investigations of Sea-Wave Motion*, Naukova Dumka, Kiev, 1962 (in Russian).

[21] L. M. Brekhovskikh, *Waves in Layered Media*, Academic, New York, 1980.

[22] T. A. Brody, J. Flores, J. B. French, P. A. Mello, A. Pandey, and S. S. M. Wong, *Rev. Mod. Phys.* **53**, 385 (1981).

[23] R. Burridge, G. Papanicolaou, and D. Mc Laughlin, *Commun. Pure Appl. Math*, **26**, 105 (1973).

[24] M. Buttiker, Y. Imry, R. Landauer, and S. Pinhas, *Phys. Rev. B* **31**, 6207 (1985).

[25] R. Camassa, M. G. Forest, and R. Knapp, *Nonlinearity.* **5**, 721 (1992).

[26] R. Carmona, A. Klein, and F. Martinelli, *Commun. Math. Phys.* **108**, 41 (1987).

[27] R. Carmona and J. Lacroix, *Spectral Theory of the Random Schrödinger Operator*, Birkhäuser, Boston, 1990.

[28] E. Codington and N. Levinson, *Theory of Ordinary Differential Equations*, Mc Graw Hill, New York, 1955.

[29] M. H. Cohen and E. N. Economou, *Phys. Rev. B* **5**, 293 (1971).

[30] S. M. Cohen and C. Machta, *Phys. Rev. Lett.* **54**, 2242 (1985).

[31] C. A. Condat and T. R. Kirkpatrick, *Phys. Rev. B* **33**, 3102 (1986).

[32] C. A. Condat and T. R. Kirkpatrick, *Phys. Rev. B* **36**, 6782 (1987).

[33] J. C. Dainty, M.-J. Kim, and A. J. Sant, in: *Notes for Tallinn Workshop*, Blackett Lab., Imperial Col., London 1988.

[34] F. Deylon and B. Souillard, *Commun. Math Phys.* **100**, 463 (1985); J. Stat. Phys. **42**, 375 (1985).

[35] P. Devillard, F. Dunlop, and B. Souillard, *Preprint Center de Physique Theor. Ecole Politech.*, A. 688. 10. **85**, Palaiseau, France, 1985.

[36] P. Devillard and B. Souillard, *J. Stat. Phys.* **43**, 423 (1986).

[37] K. A. O'Donnel and E. R. Mendez, *J. Opt. Soc. Am. A* **4**, 1194 (1987).

[38] B. Ducot and R. Rammal, *J. Phys. (Paris)* **48**, 527 (1987).

[39] V. N. Dutyshev, S. Yu. Potapenko, and A. M. Satanin, *Zh. Eksp. and Teor. Phys.* **89**, 298 (1985) [*Sov. Phys. JETP* **62**, 168 (1985)].

[40] F. J. Dyson and M. L. Mehta, *J. Math. Phys.* **4**, 701 (1963).

[41] A. L. Efros and M. Pollak (Eds.), *Electron–Electron Interactions in Disordered Systems*, North-Holland, Amsterdam, 1985.

[42] D. Escande and B. Souillard, *Phys. Rev. Lett.* **52**, 1296 (1985).

[43] S. Estemad, R. Thompson, and M. J. Andreico, *Phys. Rev. Lett.* **57**, 575 (1986).

[44] G. Farias and A. A. Maradudin, *Phys. Rev. B* **28**, 5675 (1983).

[45] C. Flytzanis, in: *Nonlinear Phenomena in Solids – Modern Topics* (Lecture Notes from the Proceedings of the Third International School on Condensed Matter Physics, M. Borisov, Ed.), World Scientific, Singapore, 1984.

[46] D. Fox and P. B. Kahn, *Phys. Rev.* **134**, B1151 (1964).

[47] V. D. Freylikher and S. A. Gredeskul, *J. Opt. Soc. Am. B* **7**, 868 (1990).

[48] V. D. Freylikher and S. A. Gredeskul, *Radio Sci.* **26**, 375 (1991).

[49] V. D. Freylikher and S. A. Gredeskul, *Progr. in Optics* **XXX**, 137 (1992).

[50] H. L. Frish and S. R. Lloyd, *Phys. Rev.* **120**, 1179 (1960).

[51] J. Frolich and T. Spencer, *Commun. Math. Phys.* **88**, 151 (1983).

[52] S. V. Gaponov, *Vest. AN USSR*, No. 12, 3 (1984) (in Russian).

[53] V. M. Gasparyan, B. L. Altshuler, A. G. Aronov and Z. A. Kasamanian, *Phys. Lett. A* **132**, 201 (1988).

[54] V. M. Gasparyan, *Fiz. Tverd. Tela* **31**, 162 (1989) [*Sov. Phys. Solid State* **31**, 266 (1989)].

[55] P. G. de Gennes, *Superconductivity of Metals and Alloys*, Benjamin, New York, 1966.

[56] M. E. Gertsenstein and V. B. Vasil'ev, *Teor. Veroyatnost. Primen.* **4**, 424 (1958) (in Russian).

[57] I. Ya. Goldsheidt, S. A. Molchanov, and L. A. Pastur, *Funct. Anal. Appl.* **11**, 1 (1977) (in Russian).

[58] S. A. Gredeskul and V. D. Freylikher, *Usp. Fiz. Nauk* **120**, 239 (1990) [*Sov. Phys. Uspekhi* **33**, 134 (1990)].

[59] S. A. Gredeskul and Yu. S. Kivshar, *Phys. Rev. Lett.* **62**, 977 (1989).

[60] S. A. Gredeskul, Yu. S. Kivshar, L. K. Maslov, A. Sanchez, and L. Vazquez, *Phys. Rev. A* **45**, 8867 (1992).

[61] S. A. Gredeskul, Yu. S. Kivshar, and M. V. Yanovskaya, *Phys. Rev. A* **41**, 3994 (1990).

[62] S. A. Gredeskul, L. A. Pastur, and P. Seba, *J. Stat. Phys.* **58**, 795 (1990).

[63] S. A. Gredeskul and V. S. Shumeiko, in: *XIII All-Union Conference on the Theory of Semiconductors*, Yerevan, 1987, p. 104 (in Russian).

[64] G. Grinstein and G. Mazenko (Eds.), *Directions in Condensed Matter Physics*, World Publishing, Singapore, 1986.

[65] A. E. Grudinin, E. M. Dianov, A. M. Prokhorov, and D. V. Khaidarov, *Pis'ma Zh. Tekhn. Fiz.* **14**, 1010 (1988) (in Russian).

[66] E. Guazzelli, E. Guyon, and B. Souillard, *J. Phys. (Paris) Lett.* **44**, 837 (1983).

[67] V. N. Gusyatnikov and M. E. Raikh, *Fiz. Tekhn. Poluprovod.* **18**, 1077 (1984) [*Sov. Phys. Semiconductors* **18**, 670 (1984)].

[68] P. Hall and C. C. Heide, *Martingale Limit Theory and its Applications*, Academic, New York, 1980.

[69] A. Hasegava and F. Tappert, *Appl. Phys. Lett.* **23**, 142 (1973).

[70] A. Hasegava and F. Tappert, *Appl. Phys. Lett.* **23**, 171 (1973).

[71] C. I. Hodges, *Sound Vibr.* **82**, 411 (1982).

[72] Y. Imry, *Europhys. Lett.* **1**, 249 (1988).

[73] K. Ishii, *Progr. Theor. Phys. Suppl.* **53**, 77 (1973).

[74] S. John, H. Sompolinsky, and M. J. Stephen, *Phys. Rev. B* **27**, 5592 (1983).

[75] Yu. S. Kaganovskii, A. I. Makienko, and V. D. Freylikher, *Fiz. Met. Metalloved.* **42**, 588 (1976) [*Phys. Met. Metallography* **42**, No. 3, 121 (1976)].

[76] Yu. S. Kaganovskii, V. D. Freylikher, and S. P. Yurchenko, *Opt. Spectrosc.* **56**, 472 (1984) [*Sov. Opt. Spectrosc.* **56**, 289 (1984)].

[77] R. Kamien, H. Politzer, and M. Wise, *Phys. Rev. Lett.* **60**, 1995 (1988).

[78] F. I. Karpelevich, V. N. Tutubalin, and M. G. Shur, *Teor. Veroyatnost. Primen.* **4**, 433 (1959).

[79] T. Kato, *Perturbation Theory for Linear Operators*, Springer, Berlin, 1966.

[80] L. V. Keldysh, *Fiz. Tverd. Tela* **4**, 2265 (1962) [*Sov. Phys. Solid State* **4**, (1962)].

[81] J. S. Keller, G. C. Papanicolaou, and J. Weilenmann, *Commun. Pure Appl. Math.* **32**, 583 (1978).

[82] T. R. Kirkpatrick, *Phys. Rev. B* **31**, 5746 (1985).

[83] Yu. S. Kivshar and S. A. Gredeskul, *Opt. Commun.* **79**, 285 (1990).

[84] Yu. S. Kivshar, S. A. Gredeskul, A. Sanchez, and L. Vazquez, *Phys. Rev. Lett.* **64**, 1693 (1990).

[85] Yu. S. Kivshar, S. A. Gredeskul, A. Sanchez, and L. Vazquez, *Waves in Random Media* **2**, 125 (1992).

[86] Yu. S. Kivshar and B. Malomed, *Rev. Mod. Phys.* **61**, 763 (1989).

[87] V. I. Klyatskin, *Stochastic Equations and Waves in Randomly Inhomogeneous Media*, Nauka, Moscow, 1980 (in Russian).

[88] R. Knapp, G. C. Papanicolaou, and D. White, in: *Disorder and Nonlinearity* (A. Bishop, D. Campbell, and S. Pnevmatikos, Eds.), Springer, Berlin, 1989.

[89] R. Knapp, G. C. Papanicolaou, and D. White, *J. Stat. Phys.* **63**, 567 (1991).

[90] K. Kono and S. Nakada, *Phys. Rev. Lett.* **69**, 1185 (1992).

[91] S. Kotani, *Contemp. Math.* **50**, 277 (1986).

[92] J. M. Kowalsky and J. L. Fry, *J. Math. Phys.* **28**, 2407 (1987).

[93] D. Krökel, N. J. Halar, G. Guliani, and D. Grishkovsky, *Phys. Rev. Lett.* **60**, 29 (1988).

[94] A. V. Kukushkin, V. D. Freylikher, and I. M. Fuks, *Izv. VUZov, Radiofiz.* **30**, 811 (1987) [*Radiophysics, Quantum Electronics* **30**, 811 (1987)].

[95] L. D. Landau and E. M. Lifshits, *Statistical Physics*, Pergamon, New York, 1958.

[96] R. Landauer, *Philos. Mag.* **21**, 863 (1970).

[97] R. Landauer and J. C. Helland, *J. Chem. Phys.* **22**, 1655 (1954).

[98] P. A. Lee and A. D. Stone, *Phys. Rev. Lett.* **55**, 1622 (1985).

[99] P. A. Lee, A. D. Stone, and H. Fukuyama, *Phys. Rev. B* **35**, 1039 (1987).

[100] Q. Li, St. Pnevmatikos, E. N. Economou, and C. M. Soukoulis, *Phys. Rev. B* **37**, 3534 (1988).

[101] Q. Li, St. Pnevmatikos, E. N. Economou, and C. M. Soukoulis, *Phys. Rev. B* **38**, 11888 (1988).

[102] I. M. Lifshits, *Adv. Phys.* **13**, 483 (1964).

[103] I. M. Lifshits and V. Ya. Kirpichenkov, *Zh. Eksp. and Teor. Fiz.* **77**, 989 (1979) [*Sov. Phys. JETP* **50**, 499 (1979)].

[104] I. M. Lifshits, S. A. Gredeskul, and L. A. Pastur, *Zh. Eksp. and Teor. Fiz.* **83**, 2362 (1982) [*Sov. Phys. JETP* **56**, 1370 (1982)].

[105] I. M. Lifshits, S. A. Gredeskul, and L. A. Pastur, *Introduction to the Theory of Disordered Systems*, Wiley, New York, 1988.

[106] A. V. Marchenko and L. A. Pastur, *Teor. Matem. Fizika* **68**, 433 (1986) [*Theor. Math. Physics (USSR)* **68**, 929 (1986)].

[107] A. V. Marchenko, S. A. Molchanov, and L. A. Pastur, *Teor. Matem. Fizika* **81**, 120 (1989) [*Theor. Math. Physics (USSR)* **81**, 1096 (1989)].

[108] V. A. Marchenko, *Spectral Theory for Sturm–Liouville Operators*, Naukova Dumka, Kiev, 1986 (in Russian).

[109] F. Martinelli and E. Scoppola, *Riv. Nuovo Cimento* **10**, 1 (1987).

[110] A. Mc Gurn and A. A. Maradudin, *Phys. Rev. B* **31**, 4866 (1985).

[111] A. Mc Gurn and A. A. Maradudin, *J. Opt. Soc. Am. B* **4**, 910 (1987).

[112] P. A. Mello, *Phys. Rev. Lett.* **60**, 1089 (1988).

[113] S. A. Molchanov, *Izv. AN SSSR, Ser. Mat.* **42**, 70 (1978) [*Math. USSR Izv.* **12**, 69 (1978)].

[114] N. F. Mott and E. A. Davis, *Electronic Properties in Noncrystalline Materials*, 2-nd ed., Oxford, Clarendon Press, 1979.

[115] N. F. Mott and W. D. Twose, *Adv. Phys.* **10**, 107 (1961).

[116] K. A. Muttalib, J. L. Pichard, and A. D. Stone, *Phys. Rev. Lett.* **59**, 2475 (1987).

[117] S. P. Novikov, V. E. Zakharov, S. V. Manakov, and L. P. Pitaevsky, *Theory of Solitons: The Inverse Scattering Method*, Consultant Bureau, New York 1984.

[118] G. C. Papanicolaou and W. Kohler, *Commun. Pure Appl. Math.* **27**, 641 (1974).

[119] G. C. Papanicolaou, *// CIME //*, (J. P. Cessoni, Ed.), Liquory Editore, Napoli, 1978.

[120] L. A. Pastur, *Teor. Matem. Fizika* **6**, 415 (1971) [*Theor. Math. Physics (USSR)* **6**, 299 (1971].

[121] L. A. Pastur, *Commun. Math. Phys.* **75**, 179 (1980).

[122] L. A. Pastur, *Sov. Sci. Rev. C Maths/Phys.* **6**, 1 (1987).

[123] L. A. Pastur, *Sov. Mat. Zametki*, **46**, No. 3, 50 (1989) (in Russian).

[124] L. A. Pastur and E. P. Feldman, *Zh. Eksp. and Teor. Phys.* **67**, 487 (1974) [*Sov. Phys. JETP* **40**, 241 (1975)].

[125] L. A. Pastur and A. L. Figotin, *Random and Almost Periodic Operators*, Springer, Heidelberg, 1992.

[126] J. B. Pendry, *J. Phys. C* **15**, 3493 (1982).

[127] V. I. Perel and D. G. Polyakov, *Zh. Eksp. and Teor. Fiz.* **86**, 352 (1984) [*Sov. Phys. JETP* **59**, 204 (1984)].

[128] Ping Sheng, B. White, Zhao-Qing Zhang, and G. Papanicolaou, *Phys. Rev. B* **34**, 4757 (1986).

[129] G. Polya and G. Szegö, *Aufaben und Lehrsätze aus der Analysis*, Springer, Berlin, 1925.

[130] M. E. Raikh and I. M. Ruzin, *Fiz. Tekhn. Poluprovod.* **19**, 1217 (1985) [*Sov. Phys. Semiconductors* **19**, 745 (1985)].

[131] M. E. Raikh and I. M. Ruzin, *Zh. Eksp. and Teor. Fiz.* **92**, 2257 (1987) [*Sov. Phys. JETP* **65**, 1273 (1987)].

[132] M. E. Raikh and I. M. Ruzin, in: *Mesoscopic Phenomena in Solids* (B. L. Altshuler, P. A. Lee, and R. A. Webb, Eds.), North-Holland, Amsterdam, 1991.

[133] T. Rosenbaum, K. Anders, G. H. Thomas, and R. N. Bhatt, *Phys. Rev. Lett.* **45**, 1723 (1980).

[134] D. Ruelle, *Statistical Mechanics: Rigorous Results*, Benjamin, New York, 1969.

[135] J. Sak and B. Kramer, *Phys. Rev. B* **24**, 1761 (1981).

[136] B. I. Shklovsky and A. L. Efros, *Electronic Properties of Doped Semiconductors*, Springer (Series in Solid State Sciences, v. 45), Berlin, 1981.

[137] B. Simon and T. Wolff, *Commun. Pure Appl. Math.* **39**, 79 (1986).

[138] Yu. S. Sokolovsky and L. N. Cherkashina, *Radiotekhnika Electronika* **16**, 1391 (1971) (in Russian).

[139] A. D. Stone and A. Szafer, *IBM J. Res. Develop.* **32**, 384 (1988).

[140] V. I. Tatarsky, *Propagation of Waves in the Turbulent Atmosphere*, Nauka, Moscow, 1967 (in Russian).

[141] G. E. Uhlenbeck and G. W. Ford, *Lectures in Statistical Mechanics*, AMS, Providence, R. I., 1963.

[142] A. Ya. Vul', V. N. Karayev, P. G. Petrosyan, T. A. Polyanskaya, I. I. Saidashev, and Yu. V. Shmartsev, *Fiz. Tekhn. Poluprovod.* **16**, 1838 (1982) [*Sov. Phys. Semiconductors* **16**, 1179 (1982)].

[143] A. Ya. Vul', T. A. Polyanskaya, I. G. Savel'ev, I. I. Saidashev, and Yu. V. Shmartsev, *Fiz. Tekhn. Poluprovod.* **17**, 134 (1983) [*Sov. Phys. Semiconductors* **17**, 84 (1983)].

[144] A. M. Weiner, J. P. Heritage, R. J. Hawkins, R. N. Hutson, E. M. Kirschner, D. E. Leaird and W. J. Tomlinson, *Phys. Rev. Lett.* **61**, 245 (1988).

[145] D. Würtz, T. Schneider, and M. P. Soerensen, *Physica A* **148**, 343 (1988).

[146] A. G. Zabrodsky, *Fiz. Tekhn. Poluprovod.* **14**, 1130 (1980) [*Sov. Phys. Semiconductors* **14**, 670 (1980)].

[147] V. E. Zakharov and Shabat, *Zh. Eksp. and Teor. Phys.* **64**, 1627 (1973) [*Sov. Phys. JETP* **37**, 823 (1973)].

3

Asymptotic Equations for Nonlinear Hyperbolic Waves

John K. Hunter

3.1. INTRODUCTION

Universal asymptotic equations. It is a remarkable fact that, in suitable asymptotic limits, the behavior of almost all nonlinear waves is described by a few rather simple looking nonlinear equations. Examples include Burgers equation, the Korteweg–de Vries (KdV) equation, and the nonlinear Schrödinger equation. These equations are "universal" or "canonical" because their applicability depends only on a few, very general features of the wave motion, such as the form of the linearized dispersion relation and the type of nonlinearity acting on the wave. Universal equations thus provide a common theoretical core for the study of nonlinear waves in an enormous number of diverse physical systems.

The idea of universal asymptotic equations for nonlinear waves has a long and rather indistinct history. Taniuti and Wei [259] stated this idea explicitly and called the derivation of the equations by the method of multiple scales the reductive perturbation method (see also [132, 260, 261]). Similar ideas are implicit in the derivation of the Burgers equation by Cole [56] and the KdV equation by Gardner and Morikawa [85]. Some further information about the history is given in [70].

JOHN K. HUNTER • Department of Mathematics and Institute of Theoretical Dynamics, University of California, Davis, California 95616.

Surveys in Applied Mathematics, Volume 2. Edited by Mark Freidlin, Sergey A. Gredeskul, John K. Hunter, Andrew V. Marchenko, and Leonid A. Pastur, Plenum Press, New York, 1995.

John K. Hunter

Table 3.1.

	Hyperbolic	Dispersive
Weakly Nonlinear	$u_t + (\frac{1}{2}u^2)_x = 0$ Inviscid Burgers Eq.	$iA_t + A_{xx} + \sigma|A|^2 A = 0$ Nonlinear Schrödinger Eq.
Strongly Nonlinear	—	$\omega = W(k, a),\ L = L(\omega, k, a),$ $k_t + \omega_x = 0, \quad \partial_t L_\omega + \partial_x L_k = 0$ Averaged Lagrangian Eq.

The discovery of the inverse scattering method for the KdV equation [166] and the nonlinear Schrödinger equation [280] provided a tremendous impetus to the study of nonlinear waves, since it showed that a surprising number of these universal equations are exactly solvable. This connection between universality and solvability is still something of a mystery; see [41] for one discussion of the possible reasons.

Classification of nonlinear waves. To explain the scope of this article, we roughly classify nonlinear waves into four classes, depending on whether they are weakly or strongly nonlinear, and on whether they are hyperbolic or dispersive [272]. In Table 3.1, we indicate schematically the simplest form of the corresponding asymptotic equations. We will be concerned with weakly nonlinear hyperbolic waves.

Weakly nonlinear theory usually applies to small amplitude waves propagating over large distances, while strongly nonlinear theory applies to slow modulations of large amplitude waves which are close to exact solutions of the fully nonlinear wave equations.

The distinction between dispersive and hyperbolic waves is not completely sharp. Dispersive waves are defined by the property that different frequencies propagate at different velocities, while hyperbolic waves are described by strongly hyperbolic partial differential equations (see Definition 3.1). The classification of a partial differential equation as strongly hyperbolic depends only on the structure of its highest order derivatives. Lower order terms may introduce dispersion so that a wave can be both hyperbolic and dispersive. A typical example is the Klein–Gordon equation,

$$u_{tt} - c_0^2 \nabla^2 u + \omega_0^2 u = 0.$$

Waves whose frequency is comparable with ω_0 are strongly dispersive, but high frequency waves are weakly dispersive and hyperbolic in nature.

When we refer to hyperbolic waves, we will always mean waves modeled by a homogeneous system of hyperbolic partial differential equations without lower order terms.[1] Such systems are nondispersive and have traveling wave solutions with arbitrary waveforms, $u = u(x - ct)$. A simple example is d'Alembert's solution of the one-dimensional wave equation.

---

[1] More generally, the system of equations may be perturbed by lower or higher order derivatives, provided the strength of these perturbations is weak for the solutions being considered.

Waveform distortion. The fact that linear hyperbolic waves have arbitrary waveforms has a profound effect on the behavior of weakly nonlinear waves. Hyperbolic waveforms are easily distorted because they are "neutrally stable." Even small nonlinear effects can gradually distort the waveform and cause the wave to break.

Dispersive waves behave very differently. A linear dispersive wave spreads out into a wavetrain with a locally sinusoidal wave profile and slowly modulated amplitude, wavelength, and phase (this follows from the method of stationary phase [272]). Weak nonlinearity cannot oppose strong dispersion to alter the sinusoidal waveform, so waveform distortion does not occur for weakly nonlinear dispersive waves.

The nonlinear Schrödinger equation. Since a modulated dispersive wavetrain is locally sinusoidal, it is completely defined by its local amplitude and phase. These vary according to the cubically nonlinear Schrödinger equation [3, 9, 70, 251]. For a description of weakly nonlinear dispersive waves in the context of nonlinear optics, see [208].

Nonlinear Schrödinger equations with complex coefficients (Ginzburg–Landau and Newell–Whitehead equations) arise in the study of strongly dissipative, driven systems near points of marginal instability [193, 209]. This article concerns systems which, in the first approximation, are nondissipative and nondispersive wave motions. We do not discuss the nonlinear Schrödinger equation or related equations at all.

Strongly nonlinear waves. Whitham developed an elegant asymptotic theory for large amplitude dispersive waves [272]. Since he based his approach on variational principles, it is called Whitham's averaged Lagrangian method, although it can also be derived for non-variational equations [160]. This theory describes modulated periodic waves, and depends on the fact that nonlinear dispersive waves typically have families of large amplitude, periodic traveling waves. There is no analogous theory for strongly nonlinear hyperbolic waves, because in the absence of dispersive effects a large amplitude nonlinear hyperbolic wave breaks quickly, and traveling waves do not exist.

One important approximate theory for large amplitude hyperbolic waves is Whitham's shock dynamics [272]. This theory describes the propagation of strong shocks. However, there does not seem to be a systematic asymptotic derivation of shock dynamics, except in certain cases involving the propagation of curved detonation shocks in combustion [17]. See [9, 120, 194, 201, 222, 223, 228, 245] and the references given there for work related to shock dynamics.

In this article, we consider only weakly nonlinear waves. This does not always mean that the wave amplitude is small. In some cases, such as for semi-linear equations or linearly degenerate waves, nonlinear hyperbolic equations have exact traveling wave solutions which are of the same form as the linear solution $u = u(x - ct)$. In these exceptional cases, a modulation theory for large amplitude waves can be developed, but the theory is much closer in spirit to the weakly nonlinear theory than to a genuine strongly nonlinear theory like the averaged Lagrangian method.

Types of nonlinearity. It is useful to separate the nonlinearity acting on hyperbolic waves into two main types, depending on the form of the nonlinear terms in the wave equation: advective nonlinearity (for example, $u \cdot \nabla u$), and semi-linearity (for example, $g(u)$).

We will mainly — but not exclusively — consider advective nonlinearity. There are two reasons for this. The first reason is that advective nonlinearity is important for almost all waves in continuum mechanics, including waves in fluids, solids, and plasmas, and this is the main physical application of the asymptotic theory. The second reason is that advective nonlinearity is "generic," in the sense that it corresponds to problems where the speed of a wave, c, depends on the wave amplitude, a. In general, we expect a "genuine nonlinearity" condition,

$$\frac{dc}{da} \neq 0$$

to hold, leading to advective types of nonlinearity.

Semi-linear effects may be important if there are special symmetries in a problem which eliminate advectively nonlinear effects. For example, in a Lorentz invariant field theory, the speed of light cannot depend on the field amplitude. Even if both semi-linear and advectively nonlinear effects are present, semi-linear effects may dominate in certain limits, such as for long waves — see Subsection 3.3.2.3.

The characteristic effect of advective nonlinearity is its tendency to distort the wave, leading to wave breaking and, possibly, to the formation of shocks. This is simply a consequence of the fact that larger amplitude parts of the wave travel faster than smaller amplitude parts (if $dc/da > 0$, say). A common feature of semi-linearity is the generation of a mean field by a rapidly oscillating wave (see Subsection 3.4.2). This does not typically occur in the advective case, since advective nonlinearity is usually in divergence form (for example, $uu_x = (u^2/2)_x$) and therefore has zero spatial mean.

The method of deriving asymptotic equations is independent of the type of non-linearity, although, of course, the details of the scaling required, and the form of the final equations do differ.

For genuinely nonlinear, strictly hyperbolic waves, the resulting asymptotic wave equation is usually of the form

$$u_t + \left(\frac{1}{2}u^2\right)_x = \mathcal{L}[u], \tag{3.1.1}$$

where $\mathcal{L}$ is a linear differential or pseudo-differential operator. This equation combines advective nonlinearity — described by $(u^2/2)_x$ — with linear dispersion, dissipation, relaxation, diffraction, or any number of other effects — described by $\mathcal{L}[u]$. Even such a deceptively simple class of nonlinear equations can display a wide range of phenomena, including shock waves, solitons, and temporal chaos. Of course, many variations of (3.1.1) also arise.

Distinguished limits. We pay particular attention to motivating the choice of scaling used in the asymptotic expansions. Once the appropriate scaling is known, it is usually just a matter of algebra to determine the resulting asymptotic equations; the art of using singular perturbation methods is to find the right scaling.

The key idea is to look for limits in which the various effects one wants to study are weak and of the same order of magnitude. Such a dominant balance typically picks out certain special scalings, called distinguished limits [18, 154, 165]. For example, in Subsection 3.2.3.1 we show that a dominant balance between weak nonlinearity and weak nonplanar effects leads to the correct ansatz for deriving weakly nonlinear geometrical optics, and in Subsection 3.3.2.2, we show that a dominant balance between weak nonlinearity and weak long wave dispersion leads to the appropriate scaling for deriving the KdV equation. The resulting ansatz, (3.3.43)–(3.3.44), is not at all obvious.

The issue of whether a given asymptotic equation models a real physical system is a separate question from this distinguished limit argument. For example, suppose the actual values of two dimensionless parameters characterizing some physical system are $\lambda = 0.1$ and $\mu = 0.05$. It is not clear what kind of limit one should take to obtain an asymptotic theory; we might assume $\lambda = O(1)$ and $\mu \to 0$, or $\lambda \to 0$ with $\mu = O(\lambda^2)$, or one of many other possibilities. Developing an effective asymptotic theory often requires a consideration of the physics, including experimental and numerical results, in addition to the use of singular perturbation methods. Moreover, it is usually not easy to estimate how small a parameter ε must actually be in order for an asymptotic theory as $\varepsilon \to 0$ to give good qualitative or quantitative results. In some problems, $\varepsilon = 1$ is "small." In other problems, $\varepsilon = 0.01$ is not small enough.

Outline of the article. In Section 3.2, we derive the inviscid Burgers equation from a general system of hyperbolic conservation laws. We begin with a discussion of weakly nonlinear plane waves. We then show how to combine geometrical optics type expansions with the weakly nonlinear expansion to obtain a generalized inviscid Burgers equation for nonplanar waves in several space dimensions. We call the resulting theory "weakly nonlinear geometrical optics."

The name "geometrical optics" originates from the fact that ray methods for high-frequency waves were first used to study light waves. Here, we use it to denote any type of high-frequency expansion, irrespective of the physical application.

In Section 3.3, we give a systematic derivation of equations like (3.1.1) starting from two types of primitive equations: first order systems of hyperbolic partial differential equations perturbed by higher order derivatives, and general systems of first order equations. We also show how the appearance of higher order dissipative and dispersive terms for long waves modeled by first order systems can be understood using a Chapman–Enskog expansion, and give general expansions for short waves.

In Section 3.4, we consider a variety of applications, including conservation laws with internal state variables, mean field interactions, and a class of hyperbolic variational equations. In Section 3.5, we discuss multi-dimensional versions of (3.1.1), and in Section 3.6 we briefly mention some aspects not discussed here, such as the resonant interaction of hyperbolic waves.

The subject matter of this article is necessarily restricted. Our aim is to explain in a systematic way how to derive asymptotic equations from a large class of hyperbolic partial differential equations. We do not describe the properties of the resulting equations in any detail, and the physical applications are of the most basic kind. Throughout, we consider waves propagating freely in R^n. There is no discussion of wave scattering by obstacles, nonlinear waves in resonators, or other problems involving boundaries.

Numerical solutions of the asymptotic equations are frequently very useful, but we mention them only briefly or not at all. Asymptotic equations typically have significantly fewer dependent and independent variables than the original systems of equations they are derived from. This greatly reduces the amount of computation required to solve them numerically.

To illustrate the theory, we give a step by step development of its application to the compressible Euler equations. We start with linear acoustics, and show how to add weak nonlinearity, nonplanar effects, dissipation, and diffraction. These developments appear in the subsubsections on nonlinear acoustics. We also present various other applications, for example to ion acoustic waves in a plasma, sound waves in bubbly fluids, and gravitational waves.

All of the general expansions are illustrated by specific physical applications. To obtain an overview of the contents of this article, the reader can look through the subsubsections on applications.

General references. The asymptotic methods explained here are very well established and there is an enormous literature on them. It is not possible to give a comprehensive bibliography. For the most part, we refer either to earlier papers which originated some of the fundamental ideas, or to papers, reviews, and books which provide readable introductions and further references.

For a general description of nonlinear waves, see Whitham [272]. Introductions to the theory of partial differential equations are given in [77, 226, 227]. For an introduction to numerical finite difference methods see [252], and for introductions to asymptotics and singular perturbation methods, see [109, 154].

Some other reviews of asymptotic equations for nonlinear hyperbolic waves are [9, 187, 205, 231, 232].

3.2. HYPERBOLIC WAVES

The inviscid Burgers equation. In this subsection we derive the fundamental asymptotic equation for weakly nonlinear, genuinely nonlinear hyperbolic waves, namely, the inviscid Burgers equation,

$$u_t + \left(\frac{1}{2} u^2 \right)_x = 0. \tag{3.2.1}$$

This equation is the simplest example of a nonlinear hyperbolic conservation law [172, 249, 272]. Smooth solutions of (3.2.1) typically break down in finite time because their derivative blows up. They can be continued by weak solutions [227] which contain discontinuities, or shocks. The jump in u across a shock satisfies a jump relation derived from the weak form of (3.2.1). When considering weak solutions it is essential to preserve the conservation form of (3.2.1), since weak solutions are not preserved by nonlinear changes of the dependent variable [272].

An additional entropy condition is required to ensure uniqueness of weak solutions [67, 173]. This entropy condition introduces irreversibility and decay of solutions of (3.2.1). As $t \to +\infty$, compactly supported solutions of (3.2.1) approach a special weak solution called an N-wave, which consists of a linear expansion wave sandwiched between two shocks [181, 272].

Explicit solutions of (3.2.1) can be obtained using the method of characteristics and shock fitting [272].

The presence of shocks means that special care must be used in devising numerical schemes for solving (3.2.1) and other hyperbolic conservation laws. The schemes have to converge to weak solutions, even when they contain discontinuities. Moreover, the schemes must pick out the correct entropy solutions. For an introduction to numerical methods for conservation laws see [177]. A detailed comparison of various schemes for (3.2.1) is given in [274].

We mention that [246, 248] study (3.2.1) with Brownian motion initial data, and [229] shows that scalar conservation laws arise as a hydrodynamic limit of interacting particle systems.

The asymptotic expansion. The fundamental weakly nonlinear geometrical optics expansion was derived by Choquet-Bruhat [54], and appears in Subsection 3.2.3.2. Geometrical optics methods are amazingly flexible and powerful: in some sense, everything we describe in this article is a generalization or an application of this expansion.

The result of the weakly nonlinear geometrical optics expansion is a pair of equations. The phase of the wave satisfies an eikonal equation, and the amplitude of the wave satisfies an inviscid Burgers equation. The wave amplitude does not appear in the eikonal equation, so the phase is the same as in linear geometrical optics. Ostrovsky [215] therefore refers to the resulting theory as the "linear ray approximation." Descriptions of linear geometrical optics are given in [34, 59, 163].

Straightforward geometrical optics breaks down when waves focus and form a caustic. The behavior of nonlinear hyperbolic waves at caustics is not fully understood.

An asymptotic expansion of a system of differential equations tacitly assumes that the solution remains smooth. Nevertheless, the final result (3.2.1) remains valid in the weak sense even after shocks form, provided that one preserves the correct conservation form throughout the expansion. This can be shown formally by expanding the jump conditions across a shock in addition to the differential equations [118] or by including a small amount of viscosity in the expansion [46]. It should also be possible to expand the weak form of the differential equation directly, using suitable rapidly varying test functions. In one space dimension, the formal theory has been rigorously justified for weak solutions [69], even when resonant interactions occur [239]. The fact that the weakly nonlinear theory handles shocks correctly is one of its most useful features.

We begin this Section with an introduction to hyperbolic waves, including a description of our main physical example, the flow of a compressible fluid. For further information on hyperbolic waves, see [132, 272].

3.2.1. Linear Hyperbolic Waves

3.2.1.1. Linearization. We consider a first order hyperbolic system of conservation laws

$$u_t + f^\alpha(u)_{x^\alpha} = 0. \tag{3.2.2}$$

Here, $u(x,t) \in R^m$ is a vector of conserved quantities and $f^\alpha : R^m \to R^m$ is the flux vector in the αth direction. We use the summation convention over repeated spatial indices ($\alpha = 1, \cdots, n$).

Linear theory can be used to study solutions which are small amplitude perturbations of a given state,

$$u(x,t) = u_0 + \varepsilon u'(x,t;\varepsilon), \qquad \varepsilon \ll 1. \tag{3.2.3}$$

We assume that u_0 is constant, but similar results follow for high frequency waves in a non-constant medium, by "freezing" the coefficients of the corresponding variable coefficient, linearized equations. For a discussion of the connection between well-posedness of a variable coefficient equation and the associated frozen equations, see [164].

We use (3.2.3) in (3.2.2) and linearize the resulting equation for u'. This gives

$$u'_t + A^\alpha u'_{x^\alpha} = 0, \qquad (3.2.4)$$

where $A^\alpha = D_u f^\alpha(u_0)$ is the $m \times m$ Jacobian matrix of f^α evaluated at u_0.

Dispersion relations. The general solution of (3.2.4) can be obtained by superposing normal mode solutions, of the form

$$u' = \hat{a} \exp(ik \cdot x - i\omega t)r. \qquad (3.2.5)$$

Equation (3.2.5) is a solution of (3.2.4) if the wavenumber vector $k = (k_1, \cdots, k_n)$ and the frequency ω satisfy

$$\det(k_\alpha A^\alpha - \omega I) = 0, \qquad (3.2.6)$$

and the vector r in (3.2.5) is a right null vector of the matrix in (3.2.6). The arbitrary constant $\hat{a}$ measures the amplitude of the wave.

Equation (3.2.6) is the linearized dispersion relation of (3.2.2). See [108, 272] for more extensive discussions of dispersion relations. We denote the roots of (3.2.6) by

$$\omega = W_j(k), \qquad j = 1, \cdots, m.$$

We denote associated left and right eigenvectors by $r_j(k)$ and $l_j(k)$, where

$$(k_\alpha A^\alpha) r_j = W_j(k)r_j,$$
$$l_j \cdot (k_\alpha A^\alpha) = W_j(k)l_j$$

We normalize the eigenvectors so that $l_j \cdot r_j = 1$.

Definition 3.1. *The system of conservation laws (3.2.2) is strongly hyperbolic at $u = u_0$, with t as a timelike direction, if: (a) $W_j(k)$ is real for every $k \in R^n$ and if the right eigenvectors $\{r_1, \cdots, r_m\}$ form a basis of R^m; (b) the matrix $R(k) = [r_1(k), \cdots, r_m(k)]$ of normalized right eigenvectors satisfies the uniform bound*

$$\|R(k)\| + \|R^{-1}(k)\| \le C,$$

for all $k \in R^n$ with $|k| = 1$. If the roots $W_j(k)$ are distinct for every $k \in R^n \setminus \{0\}$, then (3.2.2) is strictly hyperbolic.

The conditions in (a) and (b) are necessary and sufficient conditions for the Cauchy problem for (3.2.4) to be well-posed in $L^2(R^n)$ [164]. Hyperbolicity implies that for any initial data of the form $u(x, 0) = \exp(ik \cdot x)v$, the solution is a sum of waves,

$$u(x, t) = \sum_{j=1}^m \hat{a}_j \exp(ik \cdot x - iW_j(k)t) r_j.$$

It follows that solutions of hyperbolic partial differential equations which oscillate in space also oscillate in time, so hyperbolic equations are mathematical models of wave motions.

The phase velocity c of a wave with dispersion relation $\omega = W(k)$ is defined by

$$c(k) = \frac{W(k)}{|k|^2}k.$$

Wavefronts, $k \cdot x = $ constant, propagate at the phase velocity. The group velocity C is defined by

$$C(k) = (C^1, \cdots, C^n) = \nabla_k W(k).$$

We will see below that the wave energy in a slowly modulated wave train propagates at the group velocity.

Differentiating the equation

$$(k_\alpha A^\alpha - W(k)I)\, r(k) = 0$$

with respect to k_α and taking the scalar product of the result with l shows that

$$C^\alpha = l \cdot A^\alpha r. \tag{3.2.7}$$

Scale invariance. An important symmetry of (3.2.2) and (3.2.4) is their invariance under scaling transformations, $x \to \lambda x$ and $t \to \lambda t$. This scale invariance is a consequence of the fact that the only dimensional quantities defined by (3.2.2) or (3.2.4) are velocities; there are no intrinsic length or time scales.

Physically, scale invariant equations arise as an approximate description of waves whose wavelength and period are much larger or much smaller than any other length and time scales characteristic of the wave motion.

Long waves are particularly important because they transport significant amounts of energy and because dissipative effects are typically much weaker for long waves than for short waves. Many wave motions are described by a system of hyperbolic conservation laws in a suitable long wave limit. For example, sound waves in a gas are described by the compressible Euler equations provided that their wavelength is much larger than the mean free path of the gas molecules; nonlinear elastic waves are described by the equations of finite elasticity provided that their wavelength is much larger than a typical lattice spacing; and plasma waves are described by the magnetohydrodynamic equations provided that their frequency is much lower than all characteristic frequencies of the plasma.

An example of scale invariance in a short wave limit is provided by shallow water waves on a slowly rotating fluid (see Subsection 3.3.3.1).

A large separation of scales between wave parameters and the parameters of the medium occurs very commonly— this is typical of the "asymptotic character" of physics — so scale invariant equations are of wide applicability despite their apparently special nature. Small departures from scale invariance (such as weak dispersion) can be treated perturbatively.

Scale invariance implies that the dispersion relation satisfies

$$W(\lambda k) = \lambda W(k) \tag{3.2.8}$$

for all $\lambda \in R$. As a result, the phase and group velocities of waves propagating in a given direction are independent of frequency, so the waves are nondispersive.

It does not follow from (3.2.8) that the phase and group velocities are in the same direction, unless the dispersion relation is isotropic, meaning that $\omega = W(|k|)$.

Plane waves. For nondispersive waves satisfying (3.2.8), Fourier superposing all the normal modes (3.2.5) with k in a fixed direction gives a plane wave solution of (3.2.4). Normalizing $r(k)$ so that it is independent of the magnitude of k, we obtain

$$u'(x,t) = \int_R \hat{a}(\lambda)e^{i\lambda k \cdot x - iW(\lambda k)t}r(\lambda k)\,d\lambda$$
$$= \int_R \hat{a}(\lambda)e^{i\lambda(k \cdot x - W(k)t)}r(k)\,d\lambda.$$

Denoting the inverse Fourier transform of $\hat{a}(\lambda)$ by $a(\theta)$, this gives the plane wave solution

$$u' = a(k \cdot x - \omega t)r, \tag{3.2.9}$$

where $\omega = W(k)$. It is simple to verify directly that (3.2.9) is a solution of (3.2.4) for an arbitrary scalar-valued function a. We call a the wave amplitude.

The functional dependence of $a(\theta)$ on the phase variable,

$$\theta = k \cdot x - \omega t,$$

describes the waveform of the wave. For example, if a is compactly supported, then (3.2.9) represents a pulse; if

$$a(\theta) \to a_L \qquad \text{as } \theta \to -\infty,$$
$$a(\theta) \to a_R \qquad \text{as } \theta \to +\infty,$$

then (3.2.9) represents a shock-type profile; if a is periodic, or almost periodic, then (3.2.9) represents an oscillatory wave; if

$$a(\theta, x) = \begin{cases} 0 & \text{if } \theta \le 0 \\ \sigma(x)\theta^n & \text{if } \theta > 0 \end{cases}$$

then (3.2.9) represents a wave near a wavefront, and so on.

The weakly nonlinear theory for single waves is insensitive to the type of waveform. However for interactions between several different waves (see Subsection 3.6.1), the type of waveform involved is very important. For example, oscillatory waves can interact in very complicated ways, whereas interactions between short pulses are negligible.

Superposing the plane wave solutions (3.2.9) gives the general solution of (3.2.4),

$$u'(x,t) = \sum_{j=1}^m \int_{\mathbf{S}^{n-1}} \left\{ a_j\left[k \cdot x - W_j(k)t, k\right] r_j(k)\,dS(k) \right\}.$$

Here, the integration is over the $(n-1)$-dimensional unit sphere $\mathbf{S}^{n-1}$, since we have integrated over all wavenumbers propagating in a given direction in going from (3.2.5) to (3.2.9). This solution can also be obtained using the Radon transform [137].

In one space dimension ($n = 1$), this solution is a sum, but in several space dimensions, it is necessary to integrate over all directions. This is a consequence of the fact that, in one space dimension, there is a unique characteristic curve associated with a given family of waves, passing through each point of space–time, but in several space dimensions there is a cone of characteristic surfaces passing through each point [59, 132]. This nonuniqueness of characteristic surfaces, and related phenomena like focusing, is a fundamental obstacle to understanding nonlinear hyperbolic waves in several space dimensions.

Critical phenomena and turbulence. Much of the interesting nonlinear behavior in hyperbolic waves comes from the fact that they are scale invariant (at least in a first approximation). Two other areas in which scale invariance arises are critical phenomena in second order phase transitions [26, 90], and turbulence in the inertial range [198]. These problems are much harder to treat than wave problems because there is no separation of length-scales, as there is in a modulated wave, but the idea of universality is a common thread in all three areas. It would be interesting to look for further connections. For example: can renormalization methods be used to derive universal asymptotic equations? (asymptotic equations are fixed points of the expansions used to derive them); and can multiple scale methods be extended to treat scale-invariant partial differential equations whose solutions depend on a continuum of length scales?

For an analysis of related issues in "weak wave turbulence," which concerns random ensembles of interacting weakly nonlinear waves, see [279].

3.2.1.2. Acoustics.

The gas dynamics equations. The basic physical example of a hyperbolic system of conservation laws is the gas dynamics or compressible Euler equations which describe the flow of a simple compressible fluid. We denote the density of the fluid by ρ, the pressure by p, the specific internal energy by e, and the velocity by u. Conservation of mass, momentum, and energy imply that [102, 272]

$$\rho_t + \nabla \cdot (\rho u) = 0,$$
$$(\rho u)_t + \nabla \cdot (\rho u \otimes u - \mathcal{T}) = 0, \tag{3.2.10}$$
$$\left[\rho \left(e + \frac{1}{2} u \cdot u \right) \right]_t + \nabla \cdot \left[\rho \left(e + \frac{1}{2} u \cdot u \right) u - \mathcal{T} u + q \right] = 0.$$

Here, $\mathcal{T}$ is the Cauchy stress tensor and q is the heat flux vector. For an inviscid gas with no thermal conductivity, the constitutive relations for the stress and heat flux are

$$\mathcal{T} = -pI, \qquad q = 0. \tag{3.2.11}$$

The internal energy is given in terms of ρ and p by the equation of state, $e = e(\rho, p)$. The specific entropy $s(\rho, p)$ and the temperature $T(\rho, p)$ are given functions which satisfy the thermodynamic identity

$$T ds = de + p \, d \left(\frac{1}{\rho} \right).$$

The sound speed $c(\rho, p)$ is defined by

$$c^2 = \left. \frac{\partial p}{\partial \rho} \right|_s. \tag{3.2.12}$$

For an ideal gas with constant specific heats,

$$e = \frac{p}{(\gamma - 1)\rho},$$
$$p = \kappa \exp(s/c_v)\rho^\gamma, \tag{3.2.13}$$
$$c^2 = \frac{\gamma p}{\rho},$$

where γ is the ratio of specific heats, c_v is the specific heat at constant volume, and κ is a constant. A detailed discussion of equations of state for real gases appears in [200].

For smooth solutions, the non-dissipative gas dynamics equations (3.2.10), (3.2.11) are equivalent to the following equations in non-conservative form,

$$\rho_t + u \cdot \nabla \rho + \rho \nabla \cdot u = 0,$$
$$u_t + u \cdot \nabla u + \frac{1}{\rho}\nabla p = 0, \tag{3.2.14}$$
$$s_t + u \cdot \nabla s = 0.$$

In (3.2.14), we use $(\rho, u, s)^t$ as the vector of dependent variables, and $p = p(\rho, s)$ is a given function. If shocks are present, then one must use the weak form of the conservative equations (3.2.10) rather than (3.2.14). Alternatively, for piecewise smooth solutions, one can use (3.2.14) in the regions where the solution is smooth, supplemented by the jump conditions across shocks derived from the weak form of (3.2.10).

Linear acoustics. Linearizing (3.2.14) about a constant background state, $u = u_0$, $\rho = \rho_0$, and $s = s_0$, gives the acoustics equations,

$$\rho'_t + u_0 \cdot \nabla \rho' + \rho_0 \nabla \cdot u' = 0,$$
$$\rho_0 (u'_t + u_0 \cdot \nabla u') + c_0^2 \nabla \rho' + d_0 \nabla s' = 0, \tag{3.2.15}$$
$$s'_t + u_0 \cdot \nabla s' = 0.$$

In (3.2.15), ρ' stands for the perturbation in density about ρ_0, with u' and s' defined similarly, c_0 is the sound speed in the unperturbed state, and

$$d_0 = \left.\frac{\partial p}{\partial s}\right|_\rho (\rho_0, s_0). \tag{3.2.16}$$

The normal mode solutions of (3.2.15) are

$$\begin{bmatrix} \rho' \\ u' \\ s' \end{bmatrix} = \hat{a} \exp(ik \cdot x - i\omega t) \begin{bmatrix} \hat{\rho} \\ \hat{u} \\ \hat{s} \end{bmatrix}.$$

In n space dimensions, the dispersion relation of (3.2.15) is

$$\Omega^n \left[\Omega^2 - c_0^2 |k|^2\right] = 0, \tag{3.2.17}$$

where $\Omega = \omega - u_0 \cdot k$ is the Doppler shifted frequency.

The dispersion relation of sound waves is

$$\Omega^2 - c_0^2|k|^2 = 0.$$

The associated right nullvector is

$$\begin{bmatrix} \hat{\rho} \\ \hat{u} \\ \hat{s} \end{bmatrix} = \begin{bmatrix} \rho_0 \\ c_0^2\Omega^{-1}k \\ 0 \end{bmatrix}.$$

It follows that sound waves carry longitudinal velocity perturbations and density and pressure perturbations at constant entropy.

The root $\Omega = 0$ is a multiple eigenvalue for $n > 1$. Such waves are convected by the background flow since their group velocity is $C = u_0$. The corresponding nullspace is spanned by n vectors of the form

$$\begin{bmatrix} \hat{\rho} \\ \hat{u} \\ \hat{s} \end{bmatrix} = \begin{bmatrix} -d_0 \\ 0 \\ c_0^2 \end{bmatrix} \quad \text{and} \quad \begin{bmatrix} \hat{\rho} \\ \hat{u} \\ \hat{s} \end{bmatrix} = \begin{bmatrix} 0 \\ k^\perp \\ 0 \end{bmatrix}.$$

Here, $k^\perp$ runs over $n - 1$ linearly independent vectors such that $k^\perp \cdot k = 0$. These nullvectors correspond to entropy waves at constant pressure and vorticity waves, respectively.

3.2.1.3. Breakdown of linearization. There are two conditions for solutions of the linearized equations (3.2.4) to provide a reasonable approximation of solutions of the nonlinear equations (3.2.2). The first condition is the obvious one, namely that the perturbation u'/c_0 is small. The second condition is that the propagation distance and time are not too long. The effect of even very small nonlinearities can accumulate after a sufficiently long time, so that the solution of the nonlinear equation diverges from the solution of the linearized equation. These weakly nonlinear effects will be observed in practice if the wave propagates far enough and if dissipation is sufficiently weak that the wave does not damp out before nonlinearity has a chance to influence it.

To see that linearization breaks down after long times, we will compute the first nonlinear correction to the linearized approximation and show that secular terms appear in the expansion.

We look for a straightforward asymptotic expansion of small amplitude solutions of (3.2.2),

$$u(x, t; \varepsilon) = u_0 + \varepsilon u_1(x, t) + \varepsilon^2 u_2(x, t) + O(\varepsilon^3), \tag{3.2.18}$$

as $\varepsilon \to 0$. Using (3.2.18) in (3.2.2), Taylor expanding in ε, and equating coefficients of ε and ε^2 to zero, we obtain that

$$u_{1t} + A^\alpha u_{1x^\alpha} = 0, \tag{3.2.19}$$

$$u_{2t} + A^\alpha u_{2x^\alpha} + \frac{1}{2}D_u^2 f^\alpha(u_0) \cdot (u_1, u_1)_{x^\alpha} = 0. \tag{3.2.20}$$

The component free expression for the derivative of f^α is

$$\left[D_u^2 f^\alpha(u_0) \cdot (v, w)\right]_i = \sum_{p,q=1}^m \frac{\partial^2 f_i^\alpha}{\partial u_p \partial u_q}(u_0)v_p w_q.$$

Similar component free notation for other derivatives is used below.

A solution of (3.2.19) is the plane wave (3.2.9),

$$u_1 = a(\theta)r, \qquad \theta = k \cdot x - \omega t.$$

Using this expression for u_1 in (3.2.20) and solving for u_2 by the method of characteristics gives

$$u_2 = \frac{\mathcal{G}t}{2\omega}\left(a^2\right)_\theta r + \frac{1}{2}a^2 s + v_2(x,t). \tag{3.2.21}$$

Here, v_2 is an arbitrary solution of the homogeneous equation, and

$$\mathcal{G} = l \cdot k_\alpha D_u^2 f^\alpha(u_0) \cdot (r,r). \tag{3.2.22}$$

The vector s is a solution of

$$(k_\alpha A^\alpha - \omega I)\, s = -k_\alpha D_u^2 f^\alpha(u_0) \cdot (r,r) + \mathcal{G}r.$$

This equation has a solution because the right hand side is orthogonal to the left null vector l, by (3.2.22).

If $\mathcal{G} \neq 0$, then (3.2.21) shows that u_2 contains secular terms which grow linearly in time. It follows from (3.2.18) that the linearized solution does not give a valid asymptotic approximation when $t = O(\varepsilon^{-1})$, because then the second order correction, $\varepsilon^2 u_2$, is of the same order of magnitude as the linearized solution, εu_1.

To explain the significance of $\mathcal{G}$, we introduce the nonlinear dispersion relation $\omega = W(u,k)$, and an associated right eigenvector $r(u,k)$, which satisfy

$$[k_\alpha D_u f^\alpha(u) - W(u,k)I] \cdot r(u,k) = 0.$$

We take the directional derivative of this equation with respect to u in the direction r. This gives

$$[k_\alpha D_u f^\alpha - WI]\, D_u r \cdot (r) + k_\alpha D_u^2 f^\alpha \cdot (r,r) - D_u W \cdot (r)r = 0.$$

Taking the scalar product of this equation with the left eigenvector $l(u,k)$, and using the normalization $l \cdot r = 1$, implies that

$$\begin{aligned}
\mathcal{G}(u,k) &= l(u,k) \cdot k_\alpha D_u^2 f^\alpha(u) \cdot (r(u,k), r(u,k)) \\
&= D_u W(u,k) \cdot r(u,k). \tag{3.2.23}
\end{aligned}$$

Thus, $\mathcal{G} = \mathcal{G}(u_0, k)$ in (3.2.22) is the derivative of the nonlinear frequency (or, if $|k|$ is normalized to one, the wave speed) with respect to the wave amplitude. If $\mathcal{G}(u,k) \neq 0$ for all (u,k), then the wave is genuinely nonlinear [172]. On the other hand, if $\mathcal{G}(u,k) \equiv 0$ for all (u,k), then the wave is linearly degenerate.

We have shown that the linearized solution breaks down for large times, $t = O(\varepsilon^{-1})$, when the wave is genuinely nonlinear.

3.2.2. *Weakly Nonlinear Plane Waves*

In the previous subsection we saw that the linearized solution does not provide a uniformly valid approximation for small amplitude, genuinely nonlinear waves over long times. In this subsection we use the method of multiple scales [154] to derive a weakly nonlinear solution which describes the long time behavior of small amplitude waves. The main result is an inviscid Burgers equation for the wave amplitude.

In the method of multiple scales, we introduce new independent variables, which explicitly separate the "fast" and "slow" behaviors of the solution. This extra flexibility allows us to construct uniformly valid approximations. Here, the "fast" time scale is a typical wave period and the "slow" time scale is the time taken for nonlinearity to influence the wave.

3.2.2.1. The inviscid Burgers equation. To analyze the behavior of a single small amplitude wave over large times and propagation distances, we look for a multiple scale asymptotic solution of (3.2.2) of the form

$$
\begin{aligned}
u(x,t;\varepsilon) = u_0 &+ \varepsilon u_1(k \cdot x - \omega t, \varepsilon x, \varepsilon t) \\
&+ \varepsilon^2 u_2(k \cdot x - \omega t, \varepsilon x, \varepsilon t) + O(\varepsilon^3),
\end{aligned}
\tag{3.2.24}
$$

where $\varepsilon \to 0$ with $x, t = O(\varepsilon^{-1})$, and u_0 is a constant vector.

Using (3.2.24) in the conservation law (3.2.2), Taylor expanding in ε, and equating coefficients of ε and ε^2 to zero implies that $u_1(\theta, \xi, \tau)$ and $u_2(\theta, \xi, \tau)$ satisfy

$$
(k_\alpha A^\alpha - \omega I)\, u_{1\theta} = 0,
\tag{3.2.25}
$$

$$
(k_\alpha A^\alpha - \omega I)\, u_{2\theta} + u_{1\tau} + A^\alpha u_{1\xi^\alpha}
$$
$$
+ \frac{1}{2} k_\alpha D_u^2 f^\alpha(u_0) \cdot (u_1, u_1)_\theta = 0.
\tag{3.2.26}
$$

Equation (3.2.25) has a nontrivial solution if (ω, k) satisfies the linearized dispersion relation (3.2.6). A solution is then

$$
u_1 = a(\theta, \xi, \tau) r,
\tag{3.2.27}
$$

where a is an arbitrary scalar valued wave amplitude and r is the nullvector associated with (ω, k). We use (3.2.27) in (3.2.26) and take the scalar product of the result with the left nullvector l. This shows that

$$
a_\tau + C \cdot \nabla_\xi a + \left(\frac{1}{2} \mathcal{G} a^2 \right)_\theta = 0,
\tag{3.2.28}
$$

where C is the group velocity vector defined in (3.2.7) and $\mathcal{G}$ is the genuine nonlinearity coefficient defined in (3.2.23).

If $\mathcal{G} \neq 0$, we can introduce new variables

$$
\bar{u} = \mathcal{G} a, \qquad \partial_{\bar{t}} = \partial_\tau + C \cdot \nabla_\xi, \qquad \bar{x} = \theta.
$$

This transforms (3.2.28) to the inviscid Burgers equation,

$$
\bar{u}_{\bar{t}} + \left(\frac{1}{2} \bar{u}^2 \right)_{\bar{x}} = 0.
\tag{3.2.29}
$$

Some care is needed in interpreting the independent variables in (3.2.29), since $(\bar{x}, \bar{t})$ are not the original space–time variables. In general, $\partial_{\bar{t}}$ is a "slow" derivative along a ray, which is a curve in space–time moving with the group velocity, while $\bar{x}$ is a "fast" phase variable. Usually, one can think of $\bar{x}$ as a space variable in a reference frame moving with the local group velocity of the wave, and $\bar{t}$ as a time variable. In signaling problems, $\bar{t}$ is a variable measuring the propagation distance of the wave while $\bar{x}$ is proportional to time at a fixed point in space (see the discussion at the end of Subsection 3.2.2.2).

3.2.2.2. *Nonlinear acoustics (1)*. From (3.2.10) and (3.2.11), the governing equations for the one-dimensional flow of an inviscid, non-conducting compressible fluid are

$$\rho_t + (\rho u)_x = 0,$$
$$(\rho u)_t + (\rho u^2 + p)_x = 0,$$
$$\left[\rho\left(e + \frac{1}{2}u^2\right)\right]_t + \left[\rho\left(e + \frac{1}{2}u^2\right)u + pu\right]_x = 0,$$

with $e = e(p, \rho)$ a given function.

Applying (3.2.24) and (3.2.27) to this system gives the following asymptotic solution:

$$\begin{bmatrix} \rho \\ u \\ s \end{bmatrix} = \begin{bmatrix} \rho_0 \\ 0 \\ s_0 \end{bmatrix} + \varepsilon a(kx - \omega t, \varepsilon x, \varepsilon t) \begin{bmatrix} \rho_0 \\ c_0 \\ 0 \end{bmatrix} + O(\varepsilon^2), \qquad (3.2.30)$$

where $\varepsilon \to 0$ with $x, t = O(\varepsilon^{-1})$. The frequency ω and the wavenumber k are related by

$$\omega = c_0 k,$$

where c_0 is the sound speed defined in (3.2.12), evaluated at $\rho = \rho_0$ and $s = s_0$.

The dimensionless parameter ε is a measure of the strength of the sound wave. For example, we can use the acoustic Mach number

$$\varepsilon = \frac{u_{max}}{c_0},$$

where u_{max} is the maximum velocity of the velocity perturbations carried by the sound wave. For simplicity, we suppose that there is no background flow. This does not involve any loss of generality because a constant background flow can be removed by a Galilean transformation.

The inviscid Burgers equation for the wave amplitude $a(\theta, \xi, \tau)$ is, from (3.2.28),

$$a_\tau + c_0 a_\xi + \left(\frac{1}{2}\omega \mathcal{G} a^2\right)_\theta = 0. \qquad (3.2.31)$$

Here, we write $\omega \mathcal{G}$ in (3.2.31) instead of $\mathcal{G}$ in (3.2.28). The dimensionless coefficient of the nonlinear term is the fundamental derivative [200],

$$\mathcal{G} = \frac{1}{c}\frac{\partial(c\rho)}{\partial \rho}\bigg|_s \ (\rho_0, s_0). \qquad (3.2.32)$$

For an ideal gas with constant specific heats,

$$\mathcal{G} = \frac{\gamma + 1}{2}.$$

Thus, sound waves are genuinely nonlinear when $\gamma \neq -1$.

Ultrasonic waves Equation (3.2.31) describes shock formation in intense ultrasonic waves. Ultrasonic waves are sound waves whose frequency is higher than 20kHz, the upper limit of human hearing [247]. In a typical experiment, an ultrasonic wave is generated at one end of a tube of fluid by a transducer, and the pressure profile is measured at different distances down the tube. After a sufficient time a steady wave pattern is set up in which the wave amplitude is modulated in space but not in time. The individual wavefronts propagate down the tube, but the wave envelope is steady.

To give a rough idea of the parameter values in such an experiment ([247], p. 93), consider an ultrasonic wave with a frequency of 0.5 MHz in water at standard conditions. The sound speed in water is $c_0 \approx 1500$ m/s, so the wavelength is of the order $\lambda \approx 3$ mm. A very strong ultrasonic wave has an acoustic Mach number of $\varepsilon = 5.0 \times 10^{-4}$. Water is highly incompressible, so this Mach number corresponds to a maximum pressure of over ten atmospheres; nevertheless the sound wave is weakly nonlinear. The propagation distance for nonlinearity to become important is of the order $L_n = \lambda/\varepsilon \approx 6$ m. Taking numerical factors into account ($\mathcal{G} \approx 4.3$ for water, and the maximum initial slope of a sine wave with wavelength λ is $2\pi/\lambda$) shows that shocks form over a significantly shorter distance, $L_* \approx 25$ cm.

To use the asymptotic equation in this case, we write $a = a(\sigma, \xi)$, where

$$\sigma = -\frac{\theta}{\omega} = t - \frac{x}{c_0}$$

is the retarded time. Equation (3.2.31) becomes

$$c_0 a_\xi - \left(\frac{1}{2}\mathcal{G}a^2\right)_\sigma = 0.$$

Experimental observations agree very closely with the predictions of this equation. Photographs of measured pressure profiles showing the formation of shocks appear in [24, 25, 247]. Note that for this signaling problem, the wave profile is measured as a function of time at a fixed spatial location. As a result, the nonlinear coefficient changes sign, and the wave is distorted "backwards" in comparison with its behavior for an initial value problem.

Experimental observations of the generation of higher harmonics in ultrasonic waves, which is associated with waveform distortion, gives a method of measuring $\mathcal{G}$. The value of $\mathcal{G}$ for sound waves in elastic solids provides information about nonlinear elastic constants, which are important in solid state physics [33].

For reviews of nonlinear acoustics see [63, 64, 103, 205, 235].

3.2.3. Weakly Nonlinear Nonplanar Waves

In this subsection we combine the weakly nonlinear expansion derived in Subsection 3.2.2 with a geometrical optics expansion. This gives an asymptotic solution for weakly nonlinear, nonplanar waves.

We begin with a heuristic discussion to explain why the ansatz used in Subsection 3.2.3.2 is the correct one for combining nonlinear and nonplanar effects.

3.2.3.1. Modulated waves. A small amplitude, slowly varying nonplanar wave is described by an asymptotic solution of the form

$$u(x, t; \delta, \varepsilon) = u_0 + \varepsilon u' \left(\delta^{-1} \varphi(\delta x, \delta t), \delta x, \delta t; \delta, \varepsilon \right). \qquad (3.2.33)$$

Here, $\varepsilon \ll 1$ measures the dimensionless wave amplitude and $\delta \ll 1$ measures the ratio of the wavelength to the radius of curvature of the wavefront (see equation (3.2.34) below).

To show that (3.2.33) represents a slowly varying wave, we expand the right-hand side of (3.2.33) about a fixed point, $x = x_0$ and $t = t_0$. Assuming that $|x - x_0|$, $|t - t_0| \ll \delta^{-1/2}$, this gives

$$u \sim u_0 + \varepsilon u_0' \left(k_0 \cdot (x - x_0) - \omega_0(t - t_0) \right),$$

where

$$
\begin{aligned}
u_0'(\theta) &= u' \left(\theta + \theta_0, \xi_0, \tau_0; 0, 0 \right), \\
k_0 &= \nabla_\xi \varphi(\xi_0, \tau_0), \qquad \omega_0 = -\varphi_\tau(\xi_0, \tau_0), \\
\xi_0 &= \delta x_0, \qquad \tau_0 = \delta t_0, \qquad \theta_0 = \delta^{-1} \varphi(\xi_0, \tau_0).
\end{aligned}
$$

Thus, locally (3.2.33) is approximated by a plane wave. The local wavenumber and frequency of the plane wave are

$$
\begin{aligned}
k &= \nabla_\xi \varphi = \nabla_x \theta, \\
\omega &= -\varphi_\tau = -\theta_t.
\end{aligned}
$$

Here, $\theta(x, t) = \delta^{-1} \varphi(\delta x, \delta t)$ is the fast phase variable.

In weakly nonlinear geometrical optics, we consider a limit in which nonlinear effects and nonplanar effects are small and of the same order of magnitude. This limit corresponds to taking $\delta = \varepsilon$ in (3.2.33). To show this fact, we make some rough order of magnitude estimates.

We consider a genuinely nonlinear wave of dimensionless amplitude ε and wavelength λ.[2] Advectively nonlinear terms like $u \cdot \nabla u$ are then of the order $\varepsilon^2 \lambda^{-1}$.

We can think of these advectively nonlinear terms as a small forcing term on the linearized equations. This forcing term can be approximated by evaluating it at the linearized solution. Since the linearized solution propagates with the linearized wave speed, the nonlinear forcing term does also. As a result, the nonlinear forcing resonates with the linearized wave and the cumulative effect of nonlinearities grows linearly in time or propagation distance (see (3.2.21) in Subsection 3.2.1.3).

To estimate the cumulative nonlinear effect, we therefore multiply the size of the nonlinear terms by the propagation distance. After a propagation distance L, the cumulative nonlinear effect is thus of the order $\varepsilon^2 L \lambda^{-1}$. Nonlinearity alters the wave significantly when the cumulative nonlinear effect is of the same order ε as the wave amplitude. This occurs when the propagation distance is of the order

$$L_n = \frac{\lambda}{\varepsilon}.$$

---

[2]In the nondimensionalization implicitly adopted in (3.2.33), we have $\lambda = O(1)$.

Nonplanar effects, like focusing, are important over lengthscales L_f such that the local wavenumber changes by a significant fraction. This lengthscale is of the order

$$L_f = O\left(\frac{|k|}{|\nabla_x k|}\right) = O\left(\frac{\lambda}{\delta}\right).$$

Neglecting order one constants, we therefore have

$$L_f = \frac{\lambda}{\delta}. \tag{3.2.34}$$

A balance between nonlinear and nonplanar effects occurs when $L_f = L_n$. This gives $\delta = \varepsilon$. If $\varepsilon \ll \delta$ then nonplanar effects dominate, and they are described by linear geometrical optics. If $\delta \ll \varepsilon$ then nonlinear effects dominate, and they are described by the weakly nonlinear plane wave expansion in Subsection 3.2.2.

Similar arguments apply to the influence of nonuniformities on the wave. The effects of nonuniformities and nonlinearity are both significant provided the lengthscale over which the medium varies is of the order L_n.

Sonic booms. A concrete example of these scalings occurs in the propagation through the atmosphere of the sonic boom generated by a supersonic aircraft. The typical width of a sonic boom signal is of the order of $\lambda = 100$ m. Suppose the strength of the boom — as measured for example by the pressure rise over the ambient pressure — is of the order 10^{-2} in the immediate far field of a supersonic aircraft. (This is actually stronger than most booms, whose strength when they hit the ground is usually no more than 10^{-3}.) In that case, $L_n = 10km$. The density of the atmosphere decreases by a factor of e over a height of about $7km$. Thus, for an aircraft flying at $10km$, we can use weakly nonlinear geometrical optics to account for the combined effects of nonlinearity, nonplanar wavefronts, and refraction by the atmosphere, on the boom as it propagates down to the ground.

Whitham developed his "nonlinearization technique" [272] to study sonic boom propagation in the atmosphere. This was one of the first applications of weakly nonlinear geometrical acoustics. The asymptotic expansion below gives a formal justification of Whitham's results.

Other scalings. One can also imagine situations where other scalings are of interest. For example, suppose that a sound wave is propagating in an oceanic wave-guide. At high frequencies the rays are trapped, and bounce back and forth between caustics. If nonlinear effects are important only after many such reflections, then $L_n \gg L_f$, and one could study the propagation of waves over distances of the order L_n along the waveguide ("weakly-weakly nonlinear" geometrical optics).

Far-field versus geometrical optics expansions. We do not need to explicitly distinguish time t in the next subsection, so we will write $t = x^0$ and use $x = (x^0, \cdots, x^n)$ to stand for space–time variables. We also change variables from x to x/ε. In the old variables (the "far-field" expansion), the wavelength is of the order one and the nonlinear lengthscale is of the order ε^{-1}. In the new variables (the "high-frequency" or "geometrical optics" expansion), the wavelength is of the order ε and the nonlinear lengthscale is of the order one. The resulting expansions are equivalent; physically they differ only in the choice of lengths used to nondimensionalize the space variables.

In most cases, we use "far-field" variables for plane waves and "geometrical optics" variables for nonplanar waves. To indicate which variables are in use, we give the region of validity for each ansatz. Typically:

$$\varepsilon \to 0 \quad x = O(\varepsilon^{-1}), \qquad \text{far-field};$$
$$\varepsilon \to 0 \quad x = O(1), \qquad \text{geometrical optics}.$$

3.2.3.2. The asymptotic expansion. In this subsection, we derive the equations of weakly nonlinear geometrical optics [54, 118]. The phase of the wave satisfies the usual eikonal equation of linear geometrical optics. The amplitude of the wave satisfies an inviscid Burgers equation which generalizes the transport equation of linear geometrical optics. We discuss these equations further in Subsections 2.3.3 and 2.3.4.

We consider a hyperbolic system of conservation laws,

$$f^\alpha(x, u)_{x^\alpha} + g(x, u) = 0. \tag{3.2.35}$$

In equation (3.2.35) the implied summation on space–time variables runs over $\alpha = 0, \cdots, n$. We look for an asymptotic solution of the form

$$u(x; \varepsilon) = u_0(x) + \varepsilon u_1 \left(\varepsilon^{-1}\varphi(x), x\right) + \varepsilon^2 u_2 \left(\varepsilon^{-1}\varphi(x), x\right) + O(\varepsilon^3), \tag{3.2.36}$$

where $\varepsilon \to 0$ with $x = O(1)$. Here, u_0 is a given smooth solution of (3.2.35).

We use (3.2.36) in (3.2.35) and equate coefficients of ε^0 and ε^1 to zero. This implies that $u_1(\theta, x)$ and $u_2(\theta, x)$ satisfy

$$\kappa_\alpha A^\alpha u_{1\theta} = 0, \tag{3.2.37}$$
$$\kappa_\alpha A^\alpha u_{2\theta} + (A^\alpha u_1)_{x^\alpha}$$
$$+ \frac{1}{2}\kappa_\alpha D^2 f^\alpha(u_0) \cdot (u_1, u_1)_\theta + G u_1 = 0. \tag{3.2.38}$$

Here, $\kappa = (\kappa_0, \cdots, \kappa_n)$ is the frequency-wavenumber vector associated with the phase φ,

$$\kappa_\alpha = \varphi_{x^\alpha},$$

and

$$A^\alpha(x) = D_u f^\alpha(x, u_0(x)),$$
$$G(x) = D_u g(x, u_0(x)).$$

Equation (3.2.37) has a nontrivial solution provided that κ satisfies the local linearized dispersion relation of (3.2.35),

$$D(\kappa; x) = \det(\kappa_\alpha A^\alpha(x)) = 0. \tag{3.2.39}$$

A solution for u_1 is then

$$u_1 = a(\theta, x) r(x), \tag{3.2.40}$$

where a is an arbitrary scalar amplitude and r is a right nullvector, satisfying

$$\kappa_\alpha A^\alpha r = 0.$$

We assume that this matrix has a nullspace with algebraic multiplicity one, and we denote an associated left nullvector by $l(x)$.

Equation (3.2.38) can be solved for u_2 provided that the nonhomogeneous term involving u_1 is orthogonal to l. Taking the scalar product of (3.2.38) with l and using (3.2.40) in the result implies that a satisfies

$$C^\alpha a_{x^\alpha} + \left(\frac{1}{2}\mathcal{G}a^2\right)_\theta + \mathcal{H}a = 0, \tag{3.2.41}$$

where

$$\begin{aligned}
C^\alpha &= l \cdot A^\alpha r, \\
\mathcal{G} &= l \cdot \kappa_\alpha D_u^2 f^\alpha \cdot (r, r), \\
\mathcal{H} &= l \cdot \frac{\partial}{\partial x^\alpha}(A^\alpha r) + l \cdot Gr.
\end{aligned} \tag{3.2.42}$$

Equation (3.2.41) is a generalized inviscid Burgers equation; it is the main result of the expansion. As we will see, the normalized form of (3.2.41) is the constant coefficient inviscid Burgers equation (3.2.1).

We follow the usual terminology in the applied literature, and use "generalized" to denote variable coefficient versions of an equation, and "modified" to denote changes in the nonlinearity, for example from quadratic to cubic. However, note that in the PDE literature, "generalized" is often used to mean modified nonlinearity!

Multiple characteristics. If the original system is not strictly hyperbolic at $u = u_0$ we obtain a system of inviscid Burgers equations instead of (3.2.41). Suppose there are m right null vectors $\{r_1, \cdots, r_m\}$ and m left null vectors $\{l_1, \cdots, l_m\}$ such that

$$\begin{aligned}
\kappa_\alpha A^\alpha r_k &= 0, \\
l_j \cdot \kappa_\alpha A^\alpha &= 0.
\end{aligned}$$

We assume that the algebraic multiplicity of $\kappa_\alpha A^\alpha$ is equal to its geometric multiplicity m and is independent of x. A solution of (3.2.37) for u_1 is then

$$u_1 = \sum_{k=1}^m a^k(\theta, x) r_k(x).$$

Imposing solvability conditions for u_2 leads to a system of equations for the wave amplitudes $\{a^1, \cdots, a^m\}$,

$$\sum_{k=1}^m C_{jk}^\alpha a_{x^\alpha}^k + \sum_{p,q=1}^m \left(\frac{1}{2}\mathcal{G}_{jpq}a^p a^q\right)_\theta + \sum_{k=1}^m \mathcal{H}_{jk}a^k = 0,$$

where

$$\begin{aligned}
C_{jk}^\alpha &= l_j \cdot A^\alpha r_k, \\
\mathcal{G}_{jpq} &= l_j \cdot \kappa_\alpha D_u^2 f^\alpha \cdot (r_p, r_q), \\
\mathcal{H}_{jk} &= l_j \cdot \frac{\partial}{\partial x^\alpha}(A^\alpha r_k) + l_j \cdot Gr_k.
\end{aligned}$$

A classification of two by two systems of conservation laws with quadratic fluxes appears in [129, 241].

Eckhoff [73] has used linear geometrical optics for nonstrictly hyperbolic systems to study the stability of the background state. Weakly nonlinear generalizations of this stability analysis for strictly hyperbolic systems are considered in [51].

Change of type. If the geometric multiplicity of the eigenvalue is strictly less than its algebraic multiplicity, one can still carry out an asymptotic expansion. The resulting equations change type. The simplest case is when the algebraic multiplicity is two and the geometric multiplicity is one. Under an appropriate "genuine nonlinearity" condition, the normalized asymptotic equations are the steady transonic small disturbance equations [36, 99],

$$u_t + v_x = 0,$$
$$v_t + \left(\frac{1}{2}u^2\right)_x = 0.$$

These equations are elliptic when $u < 0$, so the initial value problem is ill-posed and the consistency of the asymptotics is very doubtful. In the "non-genuinely nonlinear" case some interesting cubically nonlinear equations arise which are parabolic on a line in state space and hyperbolic elsewhere [36].

One application of parabolically degenerate systems of conservation laws arises in combustion [74, 75], where the resonance associated with the degeneracy provides a mechanism for the transition to detonation in a reactive granular medium.

3.2.3.3. The eikonal equation. In Subsection 2.3.2, we showed that the local frequency and wavenumber of a modulated wave satisfy the local linearized dispersion relation (3.2.39) of the wave motion. Returning to physical notation (x, t) for the space–time variables, we have

$$\omega = W(k; x, t), \tag{3.2.43}$$
$$\omega = -\varphi_t, \tag{3.2.44}$$
$$k = \nabla\varphi. \tag{3.2.45}$$

The phase φ therefore satisfies an eikonal equation,

$$\varphi_t + W(\nabla\varphi; x, t) = 0.$$

Ray tracing. Equations (3.2.43) – (3.2.45) can be solved using the method of characteristics [272]. From (3.2.44)–(3.2.45) and the equality of mixed partial derivatives, it follows that

$$k_t + \nabla\omega = 0.$$

This equation expresses "conservation of wave crests." Using (3.2.43) in this equation gives

$$k_t + C \cdot \nabla k + \nabla_x W = 0,$$

where $C = \nabla_k W$ is the group velocity. Introducing the rays $x = x(t)$ with $dx/dt = C$, then gives a system of ordinary differential equations on each ray:

$$\frac{dx}{dt} = \nabla_k W, \tag{3.2.46}$$
$$\frac{dk}{dt} = -\nabla_x W. \tag{3.2.47}$$

These equations determine the location of a ray and the evolution of the wavenumber along a ray.

To obtain the phase, we use the equation

$$\frac{d\varphi}{dt} = \varphi_t + C \cdot \nabla\varphi = -\omega + C \cdot k.$$

For a scale invariant wave, differentiating (3.2.8) with respect to λ, and setting $\lambda = 1$, shows that the right hand side of this equation is zero. Therefore

$$\frac{d\varphi}{dt} = 0,$$

and the phase is constant along a ray.

Now suppose that we are given initial data for the phase, $\varphi(x, 0) = \varphi_0(x)$. Differentiating this initial data with respect to x gives initial data for k. (More generally, the phase may be given on an n-dimensional submanifold of space–time.) Equations (3.2.46) and (3.2.47) with initial data

$$x(0; \beta) = \beta, \tag{3.2.48}$$

$$k(0; \beta) = \nabla_\beta \varphi_0(\beta), \tag{3.2.49}$$

then determine an n-parameter family of rays, $x = x(t; \beta)$ with $k = k(t; \beta)$.

In many applications, the initial data for the ray equations is not given a priori, but must be obtained by matching with some kind of inner problem. For examples of this, and for a discussion of the numerical implementation of ray tracing, see [27].

Quantum mechanics. Equations (3.2.46)–(3.2.47) are Hamiltonian with momentum k and Hamiltonian W. The fact that we obtain a system of Hamiltonian equations for the rays is not a coincidence. Classical mechanics is the high-frequency limit of quantum mechanics, and the classical particle paths are just the rays associated with the quantum mechanical waves [151].

An interesting question connected with semi-classical mechanics is "quantum chaos." That is, what are the properties of a quantum system (such as the semi-classical asymptotics of its spectrum and eigenfunctions) when the corresponding classical Hamiltonian system is chaotic [88]. We do not consider this question here. Our interest is in using high-frequency expansions for nonlinear partial differential equations. The differential equations of quantum mechanics, such as the Schrödinger equation, are always linear. It is only the limiting classical equations are nonlinear. Moreover, quantum chaos requires the study of semi-classical solutions for arbitrarily large times. Our expansions for nonlinear equations are only valid for finite times, such as $t = O(1)$ or $t = O(\varepsilon^{-1})$, depending on the nondimensionalization.

Caustics. Taking the union of the solutions $(x(t; \beta), k(t; \beta), t)$ of the ray equations (3.2.46)–(3.2.49), where $\beta \in R^n$, gives an $(n + 1)$-dimensional submanifold $\mathcal{L}$ of $(2n + 1)$-dimensional (x, k, t) space. We denote space–time by $\mathcal{M}$. If the projection

$$\pi : \mathcal{L} \ni (x, k, t) \mapsto (x, t) \in \mathcal{M},$$

is nonsingular, we obtain a global solution for $\varphi(x, t)$. More frequently, the solution for φ is single-valued only for short times, because the submanifold $\mathcal{L}$ "folds" over.

In some regions ("shadow regions"), $\mathcal{L}$ may not cover $\mathcal{M}$ at all. In other regions ("illuminated regions"), several points of $\mathcal{L}$ may project to a single point of $\mathcal{M}$. This means that several waves propagate from different initial locations to the same point at a later time.

In linear geometrical optics, the presence of several waves at the same point does not present any new difficulties, since their amplitudes superpose. In nonlinear theories, one has to deal with interactions between the waves. For oscillatory waves, this issue is not fully resolved, since the multi-phase waves on the illuminated side of a caustic are necessarily "incoherent" (see Subsection 3.6.1). For pulses, the interactions are negligible, since the different pulses only overlap for a short time.

Curves or surfaces in $\mathcal{M}$ on which the projection π is singular are called caustics. Caustics can be classified using the methods of singularity theory [11].

The map from space–time coordinates to ray coordinates $(x, t) \mapsto (\beta, t)$ does not have a smooth local inverse on a caustic, and the straightforward geometrical optics solution breaks down near caustics because the coefficient $\mathcal{H}$ in (3.2.41) becomes infinite. This fact can be seen from (3.2.52) below: $\mathcal{H}$ contains a term proportional to the divergence of the group velocity vector, and, by a standard result,

$$C^{\alpha}_{x^{\alpha}} = \frac{C^{\alpha} J_{x^{\alpha}}}{J},$$

where J is the Jacobian of the transformation from ray to space–time coordinates. This Jacobian is zero on a caustic.

For linear waves, the only effect of propagating through a smooth convex caustic is a phase shift of $-\pi/2$ in each Fourier component of the wave amplitude. This implies that the outgoing waveform is the Hilbert transform of the incoming waveform. The number of such phase shifts is related to the Maslov index of the ray [11].

The linear theory of caustics is described in [15, 185].

3.2.3.4. The transport equation. We now restrict our discussion to regions of space–time which do not include caustics and which contain a single wave. The evolution of the amplitude of a weakly nonlinear wave is then given by the transport equation (3.2.41).

Using ray coordinates in which $a = a(\theta, t, \beta)$, equation (3.2.41) becomes

$$a_t + \left(\frac{1}{2} \mathcal{G} a^2 \right)_{\theta} + \mathcal{H} a = 0, \tag{3.2.50}$$

where $\mathcal{G} = \mathcal{G}(t, \beta)$ and $\mathcal{H} = \mathcal{H}(t, \beta)$. Equation (3.2.50) is an n-parameter family of equations, one for each ray. The equation gives the rate of change of the wave amplitude along a ray due to nonlinear effects (the second term), and due to nonuniformities, focusing, and sources (the third term).

If $\mathcal{G} \neq 0$, equation (3.2.50) can be put in a normalized form by the change of variables,

$$\bar{u}(\bar{x}, \bar{t}) = \mathcal{E}(t) a(\theta, t), \qquad \bar{t} = \mathcal{T}(t), \qquad \bar{x} = \theta,$$

$$\mathcal{E} = \int e^{\mathcal{H}} \, dt, \qquad \mathcal{T} = \int \frac{\mathcal{G}}{\mathcal{E}} \, dt. \tag{3.2.51}$$

Using these expressions in (3.2.50) implies that $\bar{u}(\bar{x}, \bar{t})$ satisfied (3.2.29). Thus the normalized asymptotic equation for a nonplanar hyperbolic wave is the usual inviscid Burgers equation.

If additional terms, such as dissipative or dispersive terms, are present, it is usually not possible to transform the variable coefficient equations to the corresponding constant coefficient version (see Section 3.3).

The energy equation. To explain the physical meaning of the transport equation (3.2.50) in more detail, we rewrite (3.2.42) for $\mathcal{H}$ as

$$\mathcal{H} = \frac{1}{2}\frac{\partial C^\alpha}{\partial x^\alpha} + \tilde{\mathcal{H}},$$

$$\tilde{\mathcal{H}} = \frac{1}{2}\left(l \cdot A^\alpha \frac{\partial r}{\partial x^\alpha} - \frac{\partial l}{\partial x^\alpha} A^\alpha r \right) + \frac{1}{2} l \cdot \frac{\partial A^\alpha}{\partial x^\alpha} r + l \cdot Gr, \qquad (3.2.52)$$

where have written x^0 instead of t again and used (3.2.7). The first term in $\mathcal{H}$ is proportional to the divergence of the group velocity vector. It therefore describes changes in the wave amplitude due to focusing of the rays. The terms in $\tilde{\mathcal{H}}$ arise from the departure of the system from a symmetric hyperbolic system, nonuniformities in the medium, and sources.

The system is symmetrizable at $u = u_0$ if there is a nonsingular matrix $S(x)$ and symmetric matrices $B^\alpha(x)$ such that

$$A^\alpha = S^{-1}B^\alpha S.$$

In this case, it follows that

$$\tilde{\mathcal{H}} = \frac{1}{2}l \cdot \left(\frac{\partial A^\alpha}{\partial x^\alpha} + A^\alpha \frac{\partial S^{-1}}{\partial x^\alpha}S + \frac{\partial S^{-1}}{\partial x^\alpha}SA^\alpha \right) r + l \cdot Gr. \qquad (3.2.53)$$

The two terms in this expression describe the effects of nonuniformities, and source terms, respectively.

For smooth solutions, the transport equation (3.2.41) implies that

$$\frac{\partial}{\partial x^\alpha}\left(\frac{1}{2}C^\alpha a^2 \right) + \frac{\partial}{\partial \theta}\left(\frac{1}{6}\mathcal{G}a^3 \right) + \tilde{\mathcal{H}}a^2 = 0. \qquad (3.2.54)$$

When shocks form, this equation does not follow from (3.2.41), since nonlinear transformations of dependent variables are not valid for weak solutions. Rosales [231] explains how to modify (3.2.54) when there are shocks.

For symmetrizable systems, we can interpret (3.2.54) as a balance of energy (or, for variational equations, as a balance of wave action [94]). The linearized system,

$$\frac{\partial}{\partial x^\alpha}\left(A^\alpha u \right) + Gu = 0,$$

has the energy equation

$$\frac{\partial}{\partial x^\alpha}\left(\frac{1}{2}u \cdot TA^\alpha u \right) + u \cdot T\left(\frac{1}{2}\frac{\partial A^\alpha}{\partial x^\alpha} + \frac{1}{2}A^\alpha \frac{\partial S^{-1}}{\partial x^\alpha}S + \frac{1}{2}\frac{\partial S^{-1}}{\partial x^\alpha}SA^\alpha + G \right) u = 0,$$

where $T = S^t S$. We substitute the geometrical optics solution $u = a(\theta, x)r(x)$ into this equation and average over θ. After using the fact that $l = Tr$, we obtain the averaged energy equation,

$$\frac{\partial}{\partial x^\alpha}\left(\frac{1}{2}C^\alpha \langle a^2\rangle\right) + \tilde{\mathcal{H}}\langle a^2\rangle = 0,$$

where $\tilde{\mathcal{H}}$ is defined in (3.2.53). This equation coincides with the phase average of the energy equation (3.2.54) obtained directly from the geometrical optics transport equation. Note that the mean energy flux is in the direction of the group velocity.

The analytical method of characteristics. A second interpretation of the transport equation (3.2.50) can be obtained by writing it in the characteristic form

$$\frac{da}{dt} + \mathcal{H}a = 0,$$
$$\frac{d\theta}{dt} = \mathcal{G}a.$$

The first equation is identical to the transport equation in linear geometrical optics. Thus the wave amplitude has the same form as in linear geometrical optics, when written as a function of the characteristic variable defined by the second equation. Nonlinearity therefore changes the location of the rays slightly in comparison with linearized theory. (This point of view is adopted by Whitham in his nonlinearization technique.)

There have been attempts to generalize this idea by giving a "rule" for changing the location of characteristic surfaces instead of the rays (the "analytical method of characteristics" [156]), but it does not seem possible to provide an asymptotic derivation of such generalizations in several space dimensions without making an additional high-frequency assumption. (In one space dimension, the rays and the characteristic curves coincide.)

Wavefront expansions. A weak discontinuity in a solution of a first order hyperbolic partial differential equation is a surface across which the first (or higher order) derivatives are discontinuous [59]. Such a discontinuity can only occur across a characteristic surface. The evolution of the jump in the derivatives can be calculated exactly using the method of wavefront expansions [50, 132]. The same results can be obtained directly from the transport equation (3.2.41) of weakly nonlinear geometrical optics.

We consider a solution

$$u = \begin{cases} u_0(x) & \text{if } \varphi(x) \leq 0 \\ u_0(x) + \varphi(x)u_1(x) + O(\varphi^2) & \text{if } \varphi(x) > 0. \end{cases}$$

Near the wavefront, when $\varphi = O(\varepsilon) \ll 1$, we can write this solution as

$$u = u_0(x) + \varepsilon a(\theta, x)r(x) + O(\varepsilon^2),$$

where $\theta = \varepsilon^{-1}\varphi$ and

$$a(\theta, x) = \begin{cases} 0 & \text{if } \theta \leq 0 \\ \sigma(x)\theta & \text{if } \theta > 0. \end{cases} \tag{3.2.55}$$

The jumps in the derivatives of u across the wavefront are given by

$$\lim_{\varphi \to 0+} u_{x^\alpha} - \lim_{\varphi \to 0-} u_{x^\alpha} = \sigma \kappa_\alpha r.$$

Using (3.2.55) in (3.2.41) gives a closed equation for the evolution of the jump σ along a ray,

$$C^\alpha \sigma_{x^\alpha} + \mathcal{G}\sigma^2 + \mathcal{H}\sigma = 0.$$

This equation is a Ricatti equation. If the damping coefficient $\mathcal{H}$ is not too large, and if the initial value of σ has the appropriate sign for a compressive wave, the solution blows up in finite time. This blow-up corresponds to the formation of a shock. After blow-up, the wavefront expansion breaks down.

It is possible to derive generalized wavefront expansions for shocks [9, 10]. One difficulty with these generalized wavefront expansions is that they lead to an infinite non-closed system of evolution equations for the jumps in derivatives of all orders across the shock. This is because of the fact that motion of the shock is affected by its interaction with the flow behind it. This flow is computed by means of a Taylor series expansion behind the shock. Generalized wavefront expansions therefore do not describe situations where the flow behind the shock is nonanalytic (for example, when one shock catches up with another).

3.2.3.5. Nonlinear acoustics (2). We apply the above theory to a sound wave propagating through a smooth medium with velocity $u = u_0(x, t)$, density $\rho = \rho_0(x, t)$, and sound speed $c = c_0(x, t)$. The phase $\varphi(x, t)$ satisfies the local linearized dispersion relation,

$$\Omega^2 = c_0^2 |k|^2,$$

where $\omega = -\varphi_t$ and $k = \nabla\varphi$ are the local frequency and wavenumber and $\Omega = \omega - u_0 \cdot k$ is the Doppler shifted frequency. The group velocity,

$$C = u_0 + \frac{c_0^2}{\Omega} k,$$

is the sum of the background flow velocity and a velocity normal to the wavefronts with magnitude equal to the sound speed. See [180] for a discussion of ray tracing in a wind.

The asymptotic solution for the sound wave is

$$\begin{bmatrix} \rho \\ u \\ s \end{bmatrix} = \begin{bmatrix} \rho_0 \\ u_0 \\ s_0 \end{bmatrix} + \varepsilon a(\varepsilon^{-1}\varphi, x, t) \begin{bmatrix} \rho_0 \\ c_0^2 \Omega^{-1} k \\ 0 \end{bmatrix} + O(\varepsilon^2), \qquad (3.2.56)$$

where $\varepsilon \to 0$ with $x, t = O(1)$. The transport equation for the wave amplitude $a(\theta, x, t)$, is

$$a_t + C \cdot \nabla a + \left(\frac{\gamma+1}{4} \Omega a^2 \right)_\theta$$
$$+ \frac{\Omega}{2\rho_0 c_0^2} \left[\left(\frac{\rho_0 c_0^2}{\Omega} \right)_t + \nabla \cdot \left(\frac{\rho_0 c_0^2}{\Omega} C \right) \right] a = 0. \qquad (3.2.57)$$

The last term describes effects due to focusing and nonuniformities.

For smooth solutions, equation (3.2.57) implies the conservation law

$$\mathcal{A}_t + \left[\frac{\gamma + 1}{3} \Omega a \mathcal{A} \right]_\theta + \nabla \cdot (\mathcal{A} C) = 0,$$

where

$$\mathcal{A} = \frac{\rho_0 c_0^2 a^2}{\Omega} = \frac{\text{energy density}}{\text{frequency}}$$

is the wave action density. When shocks are present, the wave action decreases.

Spherical waves. A simple but interesting special case is provided by outgoing spherical waves in n space dimensions. We assume the background state is constant and $u_0 = 0$. The eikonal equation is then

$$\varphi_t^2 = c_0^2 |\nabla \varphi|^2.$$

This has the spherically symmetric solution

$$\varphi(x) = r - c_0 t,$$

where $r = |x|$. The transport equation is

$$a_t + c_0 a_r + \left(\frac{\gamma + 1}{4} c_0 a^2 \right)_\theta + c_0 \left(\frac{n - 1}{2r} \right) a = 0.$$

The change of variables

$$\bar{u}(\bar{x}, \bar{t}; \beta) = c_0 \frac{\gamma + 1}{2} a(\theta, r, t),$$

$$\bar{t} = \frac{2}{3 - n} r^{(3-n)/2} \quad \text{if} \quad n \neq 3,$$

$$\bar{t} = \log r \quad \text{if} \quad n = 3,$$

$$\bar{x} = \theta, \qquad \beta = r - c_0 t,$$

puts the transport equation in standard Burgers form (3.2.29).

We note that $\bar{t} \to +\infty$ as $r \to +\infty$ when $n \leq 3$, but $\bar{t}$ is bounded above by zero as $r \to +\infty$ when $n > 3$. This fact shows that compactly supported spherically symmetric solutions of the Euler equations approach outgoing spherical N-waves as $t \to +\infty$ for $n \leq 3$, and outgoing solutions of the linear wave equation for $n > 3$.

If the shock formation time for (3.2.29) is greater than $\bar{t} = 0$, then shocks will never form in the solution of the spherical equation. Thus, geometrical attenuation can prevent shock formation when $n > 3$ but not when $n \leq 3$, however small the initial data. In the borderline case of $n = 3$, the time for shock formation is exponentially large in the derivative of the initial data (as the derivative tends to zero). Since the solution of the wave equation with compactly supported initial data approaches an outgoing spherical wave as $t \to +\infty$, with slow variations in the spherical angle, this argument gives a simple formal derivation of a number of rigorous blow up results for nonlinear wave equations [138].

If the coefficient of a^2 in the transport equation is zero (corresponding to a loss of genuine nonlinearity, or "Klainerman's null condition" [138]), then, as we show in the next subsection, the transport equation is typically cubically nonlinear,

$$a_t + c_0 a_r + \left(\frac{1}{3}c_0 \mathcal{G}_3 a^3\right)_\theta + c_0 \left(\frac{n-1}{2r}\right) a = 0.$$

A similar change of variables shows that this equation has globally smooth solutions for $n > 2$ when the initial data is small enough. The borderline case, in which smooth solutions have exponentially long life spans, is now $n = 2$ space dimensions.

3.2.4. Waves with Cubic Nonlinearity

The inviscid Burgers equation is the "generic" asymptotic equation for nonlinear hyperbolic waves. If the genuine nonlinearity coefficient happens to vanish, meaning that

$$\mathcal{G} = l \cdot \kappa_\alpha D_u^2 f^\alpha \cdot (r, r) = 0, \tag{3.2.58}$$

then the nonlinear term drops out of the transport equation (3.2.41).

When $\mathcal{G} = 0$, the dominant nonlinear effects are typically cubic. For small amplitude waves, cubic nonlinearities are weaker than quadratic nonlinearities, so they only become important over longer propagation distances than those described by the expansion in (3.2.36). Similar estimates to the ones in Subsection 3.2.3.1 show that the nonlinear lengthscale for cubically nonlinear effects is of the order

$$L_n = \frac{\lambda}{\varepsilon^2}.$$

If we nondimensionalize the space–time variables so that $L_n = 1$, then the orders of magnitude of the wavelength, λ, and the amplitude, ε, are related by $\lambda = \varepsilon^2$.

The appropriate asymptotic expansion of (3.2.35) to pick up cubically nonlinear effects therefore has the form

$$u = u_0(x) + \varepsilon u_1(\theta, x) + \varepsilon^2 u_2(\theta, x) + \varepsilon^3 u_3(\theta, x) + O(\varepsilon^4),$$
$$\theta = \varepsilon^{-2}\varphi(x),$$

where $\varepsilon \to 0$ with $x = O(1)$. The local wavenumber vector, $\nabla_x \theta$, is of the order ε^{-2} and varies over distances of the order one, as required.

Using the same notation as in Subsection 3.2.3.2, the perturbation equations are

$$\kappa_\alpha A^\alpha u_{1\theta} = 0, \tag{3.2.59}$$

$$\kappa_\alpha A^\alpha u_{2\theta} + \frac{1}{2}\kappa_\alpha D^2 f^\alpha(u_0) \cdot (u_1, u_1)_\theta = 0, \tag{3.2.60}$$

$$\kappa_\alpha A^\alpha u_{3\theta} + \kappa_\alpha D^2 f^\alpha(u_0) \cdot (u_1, u_2)_\theta$$
$$+ (A^\alpha u_1)_{x^\alpha} + \frac{1}{6}\kappa_\alpha D^3 f^\alpha(u_0) \cdot (u_1, u_1, u_1)_\theta + G u_1 = 0. \tag{3.2.61}$$

A solution of (3.2.59) for u_1 is

$$u_1 = a(\theta, x)r + \bar{u}(x).$$

Here, it is useful to include a mean field $\bar{u}$, to allow for small perturbations away from the state u_0 where the coefficient of genuine nonlinearity vanishes.

The solvability condition for (3.2.60) is satisfied because of the failure of genuine nonlinearity (3.2.58). A solution is

$$u_2 = \frac{1}{2}a^2(\theta, x)s(x) + a_2(\theta, x)r + \bar{u}_2(x)$$

where the vector s satisfies

$$\kappa_\alpha A^\alpha s + \kappa_\alpha D^2 f^\alpha(u_0) \cdot (r, r) = 0.$$

The functions a_2 and $\bar{u}_2$ are arbitrary, but do not appear in the final equation for a.

There are two solvability conditions for (3.2.61). The first is obtained by averaging with respect to the phase variable θ. Assuming that u_3 is a sublinear function of θ, this condition implies that $\bar{u}$ is a solution of the linearized equations,

$$(A^\alpha \bar{u})_{x^\alpha} + G\bar{u} = 0.$$

The second solvability condition is obtained by subtracting the mean from (3.2.61) and taking the scalar product of the result with the left null vector l. This gives

$$C^\alpha a_{x^\alpha} + \left(\frac{1}{3}\mathcal{G}_3 a^3 + \frac{1}{2}\mathcal{G}_2 a^2\right)_\theta + \mathcal{H}a = 0, \tag{3.2.62}$$

where C^α and $\mathcal{H}$ are given in (3.2.42) and

$$\mathcal{G}_3 = \frac{1}{2}l \cdot \kappa_\alpha D^3 f^\alpha(u_0) \cdot (r, r, r) + \frac{3}{2}l \cdot \kappa_\alpha D^2 f^\alpha(u_0) \cdot (r, s),$$
$$\mathcal{G}_2 = 2l \cdot \kappa_\alpha D^2 f^\alpha(u_0) \cdot (s, \bar{u}).$$

The coefficient $\mathcal{G}_3$ is equal to the second derivative of the wave speed with respect to the wave amplitude, while $\mathcal{G}_2$ is the derivative of the wave speed with respect to u in the direction $\bar{u}$. If both $\mathcal{G}$ and $\mathcal{G}_3$ are zero, one can derive an equation with a higher degree of nonlinearity [36].

Assuming that $\mathcal{G}_3 \neq 0$, the normalized form of (3.2.62) is a generalized modified Burgers equation,

$$u_t + \left(\frac{1}{3}u^3 - \frac{1}{2}\alpha(t)u^2\right)_x = 0. \tag{3.2.63}$$

This equation fails to be genuinely nonlinear when $u = \alpha/2$. Suitable entropy conditions for (3.2.63) were formulated by Oleinik and Liu — see [249] for a detailed discussion.

Loss of genuine nonlinearity at a single point. In the nonplanar case, it may happen that $\mathcal{G}$ vanishes at only one point along a ray, at $t = t_0$ say. The wave amplitude then satisfies a quadratically nonlinear generalized inviscid Burgers equation (3.2.41),

$$a_t + \left(\frac{1}{2}\mathcal{G}(t)a^2\right)_x + \mathcal{H}(t)a = 0,$$

with $\mathcal{G}(t_0) = 0$. This in itself leads to some quite interesting effects. For example if $\mathcal{G}(t)$ changes sign, a shock becomes inadmissible as t evolves through $t = t_0$, and will therefore spread out into a rarefaction wave.

Reasons for a loss of genuine nonlinearity. There are two main reasons for a loss of genuine nonlinearity. The first reason is that if the value of $\mathcal{G}$ depends on a parameter, then $\mathcal{G}$ may vanish at a particular value of the parameter.

For example, the fundamental derivative $\mathcal{G}$ of a gas, given in (3.2.32), is a function of the thermodynamic variables of the unperturbed state of the gas. Some gases — including certain hydrocarbons and fluorocarbons [264] — have the property that their fundamental derivative changes sign, so there is a curve $p_0 = p_*(\lambda)$, $s_0 = s_*(\lambda)$ on which $\mathcal{G} = 0$. Another physical example of this loss of genuine nonlinearity occurs in sound waves in a superfluid near the "lambda-line." Further references and several reviews of the derivation and applications of modified Burgers equations to nonlinear sound waves, appear in [157].

The second main reason for a loss of genuine nonlinearity is the presence of special symmetries. In particular, if the equations of motion are invariant under reflections $u \rightarrow -u$ of some set of dependent variables, then the wave speed must be an even function of u, and equation (3.2.23) shows that $\mathcal{G} = 0$ when $u = 0$.

Rotationally invariant conservation laws. An interesting example of systems of conservation laws with symmetries is provided by rotationally invariant conservation laws. Such laws describe transverse waves in an isotropic medium, and are invariant under rotations of the transverse dependent variables about the direction of propagation. Physical examples include transverse elastic waves in an isotropic solid and Alfvén waves in magnetohydrodynamics.

In three space dimensions, rotational invariance forces a loss of strict hyperbolicity, as well as a loss of genuine nonlinearity, because waves with different polarizations propagate with the same velocity. When dissipation or dispersion is included, the normalized asymptotic equation (see [35] and the references given there) is

$$u_t + (|u|^2 u)_x = D u_{xx}, \tag{3.2.64}$$

where $u(x,t) \in R^2$ and D is a 2×2 matrix. When D is symmetric, the higher order term is dissipative; when D is antisymmetric, the higher order term is dispersive.

The issue of which shocks are admissible for (3.2.64) is rather subtle [80]. For example, shocks across which u reverses direction (which are analogous to "intermediate shocks" in MHD [81, 152]) have a one parameter family of viscous profiles. The viscous traveling waves are stable with respect to sufficiently small disturbances, but the instability threshold tends to zero as the viscosity μ tends to zero [184]. Intermediate shocks are therefore unstable for the purely hyperbolic system, tending to split into a "rotational" contact discontinuity and a normal Lax shock.

3.3.   DISSIPATION AND DISPERSION

The weakly nonlinear expansion described in Section 3.2 is the foundation for deriving many other asymptotic equations. Other weak effects on the wave — such as

Table 3.2.

	Dissipative	*Dispersive*
Long waves	$u_t + uu_x = u_{xx}$ Burgers eq.	$u_t + uu_x + u_{xxx} = 0$ Korteweg–de Vries Eq.
Relaxing waves	$(\partial_x + 1)\left(u_t + (u^2/2)_x\right)$ $+u_x = 0$ Relaxing gas Eq.	$(\partial_x^2 + 1)\left(u_t + (u^2/2)_x\right)$ $= u_x$
Short waves	$u_t + (u^2/2)_x + u = 0$ Varley–Rodgers Eq.	$\partial_x\left(u_t + (u^2/2)_x\right) = u$ Ostrovsky Eq.
Higher order long waves	$u_t + uu_x + \alpha u_{xx}$ $+u_{xxxx} = 0$ Kuramoto–Sivashinsky Eq.	$u_t + uu_x + \alpha u_{xxx}$ $+u_{xxxxx} = 0$
Nonlocal long waves	$u_t + uu_x = \mathcal{H}[u_x]$	$u_t + uu_x = \mathcal{H}[u_{xx}]$ Benjamin–Ono Eq.
Long wave/ short wave	$u_t + uu_x + \alpha u = u_{xx}$ Burgers–Sivashinsky Eq.	$\partial_x\left(u_t + uu_x + u_{xxx}\right)$ $+\alpha u = 0$

dissipation, dispersion, and diffraction — can be incorporated into the expansion by choosing an appropriate limit in which these effects balance with nonlinearity. Table 3.2 summarizes some of the resulting asymptotic equations for weakly dissipative and dispersive plane waves.

In this Section, we show how to derive these equations from rather general primitive systems of partial differential equations. We consider two main types of primitive system: hyperbolic conservation laws perturbed by higher order derivatives; and general first order systems.

We also use a Chapman–Enskog expansion to connect these two types of systems. This expansion shows that long wave solutions of a general first order system satisfy a smaller first order system perturbed by higher order derivatives.

Asymptotic equations can be derived in a similar way starting from more complicated types of primitive equations, such as pseudo-differential equations or initial-boundary value problems describing wave propagation in a waveguide [175]. For example, the original derivation of the KdV equation was for water waves [162]. The water wave equations are a nonlinear free boundary value problem [272]. This free boundary value problem can also be formulated as a pseudo-differential equation on the free surface [60].

Long wave equations. Two of the most commonly occurring and most important asymptotic equations are Burgers equation,

$$u_t + u u_x = u_{xx}, \tag{3.3.1}$$

and the Korteweg–de Vries (KdV) equation,

$$u_t + u u_x + u_{xxx} = 0. \tag{3.3.2}$$

These equations describe the simplest kind of long wave dissipation ($\omega = -ik^2$) and dispersion ($\omega = -k^3$), respectively.

Equations which contain both dissipative and dispersive effects are also possible. The simplest example is the KdV–Burgers equation for long waves,

$$u_t + u u_x - \mu u_{xx} + \nu u_{xxx} = 0. \tag{3.3.3}$$

This equation, and some other long wave equations, are briefly discussed in Subsection 3.3.2.5.

Burgers equation. Burgers equation was first proposed as a model for turbulence [38, 100], for which purpose it is not in fact very useful. However, it does provide an excellent model for the viscous structure of weak shock waves. The initial value problem for (3.3.1) can be solved exactly by the Cole–Hopf transformation [272], which is a Backlünd transform between Burgers equation and the linear heat equation [207]. Kreiss and Lorentz [164] give an existence theorem for Burgers equation which does not depend on the Cole–Hopf transform.

The large time behavior of rapidly decaying solutions of Burgers equation is a triangular "nonlinear diffusion wave" [182]. This differs both from the N-wave limit of solutions the inviscid Burgers equation and the diffusive decay of solutions of the linear heat equation. (The zero-viscosity and the infinite time limits do not commute.)

The zero viscosity limit, $\mu \to 0+$, of solutions of Burgers equation,

$$u_t + u u_x = \mu u_{xx},$$

is a weak entropy solution of the inviscid Burgers equation (2.1).

The KdV equation. The KdV equation is exactly solvable, since it can be linearized by means of the inverse scattering transform (IST). This transform can be regarded as a change to action-angle variables, so that the KdV equation is a completely integrable, infinite dimensional Hamiltonian system [11]. In particular, the KdV equation has localized traveling waves, called solitons, which preserve their identity under interactions with other solitons, thus behaving like particles. As $t \to \infty$, solutions of the KdV equation which decrease sufficiently quickly at infinity, approach the sum of a number of solitons and small amplitude, dispersively decaying oscillations.

A description of completely integrable nonlinear partial differential equations and the inverse scattering transform is outside the scope of this article; for introductions, see [3, 39, 70, 71, 214]. These ideas have also had a significant impact on many other fields of mathematics, such as algebraic geometry [203], and integrable systems in quantum field theory and statistical mechanics [161].

For numerical solutions of the KdV equation, see [70, 256]. For a comparison of the KdV theory with experimental observations of water waves, see [105].

The zero dispersion limit, $\nu \to 0$, of the KdV equation,

$$u_t + uu_x + \nu u_{xxx} = 0,$$

is studied in [174, 266]. After shocks form, the zero dispersion limit of solutions of the KdV equation is not a solution of the inviscid Burgers equation (3.2.1), due to the appearance of rapid, large amplitude oscillations which are described by Whitham's averaged Lagrangian theory.

The averaged Lagrangian modulation equations for large amplitude, multi-phase wave trains take a particularly elegant form for the KdV equation, since they inherit some of the KdV equation's integrable structure [78].

Existence theorems and qualitative properties for the KdV equation, and related dispersive equations, are discussed in [251]. One interesting effect for the KdV equation is "dispersive smoothing." The solution of the Cauchy problem with rapidly decreasing initial data is infinitely differentiable for any $t > 0$ [148]. The heuristic explanation of this property is that the linearized group velocity,

$$C(k) = -3k^2,$$

tends to infinity as $k \to \infty$. As a result, the high frequency components of the solution propagate out to $x = -\infty$ arbitrarily quickly, leaving behind the smooth low frequency components. Conversely, if the initial data does not decay sufficiently rapidly, "dispersive blow-up" can occur. The constructive interference of oscillations propagating in from infinity can lead to blow-up of the KdV solution at one or more space–time points [29].

The BBM equation. The BBM equation [22], or regularized long wave equation, is a nonintegrable equation which is closely related to the KdV equation. The BBM equation is

$$u_t + u_x + uu_x - u_{xxt} = 0. \tag{3.3.4}$$

For small amplitude long waves, $u_t \sim -u_x$, and (3.3.4) reduces to a KdV equation,

$$u_t + u_x + uu_x + u_{xxx} = 0.$$

The short wave behavior of the BBM equation is better than that of the KdV equation, in the sense that the linearized group velocity for BBM,

$$C(k) = \frac{1 - k^2}{(1 + k^2)^2},$$

is uniformly bounded as $k \to \infty$. However, the BBM equation does not arise from systematic long wave asymptotics, which picks out the KdV equation uniquely (see the discussion in [166], for example). Since the KdV variables are correctly scaled, solutions with wavelength of the order one describe general long wave solutions of other equations. In contrast, only the long wavelength solutions of the BBM equation describe other long waves.

Nonplanar waves. All of the asymptotic equations above can be generalized to describe nonplanar waves and slowly varying media. This typically leads to a normalized equation in which the coefficients depend on t, but are independent of x. The fact that the coefficients depend on "time" and not "space" is a consequence of the assumption that variations in the medium and the wavefront geometry occur on the slow nonlinear timescale, and not on the fast wave period timescale.

These generalized equations are usually not integrable even if their constant coefficient versions are. For example, [211] shows that there are no Bäcklund transforms for the generalized Burgers equation. An exceptional variable coefficient equation which is integrable is the cylindrical KdV equation (see Subsection 3.3.2.4).

When weak diffraction effects are included, we obtain multidimensional versions of these equations such as the KP equation (see Section 3.5).

3.3.1. Higher Order Conservation Laws

As discussed in Subsection 3.2.1.1, long waves are often modeled by a hyperbolic system of conservation laws (3.2.2) with fluxes depending on u. At the next order of approximation in the wavelength, one typically obtains a system of conservation laws in which the the fluxes depend on u and ∇u, where ∇ is the gradient with respect to x. The balance equations then take the form

$$u_t + f^\alpha(u, \delta \nabla u)_{x^\alpha} = 0. \tag{3.3.5}$$

Assuming that u is nondimensionalized appropriately, δ has dimensions of length.

For example, the Chapman–Enskog expansion of the Boltzmann equations leads to the Euler equations in the leading order approximation and to the Navier–Stokes equations in the next approximation [47]. In that case, δ is the mean free path of the gas. Long waves, with wavelength λ have small Knudsen number,

$$Kn = \frac{\delta}{\lambda} \ll 1.$$

The effect of the gradient terms in (3.3.5) is to introduce dissipation.[3] This dissipation is weak for long waves. The cumulative effect of dissipation on a wave is significant over propagation distances of the order

$$L_d = \frac{\lambda^2}{\delta}.$$

As before, nonlinear effects are important over propagation distances of the order

$$L_n = \frac{\lambda}{\varepsilon},$$

where $\varepsilon \ll 1$ is the dimensionless wave amplitude. A balance between nonlinearity and dissipation occurs when L_d and L_n are of the same order, which implies that

$$\frac{\delta}{\lambda} = O(\varepsilon).$$

---

[3]In situations where long waves are unstable, the gradient terms lead to growth, and then the initial value problem for (3.3.5) is linearly ill-posed. In that case, including higher order derivatives can sometimes give a well-posed problem — see the Kuramoto–Sivashinsky equation below.

We use nondimensional variables in which $\lambda = O(1)$, and set

$$\delta = \varepsilon \hat{\delta},$$

where $\hat{\delta}$ is an order one parameter. The asymptotic solution for plane waves is (3.2.24). There are only slight changes in the previous expansion. It is straightforward to check that u_1 is given by (3.2.27), and that the wave amplitude a satisfies a Burgers equation,

$$a_\tau + C \cdot \nabla_\xi a + \mathcal{G} a a_\theta = \mathcal{M} a_{\theta\theta}. \tag{3.3.6}$$

The diffusivity in (3.3.6) is given by

$$\mathcal{M} = -\hat{\delta} l \cdot k_\alpha D_M f^\alpha(u_0, 0) \cdot (r \otimes k).$$

Here, D_M is the derivative of $f^\alpha(u, M)$ with respect to its second argument and $[r \otimes k]_{i\alpha} = r_i k_\alpha$.

The normalized form of equation (3.3.6) is the Burgers equation (3.3.1).

Stability of viscous shock waves. Restricting (3.3.5) to one space dimension, and assuming that the flux depends linearly on u_x, gives

$$u_t + f(u)_x = (D(u)u_x)_x, \tag{3.3.7}$$

where we assume that D is a symmetric, positive definite matrix. A viscous shock wave is a traveling wave solution of (3.3.7), $u = U(x - ct)$, which approaches different constant states, U_L and U_R, as $x \to -\infty$ and $x \to +\infty$. It follows from (3.3.7) that $U(z)$ satisfies the ordinary differential equation

$$D(U)U' = f(U) - cU - f(U_L) + cU_L,$$

with boundary conditions,

$$U(z) \to U_L \quad \text{as } z \to -\infty,$$
$$U(z) \to U_R \quad \text{as } z \to +\infty.$$

Now consider (3.3.7) with initial data,

$$u(x, 0) = U(x) + \bar{u}(x), \tag{3.3.8}$$

where $\bar{u}(x)$ is a small amplitude perturbation which tends to zero sufficiently rapidly as $|x| \to \infty$. Then, under suitable hypotheses, as $t \to +\infty$, the solution of (3.3.7)–(3.3.8) approaches the sum of a translated viscous shock wave, $u = U(x - ct - \delta)$, and a number of nonlinear diffusion waves which satisfy (3.3.6) (see [182, 253], and the references given there). Since the nonlinear diffusion waves decay (in L^∞) as $t \to +\infty$, this result means that viscous shock waves are nonlinearly stable.

In this problem the width of the perturbation and the thickness of the viscous shock profile are both of the same order of magnitude, so geometrical optics does not apply. Instead, it is necessary to develop suitable energy estimates for the (variable coefficient) linearization of (3.3.7) about the traveling wave solution for the viscous shock profile.

Goodman [91] gives an analysis of viscous shock wave stability for scalar conservation laws in two space variables. In particular, he obtains a heat equation for the evolution in the translation $\delta(y, t)$ of the viscous shock wave profile as a function of time t and transverse distance y along the wave.

Nonplanar waves. It is straightforward to generalize this plane wave expansion to nonplanar waves [86, 87]. The effects of higher order derivatives can also be included if they are appropriately scaled. To be specific, suppose that

$$f^\alpha(x, u, \varepsilon^2 \nabla u, \varepsilon^3 \nabla^2 u)_{x^\alpha} + g(x, u) = 0, \tag{3.3.9}$$

where x denotes space–time variables. We use the geometrical optics scaling for nonplanar waves, so the scaling of the derivatives in the fluxes in (3.3.9) differs from (3.3.5). Looking for a weakly nonlinear solution of (3.3.9) of the form (3.2.36) and (3.2.40), leads to a generalized KdV–Burgers equation for the wave amplitude,

$$\mathcal{C}^\alpha a_{x^\alpha} + \mathcal{G} a a_\theta - \mathcal{M} a_{\theta\theta} + \mathcal{N} a_{\theta\theta\theta} + \mathcal{H} a = 0.$$

The coefficients are given in (3.2.42) and by

$$\mathcal{M} = -l \cdot \kappa_\alpha D_M f^\alpha(x, u_0, 0, 0) \cdot (r \otimes \kappa),$$
$$\mathcal{N} = l \cdot \kappa_\alpha D_N f^\alpha(x, u_0, 0, 0) \cdot (r \otimes \kappa \otimes \kappa),$$

where D_N is the derivative of $f^\alpha(x, u, M, N)$ with respect to N. The normalized form of this equation is a generalized KdV–Burgers equation, (3.3.3) with $\mu = \mu(t)$ and $\nu = \nu(t)$.

3.3.1.1. Nonlinear acoustics (3). The balance equations for mass, momentum and energy of a fluid are given in (3.2.10). The constitutive relations for a viscous, thermally conducting, Newtonian fluid are

$$T = (-p + \mu_b \nabla \cdot u)I + \mu\left(\nabla u + \nabla u^t - \frac{2}{3}\nabla \cdot u\right), \tag{3.3.10}$$

$$q = -\kappa \nabla T.$$

Here, μ is the shear viscosity, μ_b is the bulk viscosity, and κ is the thermal conductivity.

A measure of the amplitude of a sound wave is the acoustic Mach number, $\varepsilon = u_{\max}/c_0$. A measure of the importance of dissipation is the acoustics Reynolds number,

$$Re = \frac{\rho_0 c_0 \lambda}{\mu},$$

where λ is a typical wavelength of the sound wave. Weak nonlinearity and weak dissipation are important and of the same order of magnitude when the wave propagates over distance of the order

$$L = \frac{\lambda}{\varepsilon} \quad \text{with} \quad Re = O(\varepsilon^{-1})$$

as $\varepsilon \to 0$. We also assume that

$$\frac{\mu}{\mu_b} = O(1), \qquad Pr = \frac{c_p \mu}{\kappa} = O(1).$$

Here, c_p is the specific heat at constant pressure and Pr is the Prandtl number. Using nondimensional variables in which $\lambda = O(\varepsilon)$, it follows that

$$\mu = O(\varepsilon^2), \quad \mu_b = O(\varepsilon^2), \quad \frac{\kappa}{c_p} = O(\varepsilon^2).$$

In this limit one obtains the following asymptotic solution for a nonplanar, viscous sound wave propagating through an arbitrary smooth background state, $\rho = \rho_0(x,t)$, $u = u_0(x,t)$, $c = c_0(x,t)$. The fluid variables are given by (3.2.56), and the transport equation is

$$
a_t + C \cdot \nabla a + \left(\frac{\gamma+1}{4} \Omega a^2 \right)_\theta
$$
$$
+ \frac{\Omega}{2\rho_0 c_0^2} \left[\left(\frac{\rho_0 c_0^2}{\Omega} \right)_t + \nabla \cdot \left(\frac{\rho_0 c_0^2}{\Omega} C \right) \right] a = \frac{1}{2} \hat{\delta} |k|^2 a_{\theta\theta},
$$

where $\delta = \varepsilon^2 \hat{\delta}$ is the diffusivity of sound,

$$\delta = \frac{\mu}{\rho_0} \left(\frac{4}{3} + \frac{\mu_b}{\mu} + \frac{\gamma-1}{Pr} \right). \tag{3.3.11}$$

Viscous and thermal effects both contribute to the damping of sound waves. The normalized form of this equation is the generalized Burgers equation, [212, 179, 237],

$$u_t + u u_x = \mu(t) u_{xx}. \tag{3.3.12}$$

For an application of this equation to the propagation of spherical nonlinear sound waves in an exponentially stratified atmosphere, see [66].

In the special case of plane waves in a medium at rest, one obtains the usual Burgers equation,

$$a_t + C \cdot \nabla a + \left(\frac{\gamma+1}{4} \omega a^2 \right)_\theta = \frac{1}{2} \hat{\delta} |k|^2 a_{\theta\theta},$$

where $C = c_0^2 \omega^{-1} k$.

3.3.2. First Order Systems of PDEs: Long Waves

In this subsection we consider long wave solutions of a general $m \times m$ first order system of PDEs with slowly varying coefficients,

$$A^\alpha(\varepsilon x, u) u_{x^\alpha} + g(\varepsilon x, u) = 0. \tag{3.3.13}$$

Any higher order system of PDEs can be written as an equivalent first order system by treating the higher order derivatives as new dependent variables [59].

To discuss long wave solutions of (3.3.13), it is convenient to make the change of variables $\varepsilon x \to x$, corresponding to the "geometrical optics" nondimensionalization. This puts (3.3.13) in the form

$$A^\alpha(x, u) u_{x^\alpha} + \frac{1}{\varepsilon} g(x, u) = 0. \tag{3.3.14}$$

Here, $x \in R^{n+1}$ is a space–time variable, $u : R^{n+1} \to R^m$ is the vector of dependent variables, the $A^\alpha(x, u)$ are $m \times m$ real matrices, and $g : R^{n+1} \times R^m \to R^m$ is a source term.

In Subsection 3.3.2.1, we use an expansion for long wave solutions of (3.3.14) which is analogous to the Chapman–Enskog expansion of the Boltzmann equation [47]. This expansion is a generalization of the one used in [49, 183] for conservation laws with relaxation. The Chapman–Enskog expansion applies to large amplitude long wave solutions, but ultimately we will be interested in weakly nonlinear long waves.

The end result of the Chapman–Enskog expansion is a reduced system of first order equations perturbed by higher order derivatives. Depending on the system, these higher order derivatives may be dissipative or dispersive in nature. This is the usual starting point of the reductive perturbation method.

Thus, there are two equivalent ways to derive asymptotic equations for weakly nonlinear, long wave solutions of (3.3.14). The first way is to use the Chapman–Enskog expansion to derive a reduced system perturbed by higher order derivatives, and then apply the weakly nonlinear expansion described in Subsection 3.3.1. The second way is to derive the asymptotic equations directly from (3.3.14). This direct method is described in Subsection 3.3.2.2.

For applications of the Chapman–Enskog expansion, see Subsection 3.3.2.4 on ion acoustic waves (dispersive) and Subsection 3.4.1.1 on relaxing gases (dissipative). (We recommend looking at these simpler applications before examining the general expansion described next.)

3.3.2.1. The Chapman–Enskog expansion. To motivate the Chapman–Enskog expansion, we begin with a straightforward expansion of (3.3.14). This straightforward expansion is analogous to the Hilbert expansion of the Boltzmann equation.

The Hilbert expansion. We look for a power series expansion of u in (3.3.14),

$$u(x; \varepsilon) \sim \sum_{n=0}^{\infty} \varepsilon^n u_n(x), \qquad (3.3.15)$$

where $\varepsilon \to 0$ with $x = O(1)$. Using (3.3.15) in (3.3.14) and equating coefficients of ε^{-1} and ε^0 to zero implies that

$$g(x, u_0) = 0, \qquad (3.3.16)$$

$$D_u g(x, u_0) \cdot u_1 + A^\alpha(x, u_0)\frac{\partial u_0}{\partial x^\alpha} = 0, \qquad (3.3.17)$$

From (3.3.16), we see that the source term g must vanish on longwave solutions. We now make some assumptions about the zero set of g.

For each $x \in R^{n+1}$, we assume that there is a p-dimensional embedded submanifold $\mathcal{M}_x^p \subset R^m$ such that $g(x, u) = 0$ for all $u \in \mathcal{M}_x^p$. We choose coordinates $y \in R^p$ on $\mathcal{M}_x^p$ and an associated function $u_*(x, y)$ such that

$$g(x, u) = 0 \qquad \Longleftrightarrow \qquad u = u_*(x, y) \text{ for some } y \in R^p. \qquad (3.3.18)$$

(Here, we use the appropriate coordinate chart if $\mathcal{M}_x^p$ requires more than one chart to cover it.)

The simplest case is when we can use p components of the original vector of dependent variables, say $y = (u^1, \ldots, u^p)$, as coordinates. Then the solutions of $g(x, u) = 0$ are given by

$$u^{p+1} = u_*^{p+1}(x, u^1, \ldots, u^p), \ldots, u^m = u_*^m(x, u^1, \ldots, u^p). \qquad (3.3.19)$$

Our goal is to obtain a reduced $p \times p$ system of equations for $(u^1(x), \ldots, u^p(x))$ from the $m \times m$ system for $(u^1(x), \ldots, u^m(x))$. See Subsections 3.2.4 and 4.1.1 for examples.

From (3.3.18), the solution of (3.3.16) can be written as

$$u_0(x) = u_*(x, v(x)), \qquad (3.3.20)$$

where $v : R^{n+1} \to R^p$ is an arbitrary function. Equation (3.3.20) corresponds to a solution which is in "local equilibrium." The parameters v defining the local equilibrium are allowed to vary slowly in x. This solution is analogous to a local Maxwellian solution of the Boltzmann equations.

Before deriving the reduced system for $v(x)$, we introduce some notation. For each $(x, y) \in R^{n+1} \times R^p$, we define the $m \times m$ matrix $G(x, y) \in \mathcal{L}(R^m, R^m)$ and the $m \times p$ matrix $R(x, y) \in \mathcal{L}(R^p, R^m)$ by

$$G(x, y) = D_u g(x, u_*(x, y)) \qquad (3.3.21)$$

$$R(x, y) = D_y u_*(x, y) \qquad (3.3.22)$$

Here, $\mathcal{L}(R^p, R^m)$ denotes the linear maps from R^p to R^m. Differentiating the identity $g(x, u_*(x, y)) = 0$ with respect to y shows that $GR = 0$. By the Fredholm alternative, we can choose a $p \times m$ matrix $L(x, y) \in \mathcal{L}(R^m, R^p)$ such that $LG = 0$ and

$$Gu = f \quad \text{has a solution} \qquad \Longleftrightarrow \qquad Lf = 0. \qquad (3.3.23)$$

We now derive an equation for $v(x)$. Using (3.3.21), equation (3.3.17) can be written

$$Gu_1 + A^\alpha \frac{\partial u_0}{\partial x^\alpha} = 0.$$

Here and below, A^α, g, L, R and their derivatives are evaluated at $u = u_0(x)$ and $y = v(x)$. Since $u_0 = u_*(x, v)$, they are known function of x and v. From (3.3.23), this equation can be solved for u_1 if and only if

$$LA^\alpha \frac{\partial u_0}{\partial x^\alpha} = 0. \qquad (3.3.24)$$

Equation (3.3.24) is a $p \times p$ system of equations for $v(x)$, the "Euler" equations. Using (3.3.22), it can written explicitly as

$$B^\alpha(x, v)v_{x^\alpha} + h(x, v) = 0, \qquad (3.3.25)$$

where the $p \times p$ matrices B^α and the p-vector h are given by

$$B^\alpha(x, v) = L(x, v)A^\alpha(x, u_*(x, v))R(x, v),$$

$$h(x, v) = L(x, v)A^\alpha(x, u_*(x, v)) \frac{\partial}{\partial x^\alpha}\bigg|_v u_*(x, v).$$

Carrying out the expansion to higher orders shows that the higher order corrections u_n with $n \geq 1$, in (3.3.15), satisfy linearized, nonhomogeneous "Euler" equations.

A defect of the Hilbert expansion is that it is not uniformly valid. Equation (3.3.25) is a quasi-linear system of PDEs. If this system is hyperbolic (which it must be in order for the evolution of long waves to be a well-posed problem), the derivative of solutions typically blows up in finite time. This behavior is inconsistent with the long wave ansatz used to derive (3.3.25) in the first place.

After blow up it may be possible to use the Hilbert expansion to provide an outer solution, with appropriate matching to an inner shock layer governed by the full equations. This matching has been done rigorously for the Broadwell equations [273].

The Chapman–Enskog expansion. The Chapman–Enskog expansion is an attempt to avoid the failure of the Hilbert expansion when solutions of (3.3.25) loose smoothness. The main idea is as follows. We look for long wave solutions of (3.3.14) which are small perturbations of a local equilibrium solution (3.3.20),

$$u(x; \varepsilon) \sim u_* (x, v(x; \varepsilon)) + \varepsilon u'[x, v(x; \varepsilon); \varepsilon], \tag{3.3.26}$$

as $\varepsilon \to 0$ with $x = O(1)$. We want to find a closed set of equations for v and a perturbation $u'[x, v]$ such that (3.3.26) is an asymptotic solution of (3.3.14). As we will see, the perturbation u' depends on derivatives of v, so we use square brackets to indicate that it is a functional of v, not just a pointwise function.

The requirement that (3.3.26) is an asymptotic solution of (3.3.14) means that (3.3.26) should satisfy (3.3.14) up to terms of the order ε^r for some r. When $r = 0$, this yields the "Euler" equations, (3.3.25); when $r = 1$, we obtain instead a second order system of equations for v, the "Navier–Stokes" equations.

The difference between the Chapman–Enskog expansion and more usual perturbation expansions (like the Hilbert expansion) is sometimes summarized by saying that it involves an "equation expansion" rather than a "solution expansion." Since v depends explicitly on ε in a way which is not specified a priori, the final equation of the Chapman–Enskog expansion is not uniquely determined. We can always change the equation for v by lower order terms in ε without altering its formal order of accuracy.

Another well-known example of equation expansions occurs in the method of smoothing which gives effective equations for the mean of solutions of PDEs with small random coefficients [250].

Before carrying out the Chapman–Enskog expansion, we note that there is some arbitrariness in v and u' in (3.3.26), because a change in v of the order ε can be absorbed into u'. Thus, it is necessary to constrain u' in some way to obtain a unique decomposition. For example, when we use coordinates $(u^1, \ldots, u^p)$, as in (3.3.19), the natural way to specify u' uniquely is to require that $u^{1'} = \cdots = u^{p'} = 0$.

In general, suppose that

$$v = \bar{v} + \varepsilon v_1 + O(\varepsilon^2).$$

We use this expression in (3.3.26) and Taylor expand, using (3.3.22). This gives

$$u \sim u_* (x, \bar{v}) + \varepsilon \bar{u}'[x, \bar{v}; \varepsilon],$$

where

$$\bar{u}' = u'[x, \bar{v}; \varepsilon] + R(x, \bar{v})v_1 + O(\varepsilon)$$

Thus, (v, u') and $(\bar{v}, \bar{u}')$ correspond to the same solution u. In order to define u' uniquely, we choose a projection

$$P : R^m \to \ker G$$

and require that

$$P(x, v)u'[x, v; \varepsilon] = 0. \qquad (3.3.27)$$

We also choose a projection

$$Q : R^m \to \operatorname{range} G = \{f \in R^m : Lf = 0\}$$

onto the range of G. We then define a generalized inverse, $H(x, y)$, of $G(x, y)$ by

$$Hf = u \quad \Longleftrightarrow \quad Gu = Qf \quad \text{and} \quad Pu = 0. \qquad (3.3.28)$$

Now we carry out the expansion. We use (3.3.26) in (3.3.14) and Taylor expand the result in ε. We will only retain terms up to order ε^2, but the expansion can be carried out to any desired order (depending on how much algebra one is willing to do!). This gives

$$Gu' + A^\alpha \frac{\partial u_0}{\partial x^\alpha} + \varepsilon f_1[x, u_0, u'] + \varepsilon^2 f_2[x, u_0, u'] = O(\varepsilon^3), \qquad (3.3.29)$$

where the coefficients are evaluated at $y = v$ and $u = u_0$, and

$$f_1 = \frac{1}{2} D_u^2 g \cdot (u', u') + A^\alpha \frac{\partial u'}{\partial x^\alpha} + D_u A^\alpha \cdot (u') \frac{\partial u_0}{\partial x^\alpha},$$

$$f_2 = \frac{1}{6} D_u^3 g \cdot (u', u', u') + D_u A^\alpha \cdot (u') \frac{\partial u'}{\partial x^\alpha} + \frac{1}{2} D_u^2 A^\alpha \cdot (u', u') \frac{\partial u_0}{\partial x^\alpha}.$$

We rewrite (3.3.29) as the following pair of equations:

$$LA^\alpha \frac{\partial u_0}{\partial x^\alpha} + \varepsilon L f_1[x, u_0, u'] + \varepsilon^2 L f_2[x, u_0, u'] = O(\varepsilon^3), \qquad (3.3.30)$$

$$u' + HA^\alpha \frac{\partial u_0}{\partial x^\alpha} + \varepsilon H f_1[x, u_0, u'] = O(\varepsilon^2). \qquad (3.3.31)$$

The first equation is the solvability condition (3.3.23) of (3.3.29) for u', while the second equation gives the unique solution u' which satisfies (3.3.27).

Equation (3.3.31) is an implicit equation for u'. We solve it by iteration to obtain

$$u' = u_1[x, v] + \varepsilon u_2[x, v] + O(\varepsilon^2),$$

$$u_1 = -HA^\alpha \frac{\partial u_0}{\partial x^\alpha}, \qquad (3.3.32)$$

$$u_2 = -H f_1[x, u_0, u_1].$$

Using (3.3.32) in (3.3.30) and expanding, then gives a higher order $p \times p$ system of equations for $v(x; \varepsilon)$,

$$LA^\alpha \frac{\partial u_0}{\partial x^\alpha} + \varepsilon L f_1[x, u_0, u_1]$$

$$+ \varepsilon^2 \left\{ L f_2[x, u_0, u_1] + L D_{u_1} f_1[x, u_0, u_1] \cdot (u_2) \right\} = O(\varepsilon^3), \qquad (3.3.33)$$

Truncating at the order ε term and using (3.3.32) gives

$$LA^{\alpha}\frac{\partial u_0}{\partial x^{\alpha}} = \varepsilon L\left[A^{\alpha}\frac{\partial}{\partial x^{\alpha}}\left(HA^{\beta}\frac{\partial u_0}{\partial x^{\beta}}\right) + D_u A^{\alpha}\cdot\left(HA^{\beta}\frac{\partial u_0}{\partial x^{\beta}}\right)\frac{\partial u_0}{\partial x^{\alpha}}\right.$$
$$\left. -\frac{1}{2}D_u^2 g\cdot\left(HA^{\alpha}\frac{\partial u_0}{\partial x^{\alpha}}, HA^{\alpha}\frac{\partial u_0}{\partial x^{\alpha}}\right)\right]. \tag{3.3.34}$$

Equation (3.3.34) is a system of second order partial differential equations for $v(x;\varepsilon)$, analogous to the Navier–Stokes equations for the Boltzmann equations.

A further expansion in (3.3.34) is often useful. If the right hand side contains second order time derivatives (with $x^0 = t$, say), it may be possible to use the equation to express the time derivative in terms of spatial derivatives plus an $O(\varepsilon)$ correction. On replacing time derivatives in terms of spatial derivatives on the right hand side, one obtains a long wave equation which is first order in time and second order in space. The necessity of replacing time derivatives by spatial derivatives is well known in the case of Boltzmann's equation. See (3.3.74) in Subsection 3.3.2.4 or (3.4.8) in Subsection 3.4.1.1 for examples involving first order partial differential equations.

Nonuniformity of the Chapman–Enskog expansion. Solutions of the equations derived by the Chapman–Enskog expansion may remain smooth due to the higher order derivatives on the right hand side. However, the equation is no more uniformly valid in general than the "Euler" equations (3.3.25). In order to smooth out discontinuous solutions of (3.3.25), the derivatives in (3.3.34) must be of the order ε^{-1}. It is then not formally consistent to neglect the terms involving higher order derivatives. It is only when the discontinuities are weak that the first nonzero higher order terms can be consistently retained, and in that case the weakly nonlinear long wave expansions derived in the next subsection lead to much greater simplifications.

There is a similar nonuniformity in the Navier–Stokes approximation of the Boltzmann equation. Strong shocks have widths of the order of a mean free path δ, and the shock profile therefore cannot be described by a continuum limit (although the predictions of the Navier–Stokes equations turn out to be remarkably accurate for shocks of moderate strength). The width of weak shocks of strength ε is of the order δ/ε, so their profile is described by the Navier–Stokes equations, or — more simply — by Burgers equation.

The qualitative properties of the Chapman–Enskog equations and the original system may also differ. For example, it is possible that all solutions of the "Navier–Stokes" equation (3.3.34) remain globally smooth, even though solutions of the original system (3.3.14) shock up when they are sufficiently steep. This occurs in systems which have a finite threshold for shock formation, meaning that waves only shock up if their initial slope is steeper than a nonzero critical value. An example of such a system is the relaxing gas equations described in Subsection 3.4.1.1.

Straightforward long wave asymptotics cannot capture this threshold phenomenon. Other limits, such as a weak relaxation limit (Subsection 3.4.1) or a short wave limit (Subsection 3.3.3), do lead to model equations with a threshold for shock formation.

Effective diffusivity of long waves. In Subsection 3.3.2.3, we derive weakly nonlinear asymptotic equations for the original first order system (3.3.13). Equivalent results are obtained by applying the methods of the previous subsection to the long wave equation (3.3.34), and we can read off some useful quantities directly from (3.3.34).

We assume for simplicity that $u = 0$ is a solution of (3.3.14). We choose coordinates so that $u_* = 0$ when $v = 0$. Freezing coefficients in (3.3.34) and linearizing in v gives the system

$$B^\alpha \frac{\partial v}{\partial x^\alpha} = \varepsilon L D^{\alpha\beta} R \frac{\partial^2 v}{\partial x^\alpha \partial x^\beta}, \tag{3.3.35}$$

where all coefficients are evaluated at $v = 0$, and

$$D^{\alpha\beta} = A^\alpha H A^\beta.$$

The effective diffusivity matrix of long waves is $L D^{\alpha\beta} R$.

The dispersion relation of (3.3.35) at $\varepsilon = 0$ is

$$\det\left[\kappa_\alpha B^\alpha\right] = 0.$$

The geometrical optics solution v is

$$v \sim a(\kappa_\alpha x^\alpha, \varepsilon x)s \tag{3.3.36}$$

where $s \in R^p$ is a null vector of $\kappa_\alpha B^\alpha$. We denote an associated left eigenvector by $\jmath$. To express this result in terms of u, we define

$$r = Rs, \qquad l = \jmath \cdot L. \tag{3.3.37}$$

Then

$$u \sim a(\kappa_\alpha x^\alpha, \varepsilon x)r$$

The transport equation for weakly dissipative waves is obtained by using (3.3.36) in (3.3.35) and projecting onto the left null vector $\jmath$, which implies that $a(\theta, \xi)$ satisfies the heat equation

$$C^\alpha a_{\xi^\alpha} = \mu a_{\theta\theta}, \tag{3.3.38}$$

where

$$\begin{aligned}
C^\alpha &= \jmath \cdot B^\alpha s = l \cdot A^\alpha r, \\
\mu &= \kappa_\alpha \kappa_\beta \, \jmath \cdot L D^{\alpha\beta} R s = \kappa_\alpha \kappa_\beta \, l \cdot D^{\alpha\beta} r.
\end{aligned} \tag{3.3.39}$$

The coefficient μ is the effective diffusivity of the long wave.

If $\mu \neq 0$, then weakly nonlinear long waves satisfy a Burgers equation. If $\mu = 0$, and an appropriate third order coefficient is nonzero, one obtains a KdV equation instead. We treat this case in detail below.

Reaction-diffusion equations. The use of Hilbert and Chapman–Enskog expansions is not restricted to first order systems. For example, Wagner and Keizer [269] consider a system of reaction-diffusion equations (describing Calcium diffusion in cells) with fast reaction and slow diffusion of the form

$$u_t = D(u)\Delta u + \frac{1}{\varepsilon}g(u).$$

We assume a submanifold of equilibrium solutions, $u = u_*(y)$, where

$$g\left(u_*(y)\right) = 0.$$

Solutions in local equilibrium are determined by a function $v(x, t)$ where

$$u(x, t; \varepsilon) = u_*\left(v(x, t)\right) + O(\varepsilon),$$

and v satisfies a reduced diffusion equation of the form

$$L(u_*)u_{*t} = L(u_*)D(u_*)\Delta u_*.$$

Thus, the Hilbert expansion leads to an effective diffusion equations for a reduced set of variables. The Chapman–Enskog expansion leads to additional fourth order diffusion terms.

3.3.2.2. Heuristic derivation of the KdV scaling. In this subsection, we estimate when nonlinear and dispersive effects on a small amplitude long wave are of the same order of magnitude. This dominant balance argument provides the scaling which leads to the KdV equation.

Suppose that the dimensionless wave amplitude is of the order α and the wavenumber is of the order κ. We assume advective nonlinearity of the form $u \cdot \nabla u$. (Other types of nonlinearity are mentioned below.) The nonlinear terms are then of the order $\alpha^2 \kappa$. The cumulative nonlinear effect on the wave over time T and propagation distances $c_0 T$ is of the order $\alpha^2 \kappa c_0 T$. Here, c_0 is the speed of long waves. The time T_n after which nonlinear effects become important is determined by the condition:

$$\text{cumulative nonlinear effect} = O(\text{wave amplitude}),$$

or $\alpha^2 \kappa c_0 T_n = \alpha$. This gives

$$T_n = \frac{1}{\alpha c_0 \kappa}.$$

Next, we consider dispersive effects. For waves governed by systems with real coefficients, the linearized dispersion relation satisfies

$$\omega(-k) = -\omega^*(k).$$

This is because the complex conjugate of a solution is also a solution. We suppose that the wave is nondispersive in the long wave limit and that the frequency has a power series expansion in the wavenumber k. Thus, as $k \to 0$,

$$\omega = c_0 h^{-1} \left[kh - i\mu k^2 h^2 + \nu k^3 h^3 + O(k^4 h^4)\right],$$

where μ and ν are dimensionless real constants and h is a lengthscale characteristic of the motion. For example, in the case of shallow water waves, h is the mean depth of the fluid. Dissipative effects are only negligible if

$$\mu = 0, \tag{3.3.40}$$

and we assume that this is the case.

The waves are weakly dispersive when $kh \ll 1$. The group velocity of two harmonics with wavenumbers of the order κ (say κ and 2κ), differ by an amount of the order

$$C_k\kappa = \omega_{kk}\kappa = O(c_0 h^2 \kappa^2).$$

After time T, the harmonics separate by a distance $c_0 h^2 \kappa^2 T$. The timescale T_d for dispersion to be significant is determined by the condition:

$$\text{separation of harmonics} = O(\text{wavelength}),$$

or $c_0 h^2 \kappa^2 T_d = \kappa^{-1}$. This gives

$$T_d = \frac{1}{c_0 h^2 \kappa^3}.$$

Nonlinear and dispersive effects balance when T_n and T_d are the same order of magnitude. This condition implies that the wavenumber, amplitude, and propagation distance are related by

$$\kappa h = \alpha^{1/2}, \qquad T_n = T_d = \alpha^{-3/2} c_0^{-1} h.$$

To be consistent with the scaling used in (3.3.13), we choose nondimensional variables in which $c_0 = h = 1$ and $T_n = 1/\varepsilon$. It follows that

$$\alpha = \varepsilon^{2/3}, \qquad \kappa = \varepsilon^{1/3}.$$

We have gone into some detail here to explain the meaning of the scaling used in the asymptotic expansion for deriving the KdV equation. In practice, one can obtain the same scaling more quickly simply by looking at the form of the expected terms. In the case of the KdV equation,

$$u_t + c_0 u u_x + c_0 h^2 u_{xxx} = 0,$$

balancing the three terms implies that

$$T^{-1}\alpha = c_0 \alpha^2 \kappa = c_0 h^2 \kappa^3 \alpha,$$

which gives the same result as above. The linear terms can be obtained from a long wave expansion of the linearized dispersion relation, and the only other input needed is the form of the nonlinearity.

3.3.2.3. The asymptotic expansion. We consider the real $m \times m$ system,

$$A^\alpha(\varepsilon x, u)u_{x^\alpha} + g(\varepsilon x, u) = 0. \tag{3.3.41}$$

We assume that

$$g(\varepsilon x, 0) = 0, \tag{3.3.42}$$

so that $u = 0$ is a solution. This does not involve any loss of generality, since if $u = u_0(\varepsilon x)$ is a nonzero solution, we can introduce translated dependent variables $u' = u - u_0$.

Using the scaling suggested by the heuristic argument in the last subsection, we look for an asymptotic solution of (3.3.41) of the form

$$u = \varepsilon^{2/3} u_1(\theta, \xi) + \varepsilon u_2(\theta, \xi) + \varepsilon^{4/3} u_3(\theta, \xi)$$
$$+ \varepsilon^{5/3} u_4(\theta, \xi) + O(\varepsilon^2), \tag{3.3.43}$$
$$\theta = \varepsilon^{-2/3} \varphi(\varepsilon x), \quad \xi = \varepsilon x, \tag{3.3.44}$$

where $\varepsilon \to 0$ with $x = O(\varepsilon^{-1})$. The phase variable θ is the appropriate one for wavenumbers of the order

$$k = \nabla_x \theta = O(\varepsilon^{1/3})$$

which are modulated over distances of the order

$$\frac{|k|}{|\nabla_x k|} = O\left(\frac{1}{\varepsilon}\right).$$

Summary of notation. Before carrying out the expansion, we summarize some notation we will use below. We define the $m \times m$ matrix

$$G(\xi) = D_u g(\xi, 0),$$

and we assume that G has corank $p < m$ (independent of ξ). This means that there is a p-dimensional manifold of long wave solutions. The weakly nonlinear solution lies in the tangent space of this manifold at $u = 0$. We introduce a $p \times m$ matrix $L(\xi)$ and an $m \times p$ matrix $R(\xi)$ such that

$$LG = 0, \qquad GR = 0.$$

We denote a generalized inverse of $G(\xi)$ by $H(\xi)$; for example we can use the definition in (3.3.28).

We define the $p \times p$ matrices B^α by

$$B^\alpha(\xi) = L(\xi) A^\alpha(\xi, 0) R(\xi). \tag{3.3.45}$$

Below, we show that the frequency-wavenumber vector $\kappa_\alpha = \varphi_{\xi^\alpha}$ satisfies the long wave dispersion relation

$$\det(\kappa_\alpha B^\alpha) = 0. \tag{3.3.46}$$

We assume that

$$\text{zero is a simple eigenvalue of } \kappa_\alpha B^\alpha. \tag{3.3.47}$$

This means that there is only one long wave mode with wavenumber κ. We denote left and right nullvectors of $\kappa_\alpha B^\alpha$ by $j(\xi) \in R^p$ and $s(\xi) \in R^p$, with $j \cdot s = 1$, so that

$$\kappa_\alpha B^\alpha s = 0. \tag{3.3.48}$$

We denote a generalized inverse of $\kappa_\alpha B^\alpha$ by $(\kappa_\alpha B^\alpha)^\dagger$. For example, we can define

$$(\kappa_\alpha B^\alpha)^\dagger y = x \quad \Longleftrightarrow \quad \kappa_\alpha B^\alpha x = y - (j \cdot y) s \quad \text{and} \quad j \cdot x = 0.$$

Finally, we define the m-vectors $r(\xi)$ and $l(\xi)$ by

$$r = Rs, \qquad l = \jmath \cdot L,$$

and the $m \times m$ matrix M by

$$M = R \left(\kappa_\alpha B^\alpha \right)^\dagger L \kappa_\alpha A^\alpha - I_m.$$

The conditions needed for long wave solutions of (3.3.41) to satisfy a KdV equation are

$$l \cdot \kappa_\alpha A^\alpha H \kappa_\beta A^\beta r = 0, \tag{3.3.49}$$

$$L D_u^2 g \cdot (r, r) = 0. \tag{3.3.50}$$

Here and below, A^α, g, and their derivatives are evaluated at $u = 0$. Equation (3.3.49) implies that there are no second order dissipative effects on long waves — see (3.3.38)–(3.3.39) and (3.3.40). Equation (3.3.50) implies that there are no quadratically nonlinear effects of the form a^2 on a wave with amplitude a; if present, such effects dominate advective nonlinearity, of the form aa_θ. At the end of this subsection we briefly indicate how to modify the expansion if these conditions are not met.

Proposition 3.1. *Suppose that (3.3.42), (3.3.47), (3.3.49), and (3.3.50) hold. Then (3.3.41) has a formal asymptotic solution (3.3.43), with leading order term*

$$u = \varepsilon^{2/3} a \left(\varepsilon^{-2/3} \varphi(\varepsilon x), \varepsilon x \right) r(\varepsilon x) + O(\varepsilon). \tag{3.3.51}$$

The phase $\varphi(\xi)$ is a solution of the eikonal equation,

$$\det \left(\varphi_{\xi^\alpha} B^\alpha \right) = 0. \tag{3.3.52}$$

The wave amplitude a is a solution of a generalized KdV equation,

$$a_s + \mathcal{G}(s) a a_\theta + \mathcal{N}(s) a_{\theta\theta\theta} + \mathcal{H}(s) a = 0, \tag{3.3.53}$$

where

$$\partial_s = l \cdot A^\alpha r \partial_{\xi^\alpha},$$
$$\mathcal{G} = l \cdot \left[\kappa_\alpha D_u A^\alpha \cdot (r) r - \kappa_\alpha A^\alpha H D_u^2 g \cdot (r, r) - D_u^2 g \cdot (r, M H \kappa_\alpha A^\alpha r) \right],$$
$$\mathcal{N} = -l \cdot \kappa_\alpha A^\alpha H \kappa_\beta A^\beta M H \kappa_\gamma A^\gamma r,$$
$$\mathcal{H} = l \cdot A^\alpha r_{\xi^\alpha}.$$

Proof. We use (3.3.43) in (3.3.41), Taylor expand, and equate coefficients of $\varepsilon^{1/3}$ to zero. This gives

$$G u_1 = 0, \tag{3.3.54}$$

$$G u_2 + \kappa_\alpha A^\alpha u_{1\theta} = 0, \tag{3.3.55}$$

$$G u_3 + \kappa_\alpha A^\alpha u_{2\theta} + \frac{1}{2} D_u^2 g \cdot (u_1, u_1) = 0, \tag{3.3.56}$$

$$G u_4 + \kappa_\alpha A^\alpha u_{3\theta} + D_u^2 g \cdot (u_1, u_2) + A^\alpha u_{1\xi^\alpha}$$
$$+ \kappa_\alpha D_u A^\alpha \cdot (u_1) u_{1\theta} = 0. \tag{3.3.57}$$

The solution of (3.3.54) is

$$u_1 = Rw_1(\theta), \tag{3.3.58}$$

where $w_1(\theta) \in R^p$ is arbitrary. We suppress the dependence on the slow variables ξ to simplify the notation.

Left multiplying (3.3.55) by L shows that it can only be solved for u_2 if

$$L\kappa_\alpha A^\alpha u_{1\theta} = 0.$$

Using (3.3.58) and (3.3.45) in this equation implies that

$$\kappa_\alpha B^\alpha w_{1\theta} = 0. \tag{3.3.59}$$

In order to obtain a nontrivial solution, we require that κ satisfies (3.3.46), which implies that φ satisfies the eikonal equation (3.3.52).

From (3.3.48), a solution of (3.3.59) is $w_1 = as$, where $a(\theta)$ is an arbitrary scalar valued function. Using this in (3.3.58) gives

$$u_1 = a(\theta)r. \tag{3.3.60}$$

Equations (3.3.43) and (3.3.60) imply that the leading order behavior of u is given by (3.3.51).

The solution of (3.3.55) for u_2 is then

$$u_2 = -H\kappa_\alpha A^\alpha r\, a_\theta + Rw_2(\theta), \tag{3.3.61}$$

where $w_2(\theta) \in R^p$ is arbitrary.

Left multiplying (3.3.56) by L and using (3.3.58), (3.3.61), and (3.3.50) in the result gives the following solvability condition for u_3,

$$\kappa_\alpha B^\alpha w_{2\theta} = L\kappa_\alpha A^\alpha H\kappa_\beta A^\beta r\, a_{\theta\theta}. \tag{3.3.62}$$

This equation can be solved for w_2 provided that the right-hand side is orthogonal to the left null vector $\jmath$. Since $\jmath \cdot L = l$, the solvability condition for w_2 is equivalent to the condition (3.3.49) for the absence of second order dissipative effects. A solution of (3.3.62) for w_2 is

$$w_2 = (\kappa_\alpha B^\alpha)^\dagger L\kappa_\alpha A^\alpha H\kappa_\beta A^\beta r\, a_\theta + a_2 s, \tag{3.3.63}$$

where $a_2(\theta)$ is an arbitrary scalar valued function.

The solution of (3.3.56) for u_3 is then

$$u_3 = -H\kappa_\alpha A^\alpha u_{2\theta} - \frac{1}{2}HD_u^2 g \cdot (r,r)a^2 + Rw_3, \tag{3.3.64}$$

where $w_3(\theta) \in R^p$ is arbitrary.

The condition for solving (3.3.57) for both w_3 and u_4 is obtained by taking the scalar product of the equation with l. Using (3.3.60), (3.3.61), (3.3.63), and (3.3.64) in the result implies that $a(\theta, \xi)$ satisfies (3.3.53).

One can check that the solvability conditions can be satisfied at all orders by choosing the functions of integration w_n and a_n appropriately. $\square$

Dissipative and semi-linear equations. Next we briefly describe some other long wave equations without writing out the details of the derivation.

Long wave dissipation dominates dispersion whenever

$$l \cdot \kappa_\alpha A^\alpha H \kappa_\beta A^\beta r \neq 0. \qquad (3.3.65)$$

In this case, quadratic semi-linear effects are negligible provided that

$$LD_u^2 g \cdot (r, r) = 0,$$
$$l \cdot D_u^2 g \cdot (r, Rw) = 0 \text{ for all } w \in R^p.$$

The asymptotic solution of (3.3.41) which describes a balance between weak nonlinearity and weak dissipation is then

$$u = \varepsilon^{1/2} a \left(\varepsilon^{-1/2} \varphi(\varepsilon x), \varepsilon x \right) r(\varepsilon x) + O(\varepsilon),$$

where $\varepsilon \to 0$ with $x = O(\varepsilon^{-1})$. Here and below, the phase φ satisfies the eikonal equation (3.3.52), as before, and r is the associated right nullvector. The transport equation for $a(\theta, \xi)$ is

$$lA^\alpha(ar)_{\xi^\alpha} + \left[l \cdot D_u A^\alpha \cdot (r)r - l \cdot D_u^2 g \cdot (r, H\kappa_\alpha A^\alpha r) \right] aa_\theta$$
$$- \frac{1}{2} l \cdot D_u^2 g \cdot \left(r, H D_u^2 g \cdot (r, r) \right) a^3 = l \cdot \kappa_\alpha A^\alpha H \kappa_\beta A^\beta r \, a_{\theta\theta}.$$

This equation is a generalized Burgers equation, with an additional cubically nonlinear term (if the appropriate coefficient is nonzero). The normalized form of the equation is

$$u_t + \alpha(t)uu_x + \beta(t)u^3 + \gamma(t)u = \mu(t)u_{xx}.$$

For plane waves, $\gamma = 0$ and the remaining coefficients are constants.

The dominant nonlinearity is quadratically nonlinear and semi-linear if

$$LD_u^2 g \cdot (r, r) \neq 0. \qquad (3.3.66)$$

This type of nonlinearity leads to mean-field interactions (which we discuss further in Subsection 3.4.2). In the dissipative case, when (3.3.65) is satisfied, the appropriate asymptotic solution is

$$u = \varepsilon a \left(\varepsilon^{-1/2} \varphi(\varepsilon x), \varepsilon x \right) r(\varepsilon x) + R(\varepsilon x)\bar{w}(\varepsilon x) + O(\varepsilon^{3/2}),$$

where $\varepsilon \to 0$ with $x = O(\varepsilon^{-1})$. We assume the wave amplitude a has zero mean with respect to θ. The equation for the mean field $\bar{w}(\xi) \in R^p$ is

$$LA^\alpha(R\bar{w})_{\xi^\alpha} + \frac{1}{2}LD_u^2 g \cdot (R\bar{w}, R\bar{w}) + \frac{1}{2}\langle a^2 \rangle LD_u^2 g \cdot (r, r) = 0, \quad (3.3.67)$$

where the angular bracket denotes the mean with respect to θ,

$$\langle f(\theta, \xi) \rangle = \lim_{T \to \infty} \frac{1}{T} \int_0^T f(\theta, \xi) \, d\theta.$$

Equation (3.3.67) shows that the wave-field generates a mean-field.

The transport equation for the wave amplitude is a semi-linear diffusion equation

$$lA^\alpha(ar)_{\xi^\alpha} + l \cdot D_u^2 g \cdot (r, R\bar{w})\, a + \frac{1}{2} l \cdot D_u^2 g \cdot (r, r)(a^2 - \langle a^2 \rangle)$$
$$= \left(l \cdot \kappa_\alpha A^\alpha H \kappa_\beta A^\beta r\right) a_{\theta\theta}.$$

This equation has the form

$$u_t + \alpha(t)\left(u^2 - \langle u^2 \rangle\right) + \beta(t)u = \mu(t)u_{xx},$$

where the angular brackets denote an average with respect to x.

A limitation of these semi-linear equations is that their solutions may blow up in finite time, as in the classical ODE example $y' = y^2$. For example, small amplitude solutions of the semi-linear diffusion equation

$$u_t = u_{xx} + u^p$$

are known to blow up if $p \leq 3$ [82]. After the blow up time, the weakly nonlinear asymptotics is not self-consistent, and one has to study large amplitude solutions. It might, for example, be possible to match the weakly nonlinear solution to a large amplitude self-similar solution.

In the semi-linear dispersive case, when conditions (3.3.66) and (3.3.49) hold, the asymptotic solution is

$$u = \varepsilon a\left(\varepsilon^{-2/3}\varphi(\varepsilon x), \varepsilon x\right) r(\varepsilon x) + R(\varepsilon x)\bar{w}(\varepsilon x) + O(\varepsilon^{4/3}),$$

where $\varepsilon \to 0$ with $x = O(\varepsilon^{-1})$. The mean field, $\bar{w}$, satisfies (3.3.67) and

$$lA^\alpha(ar)_{\xi^\alpha} + l \cdot D_u^2 g \cdot (r, R\bar{w})\, a + \frac{1}{2} l \cdot D_u^2 g \cdot (r, r)(a^2 - \langle a^2 \rangle)$$
$$= \left(l \cdot \kappa_\alpha A^\alpha H \kappa_\beta A^\beta H \kappa_\gamma A^\gamma r\right) a_{\theta\theta\theta}.$$

This equation has the form

$$u_t + \alpha(t)\left(u^2 - \langle u^2 \rangle\right) + \beta(t)u = \nu(t)u_{xxx}.$$

We remark that if one is dealing with pulses, for which $\langle a^2 \rangle = 0$, or wavefronts, when one does not demand uniform validity of the asymptotic expansion as $\theta \to \infty$, the condition (3.3.50) in Proposition 3.1 — which rules out wave generation of a mean field as in (3.3.67) — can be replaced by the weaker condition $l \cdot D_u^2 g \cdot (r, r) = 0$, and we can set $\bar{w} = 0$.

Finally, if the strict hyperbolicity condition (3.3.47) is not satisfied, then several wave modes have the same phase velocity. In this case one typically obtains a system of coupled Burgers or KdV equations [175].

3.3.2.4. Ion acoustic waves. A plasma is a fluid of charged particles. The positively charged particles are ions and the negatively charged particles are electrons. Ion acoustic waves are low frequency plasma waves in which the ions oscillate, dragging the much lighter electrons with them. We consider the simplest set of equations which model these waves.

The two fluid equations. In the two fluid model of a plasma, the ions and electrons are treated as separate charged fluids which interact through the macroscopic electromagnetic field they generate. In a simple model of ion acoustic waves, we assume that the electromagnetic field is electrostatic (the wave velocities are much less than the speed of light), and we neglect the electron inertia and the ion fluid pressure.[4] The equations of motion are then

$$
\begin{aligned}
&n_t + \nabla \cdot (nu) = 0, \\
&(nu)_t + \nabla \cdot (nu \otimes u) = \frac{e}{m} nE, \\
&\nabla p_e = -en_e E, \\
&\operatorname{curl} E = 0, \\
&\epsilon_0 \nabla \cdot E = e(n - n_e), \\
&p_e = p_e(n_e).
\end{aligned}
\tag{3.3.68}
$$

Here, $n(x,t)$ and $u(x,t)$ are the number density and velocity of the ion fluid, $n_e(x,t)$ and $p_e(x,t)$ are the number density and pressure of the electron fluid, $E(x,t)$ is the electric field, m is the ion mass, $e > 0$ is the electronic charge, and ϵ_0 is the dielectric constant.

The first two equations are balance of mass and momentum for the ion fluid, with pressure neglected. The third equation is momentum balance for the electron fluid with the inertia term neglected. The next two equations are the electrostatic equations giving the electric field generated by the net charge density of the plasma. For simplicity, we assume an isothermal equation of state for the electron fluid,

$$
p_e = KT_e n_e,
$$

where K is Boltzmann's constant and T_e is the temperature of the electron fluid. This assumption is not essential — the method applies to arbitrary equations of state.

The linearized dispersion relation of (3.3.68) at $n = n_e = n_0$ and $u = E = 0$ is

$$
\omega^2 = \frac{c_0^2 |k|^2}{1 + \lambda_d^2 |k|^2},
$$

where c_0 is the ion acoustic speed and λ_d is the Debye length,

$$
c_0^2 = \frac{KT_e}{m}, \qquad \lambda_d^2 = \frac{\epsilon_0 KT_e}{n_0 e^2}.
$$

The Debye length is the distance over which the plasma shields a local charge imbalance. For ion acoustic waves with wavelength comparable to the Debye length, charges are only partially shielded and the waves are dispersive. For long waves, with $\lambda_d |k| \ll 1$, charges are completely shielded. The electrons and ions move in phase and the waves are nondispersive. The long wave speed c_0 is determined by the compressibility of the electron fluid, KT_e, and the inertia, m, of the ion fluid.

---

[4] Starting from more complicated primitive equations still leads to a KdV equation for long waves.

We introduce nondimensional variables, with lengths nondimensionalized by a nonlinear lengthscale, λ_d/ε, where $\varepsilon \ll 1$ is the dimensionless wave amplitude,

$$\bar{x} = \frac{\varepsilon x}{\lambda_d}, \qquad \bar{t} = \frac{\varepsilon c_0 t}{\lambda_d},$$

$$\bar{n} = \frac{n}{n_0}, \qquad \bar{n}_e = \frac{n_e}{n_0}, \qquad \bar{p}_e = \frac{p_e}{KT_e n_0},$$

$$\bar{u} = \frac{u}{c_0}, \qquad \bar{E} = \frac{e\lambda_d E}{KT_e}.$$

After dropping the bars on the nondimensionalized variables, (3.3.68) becomes

$$n_t + \nabla \cdot (nu) = 0,$$

$$(nu)_t + \nabla \cdot (nu \otimes u) = \frac{1}{\varepsilon} nE,$$

$$\nabla n_e = -\frac{1}{\varepsilon} n_e E, \qquad\qquad (3.3.69)$$

$$\operatorname{curl} E = 0,$$

$$\nabla \cdot E = \frac{1}{\varepsilon}(n - n_e).$$

The electrostatic equations are often written in terms of a potential ϕ, where $E = -\nabla\phi$. After solving for n_e, this gives

$$\Delta\phi = \frac{1}{\varepsilon}\left(e^{\phi/\varepsilon} - n\right).$$

Here, we leave (3.3.69) as a first order system.

The Chapman–Enskog expansion. The Chapman–Enskog expansion is simple to carry out for this example. Equilibrium solutions of (3.3.69) satisfy

$$n_e = n, \quad \text{and} \quad E = 0.$$

We specify a local equilibrium state by n and u, and consider solutions of the form

$$n = n(x, t; \varepsilon),$$
$$u = u(x, t; \varepsilon),$$
$$n_e = n(x, t; \varepsilon) + \varepsilon n'_e(x, t, \varepsilon),$$
$$E = \varepsilon E'(x, t; \varepsilon).$$

Using these expressions in (3.3.69) gives

$$n_t + \nabla \cdot (nu) = 0,$$
$$-nE' + (nu)_t + \nabla \cdot (nu \otimes u) = 0,$$
$$nE' + \nabla n + \varepsilon\left(\nabla n'_e + n'_e E'\right) = 0,$$
$$\varepsilon \operatorname{curl} E' = 0,$$
$$n'_e + \varepsilon\nabla \cdot E' = 0.$$

These equations can be solved for n'_e and E' provided that

$$n_t + \nabla \cdot (nu) = 0,$$
$$(nu)_t + \nabla \cdot (nu \otimes u + nI) + \varepsilon \left(\nabla n'_e + n'_e E' \right) = 0,$$
$$\varepsilon \operatorname{curl} E' = 0.$$

Here, the second equation is obtained by eliminating nE' from the full system. The remaining equations for n'_e and E' can then be written

$$n'_e = -\varepsilon \nabla \cdot E',$$
$$E' = -\nabla \log n - \varepsilon \frac{1}{n} \left(\nabla n'_e + n'_e E' \right).$$

Solving these iteratively gives

$$n'_e = \varepsilon \Delta \log n + O(\varepsilon^3),$$
$$E' = -\nabla \log n + O(\varepsilon^2).$$

Finally, using these expressions in the solvability conditions and neglecting terms of the order ε^4 gives equations for n and u,

$$
\begin{aligned}
& n_t + \nabla \cdot (nu) = 0, \\
& (nu)_t + \nabla \cdot (nu \otimes u + nI) \\
& \qquad + \varepsilon^2 \left\{ \nabla(\Delta \log n) - (\Delta \log n) \nabla \log n \right\} = 0.
\end{aligned}
\tag{3.3.70}
$$

These long wave equations contain an additional pressure term $p = n$ and small third order, nonlinear dispersive terms in the momentum equation.

The KdV expansion. The KdV expansion for (3.3.69) is

$$
\begin{bmatrix} n \\ u \\ E \\ n_e \end{bmatrix} = \begin{bmatrix} 1 \\ 0 \\ 0 \\ 1 \end{bmatrix} + \varepsilon^{2/3} a(\theta, x, t) \begin{bmatrix} \omega \\ k \\ 0 \\ \omega \end{bmatrix} + \varepsilon a_\theta(\theta, x, t) \begin{bmatrix} 0 \\ 0 \\ -\omega k \\ 0 \end{bmatrix} + O(\varepsilon^{4/3}).
$$

where $\theta = \varepsilon^{-2/3} \varphi(x, t)$, and $\varepsilon \to 0$ with $x, t = O(1)$. The same result follows from (3.3.70).

The local frequency $\omega = -\varphi_t$ and wavenumber $k = \nabla \varphi$ satisfy the linearized dispersion relation,

$$\omega^2 = |k|^2.$$

The group velocity is $C = k/\omega$. The wave amplitude solves a generalized KdV equation,

$$a_t + C \cdot \nabla a + \omega^2 a a_\theta + \frac{1}{2} \omega^3 a_{\theta\theta\theta} + \left(\frac{\omega_t + \nabla \cdot k}{2\omega} \right) a = 0.$$

In the plane wave case, this is the usual KdV equation.

A simple nonplanar example is outgoing spherical waves in n space dimensions. The phase variable is $\varphi = r - t$, where $r = |x|$. The transport equation is then

$$a_t + a_r + a a_\theta + \frac{1}{2} a_{\theta\theta} + \frac{n-1}{2r} a = 0.$$

For cylindrical waves ($n = 2$), this equation is completely integrable, since it can be transformed into the KdV equation by an appropriate change of variables [128]. For $n \geq 3$, the equation is not integrable. For further information on the cylindrical and spherical KdV equations and their application to plasma waves, see [9, 128].

Magnetized plasmas. Finally, we briefly describe the Chapman–Enskog expansion for an interesting generalization of (3.3.69) which provides a simple model of ion acoustic waves in a magnetized plasma. We suppose that there is a uniform magnetic field in the e-direction, where e is a constant unit vector. This gives [128]

$$
\begin{aligned}
& n_t + \nabla \cdot (nu) = 0, \\
& u_t + u \cdot \nabla u = \frac{1}{\varepsilon} \left(E - \Omega e \times u \right), \\
& \nabla n_e = -\frac{1}{\varepsilon} n_e E, \\
& \operatorname{curl} E = 0, \\
& \nabla \cdot E = \frac{1}{\varepsilon} (n - n_e).
\end{aligned}
\tag{3.3.71}
$$

Here, the scalar parameter Ω is a nondimensionalized magnetic field strength, which we assume is of the order one. Equation (3.3.71) differs from (3.3.69) by the addition of a Lorentz force term to the momentum equation. We have rewritten the momentum equation in nonconservative form.

The velocity vector of long wave solutions of (3.3.71) must be parallel to e in order for the Lorentz force to vanish. The long wave expansion is therefore

$$
\begin{aligned}
u &= u_\| e + \varepsilon u'_\perp, \\
n_e &= n + \varepsilon n'_e, \\
E &= \varepsilon E'.
\end{aligned}
$$

Here $u_\|$ is the scalar velocity component parallel to e, which we take to be in the x direction, and $u'_\perp$ is the velocity perturbation orthogonal to e.

Expanding (3.3.71) in the usual way leads to the solutions

$$
\begin{aligned}
n'_e &= \varepsilon \Delta N + O(\varepsilon^3), \\
E' &= -\nabla N + O(\varepsilon^2), \\
u'_\perp &= \frac{1}{\Omega} e \times \nabla N + \frac{\varepsilon}{\Omega^2} v + O(\varepsilon^2).
\end{aligned}
$$

Here, $N = \log n$, and

$$v = - \left[(\nabla N)_t + u_\| (\nabla N)_x \right]^\perp, \tag{3.3.72}$$

where the superscript $\perp$ denotes the component orthogonal to e.

The corresponding equations for n and $u_\parallel$ are, neglecting lower order terms,

$$n_t + \left(nu_\parallel\right)_x + \frac{\varepsilon^2}{\Omega^2}\nabla \cdot (nv) = 0,$$

$$u_{\parallel t} + u_\parallel u_{\parallel x} + N_x + \frac{\varepsilon}{\Omega}\left(e \times \nabla N\right) \cdot \nabla u_\parallel \qquad (3.3.73)$$

$$+\frac{\varepsilon^2}{\Omega^2}v \cdot \nabla u_\parallel + \varepsilon^2\frac{1}{n}\left(\Delta N_x - N_x\Delta N\right) = 0.$$

We can simplify these equations further by using the fact that

$$n_t + u_\parallel n_x = -nu_{\parallel x} + O(\varepsilon^2), \qquad (3.3.74)$$

to eliminate time derivatives from v in (3.3.72). Neglecting lower order terms, this gives

$$v = \frac{1}{n}\left[\left(n\nabla u_\parallel\right)_x\right]^\perp.$$

The corresponding dispersive terms in (3.3.73) then involve only spatial derivatives.

  3.3.2.5. Other long wave equations. Here, we briefly describe some other long wave equations. First, suppose that ω is not an analytic function of k. This can occur for waves modeled by initial-boundary value problems, such as internal waves in a stratified fluid. Then, as $kh \to 0$, the long wave dispersion relation takes the form

$$\omega = c_0 k\left[1 + i\mu|kh| + O(k^2h^2)\right].$$

In this case, instead of the KdV equation (3.3.2), one obtains the Benjamin–Ono (BO) equation [1, 21],

$$u_t + uu_x + \mathcal{H}[u_{xx}] = 0,$$

where $\mathcal{H}$ is the Hilbert transform,

$$\mathcal{H}[u](x, t) = \frac{1}{\pi}\text{p.v.}\int_{-\infty}^{\infty}\frac{u(y, t)}{y - x}\,dy.$$

Since the Hilbert transform of e^{ikx} is $i\,\text{sgn}\,(k)e^{ikx}$, the term $\mathcal{H}[u_{xx}]$ is dispersive with a dispersion relation of the form $\omega = k|k|$, as required. The BO equation is completely integrable [2], and has algebraically decaying soliton solutions.

  Now suppose that the wave motion depends on an additional dimensionless parameter λ, and that the long wave expansion of the dispersion relation is

$$\omega = c_0 k\left[1 - i\mu(\lambda)kh + \nu(\lambda)k^2h^2 + O(k^3h^3)\right].$$

If $\mu(\lambda_0) = 0$, then long wave dispersion and dissipation are of the same order of magnitude for λ close to λ_0. Specifically, if $\mu'(\lambda_0) \neq 0$, they are of the same order of magnitude for $\lambda - \lambda_0 = O(kh)$ as $kh \to 0$. Assuming advective nonlinearity, the same expansion that we used to derive the KdV equation now leads to the KdV–Burgers equation,

$$u_t + uu_x - \alpha u_{xx} + \beta u_{xxx} = 0. \qquad (3.3.75)$$

This is a canonical equation for the combined effects of long wave dissipation and dispersion. Depending on the relative size of α, β, and the width of the traveling wave, equation (3.3.75) has monotonic traveling waves — like the Burgers shock profile — or traveling waves with oscillatory tails [133]. The latter "collisionless shocks" are observed in plasmas and bubbly fluids [48, 204, 267]. For special values of the coefficients and wave speed, there is an explicit analytical expression for one of the monotonic traveling waves [135].

The zero dispersion and dissipation limit of (3.3.75) is studied in [30], while some numerical solutions appear in [44, 28].

Next, suppose that the wave motion is purely dissipative, so that the linearized dispersion relation takes the form

$$\omega = c_0 k \left[1 - i\mu(\lambda)kh + i\nu(\lambda)k^3 h^3 + O(k^5 h^5) \right]$$

If $\mu(\lambda_0) = 0$ and $\mu'(\lambda_0) \neq 0$, then for $\lambda - \lambda_0 = O(k^2 h^2)$ one obtains the Kuramoto–Sivashinsky (KS) equation,

$$u_t + uu_x + \alpha u_{xx} + \beta u_{xxxx} = 0.$$

We require that $\beta > 0$ for well-posedness. If $\alpha < 0$, then all Fourier modes of the linearized equations decay as $t \to +\infty$. If $\alpha > 0$, then long waves in the wavenumber band $|k| < (\alpha/\beta)^{1/2}$ are linearly unstable. This equation therefore models the loss of stability in long waves. The KS equation has many interesting properties, including the existence of an inertial manifold, that is, a finite dimensional invariant submanifold of the (infinite dimensional) state space which contains a chaotic attractor [92, 127, 210].

The KS equation does not contain any dispersive effects. Including long wave dispersion gives the KS–KdV equation,

$$u_t + uu_x + \alpha u_{xx} + \beta u_{xxx} + \gamma u_{xxxx} = 0.$$

This equation does not seem to have been studied at all.

Finally, if the wave motion is purely dispersive, so that

$$\omega = c_0 k \left[1 + \mu(\lambda)k^2 h^2 + \nu(\lambda)k^4 h^4 + O(k^6 h^6) \right],$$

and $\mu(\lambda_0) = 0$, one obtains a higher order KdV equation,

$$u_t + uu_x + \alpha u_{xxx} + \beta u_{xxxxx} = 0.$$

This equation arises in water waves with surface tension, where λ is a dimensionless surface tension called the Bond number [107, 123], and in magneto-acoustic waves [147], where λ is the angle between the propagation direction and the unperturbed magnetic field.

One interesting question concerning this higher order KdV equation is the existence or nonexistence of solitary wave solutions. If $\alpha > 0$, the equation does not have solitary wave solutions which approach the KdV soliton as $\beta \to 0$. Instead, it has traveling wave solutions which consist of a solitary wave "riding" on an infinite train of small amplitude periodic waves [7, 93, 123]. The amplitude of these waves is exponentially small in β, so it has to be computed using "asymptotics beyond all orders" [167, 221]. For other equations with similar traveling wave solutions, see [32]. An integrable equation, the Camassa–Holm equation (3.4.17), which has n-soliton solutions riding on a quasiperiodic background is studied in [4].

3.3.3. *First Order Systems of PDEs: Short Waves*

In this subsection we derive asymptotic equations for short wave solutions of a first order system of PDEs. Since shocks typically form in short wave solutions, we consider a system in conservation form,

$$f^\alpha(\varepsilon x, u)_{x^\alpha} + g(\varepsilon x, u) = 0. \tag{3.3.76}$$

We assume that $g(\varepsilon x, 0) = 0$.

For short waves, the typical nonlinearity is advective, of the form uu_x, because this is stronger than semilinear effects of the form u^2, except in special cases (for example, when (3.3.76) is semilinear).

The expansion of the linearized local dispersion relation for unidirectional short waves is of the form

$$\omega = c_0 k \left\{ 1 - i\mu(kh)^{-1} + \nu(kh)^{-2} + O(kh^{-3}) \right\} \quad \text{as } kh \to \infty.$$

If $\mu \neq 0$, the dominant linear effect on the wave is a frequency independent damping ($\mu > 0$) or growth ($\mu < 0$). If $\mu = 0$ and $\nu \neq 0$, then short waves are dispersive.

Dissipative short waves. Dissipative short waves are described by the expansion in subsection 3.2.3.2. For plane wave solutions of constant coefficient equations, the equation derived there for the wave amplitude reduces to

$$C^\alpha a_{x^\alpha} + \left(\frac{1}{2} \mathcal{G} a^2 \right)_\theta + \mathcal{H} a = 0, \tag{3.3.77}$$

where $\mathcal{H}$ is the damping rate of short waves,

$$\mathcal{H} = l \cdot Gr.$$

Here, $G = D_u g$, with f^α, g, and their derivatives evaluated at $u = 0$, as usual. The normalized form of (3.3.77) is

$$u_t + \left(\frac{1}{2} u^2 \right)_x + u = 0. \tag{3.3.78}$$

Equation (3.3.78) is sometimes called the Varley–Rodgers equation. One interesting property of (3.3.78) is that if the initial data $u(x, 0)$ is smooth and

$$u_x(x, 0) > -1,$$

then the solution $u(x, t)$ is smooth for all $t > 0$. On the other hand, if the derivative of the initial data is less than negative one at some point, then the solution shocks up in finite time [138, 272].

Dispersive short waves. To analyze weakly dispersive short waves, we assume that there is no frequency independent damping of short waves. The algebraic expression of this condition is

$$l \cdot Gr = 0. \tag{3.3.79}$$

We look for an asymptotic solution of (3.3.76) of the form

$$u = \varepsilon^2 u_1(\theta, \xi) + \varepsilon^3 u_2(\theta, \xi) + \varepsilon^4 u_3(\theta, \xi) + O(\varepsilon^5), \tag{3.3.80}$$
$$\theta = \varepsilon^{-2} \varphi(\varepsilon x), \qquad \xi = \varepsilon x,$$

where $\varepsilon \to 0$ with $x = O(\varepsilon^{-1})$. This scaling can be motivated by looking for a balance between weak nonlinearity and weak dispersion [124], just as we did for the KdV scaling above.

Using (3.3.80) in (3.3.76), Taylor expanding in powers of ε, and equating coefficients of ε, ε^2, and ε^3 to zero, implies that $u_1(\theta, x)$, $u_2(\theta, x)$, and $u_3(\theta, x)$ satisfy

$$\kappa_\alpha A^\alpha u_{1\theta} = 0, \tag{3.3.81}$$
$$\kappa_\alpha A^\alpha u_{2\theta} + G u_1 = 0, \tag{3.3.82}$$
$$\kappa_\alpha A^\alpha u_{3\theta} + G u_2 + (A^\alpha u_1)_{\xi^\alpha} + \frac{1}{2} D_u^2 f^\alpha \cdot (u_1, u_1) = 0, \tag{3.3.83}$$

where $\kappa_\alpha = \varphi_{\xi^\alpha}$, and

$$A^\alpha(\xi) = D_u f^\alpha(\xi, 0),$$
$$G(\xi) = D_u g(\xi, 0).$$

Equation (3.3.81) implies that κ satisfies the local high frequency dispersion relation,

$$\det(\kappa_\alpha A^\alpha) = 0,$$

and that

$$u_1 = a(\theta, \xi) r(\xi),$$

where r is a right null vector of $\kappa_\alpha A^\alpha$. We assume that zero is a simple eigenvalue of this matrix, and we denote a left null vector by l.

Equation (3.3.82) has a solution of the form

$$u_2 = b(\theta, \xi) s(\xi),$$

where the scalar amplitude b satisfies

$$b_\theta = a, \tag{3.3.84}$$

and the vector s is a solution of

$$\kappa_\alpha A^\alpha s + G r = 0.$$

The solvability condition for this equation is satisfied because of assumption (3.3.79) that damping is negligible.

Equation (3.3.83) can be solved for u_3 provided its scalar product with l is zero, which implies that

$$C^\alpha a_{\xi^\alpha} + \left(\frac{1}{2} \mathcal{G} a^2\right)_\theta + \mathcal{H} a + \mathcal{N} b = 0. \tag{3.3.85}$$

Here, C^α, $\mathcal{G}$, and $\mathcal{H}$ are given by (3.2.42) (with x replaced by ξ) and

$$\mathcal{N} = l \cdot Gs.$$

In the constant coefficient case, we can put equations (3.3.84)–(3.3.85) in the normalized form

$$u_t + \left(\frac{1}{2} u^2 \right)_x + v = 0,$$
$$v_x - u = 0.$$

Alternatively, eliminating v gives

$$\partial_x \left[u_t + \left(\frac{1}{2} u^2 \right)_x \right] + u = 0. \tag{3.3.86}$$

The earliest derivation of this equation we know of was by Ostrovsky [216], so we shall call (3.3.86) the Ostrovsky equation.

For long wave dispersive equations, like the KdV equation, dispersion gets stronger as a wave steepens. This prevents shock formation. In contrast, for (3.3.86), dispersion gets weaker as a wave steepens, and therefore it cannot prevent shock formation in strong waves. Thus, (3.3.86) is a canonical equation illustrating the interplay between dispersion and shocks. This is even clearer if we write (3.3.86) as an integro-differential equation. Assuming, for example, that the solution is 2π-periodic in x, when it necessarily has zero mean, we obtain

$$u_t + \left(\frac{1}{2} u^2 \right)_x = \frac{1}{2\pi} \int_0^{2\pi} S(x - y) u(y, t) \, dy$$

where $S(x)$ is the sawtooth function defined by

$$S(x) = x \text{ for } |x| < \pi,$$
$$S(x + 2\pi) = S(x).$$

Integro-differential equations of this form, but with arbitrary kernels S, were proposed by Whitham as model equations for nonlinear dispersive waves; additional analysis and numerical solutions of these equations appear in [79, 83]. Similar equations arise for resonantly interacting hyperbolic waves (see Subsection 3.6.1), for conservation laws with internal state variables (Subsection 3.4.1), and for nonlinear waves in fluctuating media [20, 101].

Equation (3.3.86) has a family of smooth periodic traveling waves. There is a limiting periodic wave of maximum amplitude which has a corner at its crest. There are no solitary wave solutions. Numerical solutions of (3.3.86) show quite complicated behavior, but little is known rigorously.

Cubically nonlinear equations. Different asymptotic equations are obtained for short wave solutions which fail to be genuinely nonlinear [257], meaning that

$$l \cdot \kappa_\alpha D_u^2 f^\alpha \cdot (r, r) = 0. \tag{3.3.87}$$

For example, suppose that, in addition to (3.3.87), equation (3.3.79) is satisfied, so that the dominant linear effects are dispersive. The asymptotic solution of (3.3.76) is then

$$u = \varepsilon a \left(\varepsilon^{-2} \varphi(\varepsilon x), \varepsilon x \right) r(\varepsilon x) + O(\varepsilon^2),$$

where $\varepsilon \to 0$ with $x = O(\varepsilon^{-1})$. The wave amplitude $a(\theta, \xi)$ satisfies

$$C^\alpha a_{\xi\alpha} + \left(\mathcal{G} a^3 + \mathcal{F} \langle a^2 \rangle a + \mathcal{E} ab \right)_\theta + \mathcal{M} \left(a^2 - \langle a^2 \rangle \right) + \mathcal{H} a + \mathcal{N} b = 0,$$

$$b_\theta = a,$$

where the angular brackets denote the average with respect to θ. Here, the coefficients are given by

$$C^\alpha = l \cdot A^\alpha r,$$

$$\mathcal{E} = l \cdot \kappa_\alpha D_u^2 f^\alpha \cdot (r, s),$$

$$\mathcal{F} = \frac{1}{4} l \cdot \kappa_\alpha D_u^2 f^\alpha \cdot (r, p) - \frac{1}{4} l \cdot \kappa_\alpha D_u^2 f^\alpha \cdot (r, q),$$

$$\mathcal{G} = \frac{1}{4} l \cdot \kappa_\alpha D_u^2 f^\alpha \cdot (r, q) + \frac{1}{6} l \cdot \kappa_\alpha D_u^3 f^\alpha \cdot (r, r, r),$$

$$\mathcal{H} = l \cdot (A^\alpha r)_{\xi\alpha},$$

$$\mathcal{M} = \frac{1}{2} l \cdot D_u^2 g \cdot (r, r) + \frac{1}{2} l \cdot \mathcal{G} q,$$

$$\mathcal{N} = l \cdot \mathcal{G} s.$$

The vectors p, q, s are solutions of

$$\kappa_\alpha A^\alpha s + \mathcal{G} r = 0,$$

$$\kappa_\alpha A^\alpha q + \kappa_\alpha D_u^2 f^\alpha \cdot (r, r) = 0,$$

$$\mathcal{G} p + D_u^2 g \cdot (r, r) = 0.$$

We can solve for s and q in view of the conditions (3.3.79) and (3.3.87). We also assume that $\mathcal{G}$ is nonsingular, or at least that the above equation for p is solvable.

 3.3.3.1. Waves on a rotating shallow fluid. The Boussinesq equations for waves on the surface of a shallow rotating layer of fluid are

$$h_t + (hu)_x + (hv)_y = 0,$$

$$u_t + uu_x + vu_y + gh_x + \frac{1}{3} c_0^2 h_0 (h_{xx} + h_{yy})_x - \Omega v = 0, \qquad (3.3.88)$$

$$v_t + uv_x + vv_y + gh_y + \frac{1}{3} c_0^2 h_0 (h_{xx} + h_{yy})_y + \Omega u = 0.$$

Here, $h(x, y, t)$ is the elevation of the free surface, and $u(x, y, t)$ and $v(x, y, t)$ are the x and y velocity components, respectively. There are three dimensioned parameters in (3.3.88): the mean depth h_0, the gravitational acceleration g, and the Coriolis frequency Ω. The shallow water speed is defined by $c_0 = (gh)^{1/2}$.

 We assume that the rotation is slow, meaning that

$$\frac{h_0 \Omega}{c_0} = \varepsilon^{1/2} f \ll 1,$$

where f is an order one dimensionless parameter. In this case, short wave dispersion (due to rotation) and long wave dispersion (due to finite depth effects) are both weak and of the same order of magnitude if

$$\frac{\lambda}{h_0} = O(\varepsilon^{1/2}),$$

where λ is a typical wavelength. Weakly nonlinear effects balance with these dispersive effects when

$$\frac{h - h_0}{h_0} = O(\varepsilon).$$

Nondimensionalizing (3.3.88) by the shallow water speed, c_0, and the Rossby deformation lengthscale, $\varepsilon^{1/2} h_0$, gives

$$h_t + (hu)_x + (hv)_y = 0,$$

$$u_t + uu_x + vu_y + h_x + \frac{1}{3}\varepsilon(h_{xx} + h_{yy})_x - fv = 0, \qquad (3.3.89)$$

$$v_t + uv_x + vv_y + h_y + \frac{1}{3}\varepsilon(h_{xx} + h_{yy})_y + fu = 0.$$

The asymptotic solution is

$$\begin{bmatrix} h \\ u \\ v \end{bmatrix} = \begin{bmatrix} 1 \\ 0 \\ 0 \end{bmatrix} + \varepsilon a(\theta, \xi, \eta, \tau) \begin{bmatrix} \omega \\ k \\ l \end{bmatrix} + O(\varepsilon^2),$$

as $\varepsilon \to 0$ with $x, t = O(\varepsilon^{-1})$, where

$$\theta = \varepsilon^{-1}\varphi(\varepsilon x, \varepsilon y, \varepsilon t),$$
$$\xi = \varepsilon x, \quad \eta = \varepsilon y, \quad \tau = \varepsilon t.$$

The local frequency and wavenumbers $(\omega, k, l) = (-\varphi_\tau, \varphi_\xi, \varphi_\eta)$ satisfy the dispersion relation,

$$\omega^2 = k^2 + l^2.$$

The wave amplitude a satisfies the transport equation

$$\frac{\partial}{\partial\theta}\left(a_\tau + ka_\xi + la_\eta + \frac{3}{2}aa_\theta + \frac{1}{6}a_{\theta\theta\theta} + \frac{\omega_\tau + k_\xi + l_\eta}{2\omega}a \right) - \frac{1}{2}f^2 a = 0.$$

For plane waves, the normalized form of this equation is

$$(u_t + uu_x + u_{xxx})_x - u = 0. \qquad (3.3.90)$$

This equation contains both weak short wave dispersion, due to the rotation, and weak long wave dispersion, due to finite depth effects. When rotation is neglected, the equation reduces to a KdV equation. When finite depth effects are neglected, the equation reduces to the Ostrovsky equation (3.3.86). Equation (3.3.90) was also derived by Ostrovsky [216] for internal waves in a rotating ocean.

One effect of rotation is to destroy the usual shallow water KdV solitary waves. The reason is that the KdV soliton propagates at the same velocity as small amplitude inertial waves (whose restoring force is the Coriolis force). Consequently, a KdV solitary wave radiates inertial waves and decays [116].

KP-type equation. The KP-version of this long wave-short wave equation, which includes weak diffractive effects (see Section 3.5), is [95]

$$(u_t + uu_x + u_{xxx})_x + u_{yy} - u = 0.$$

For further analysis and numerical solutions of this equation, see [97, 218]. There are four non-equivalent versions of this equation, depending on the choice of signs of the coefficients of u_{yy} and u.

Burgers–Sivashinsky equation. There are dissipative analogs of these dispersive long–short wave equations. For example, if we have weak short wave instability and weak long wave dissipation, the resulting asymptotic equation has the form

$$u_t + uu_x = u + u_{xx}. \tag{3.3.91}$$

This equation is studied by Goodman [92]. In view of the qualitative analogy with the Kuramoto–Sivashinsky equation, he calls (3.3.91) the Burgers–Sivashinsky equation

3.4. MISCELLANEOUS APPLICATIONS

In this Section, we derive asymptotic equations for some special types of hyperbolic equations. In Subsection 3.4.1 we analyze conservation laws with internal state variables. In Subsection 3.4.2, we derive equations for wave-mean field interactions in conservation laws with rapidly varying source terms and in semi-linear equations. In Subsection 3.4.3, we study a class of hyperbolic variational equations.

3.4.1. *Conservation Laws with Internal State Variables*

In Section 3.3, the effects of dissipation and dispersion were weak because the waves were either very long or very short. There are many other situations in which dissipation and dispersion are weak and can be treated perturbatively. In this subsection, we analyze conservation laws with internal state variables. We assume that the fluxes depend weakly on the internal state variables, and this allows us to derive asymptotic equations for weakly dissipative or dispersive waves without restricting the wavelength in any way.

The primitive conservation laws. We consider a system of equations of the form

$$f^\alpha(x, u, \varepsilon v)_{x^\alpha} = 0, \tag{3.4.1}$$

$$g^\alpha(x, u, v)_{x^\alpha} = \frac{1}{\varepsilon} h(x, u, v). \tag{3.4.2}$$

Equation (3.4.1) is a system of conservation laws for the conserved quantities u in which the fluxes depend weakly on "internal" state variables v. The internal variables satisfy a rate-type equation (3.4.2). The factor of $1/\varepsilon$ in front of the source term h is not essential — it can be removed by rescaling x. We include it so we can use the geometrical optics ansatz below (rather than the far-field ansatz we would need if the factor was omitted). We assume that there is an equilibrium state, $v = v_*(x, u)$, such that

$$h(x, u, v_*(x, u)) = 0.$$

The weakly nonlinear expansion is

$$u = u_0(x) + \varepsilon u_1\left(\varepsilon^{-1}\varphi(x), x\right) + \varepsilon^2 u_2\left(\varepsilon^{-1}\varphi(x), x\right) + O(\varepsilon^3),$$
$$v = v_0(x) + \varepsilon v_1\left(\varepsilon^{-1}\varphi(x), x\right) + O(\varepsilon^2), \tag{3.4.3}$$

where $\varepsilon \to 0$ with $x = O(1)$. In (3.4.3), we assume that u_0 and v_0 satisfy

$$f^\alpha(x, u_0, 0)_{x^\alpha} = 0,$$
$$v_0 = v_*(x, u_0).$$

We use (3.4.3) in (3.4.1) and (3.4.2) and expand in the usual way. This leads to the following results.

The frequency wavenumber vector $\kappa_\alpha = \varphi_{x^\alpha}$ satisfies the "frozen" eikonal equation,

$$\det(\kappa_\alpha A^\alpha) = 0,$$

where $A^\alpha(x) = D_u f^\alpha(x, u_0, 0)$. The first order perturbation u_1 is given by

$$u_1 = a(\theta, x) r(x),$$

where a is a scalar valued amplitude function and r is a right null vector of $\kappa_\alpha A^\alpha$. The transport equation for a and the equation for v_1 are

$$C^\alpha a_{x^\alpha} + \left(\frac{1}{2}\mathcal{G}a^2 + \mathcal{V}a\right)_\theta + \mathcal{H}a + m \cdot v_{1\theta} = 0, \tag{3.4.4}$$
$$Pv_{1\theta} + Qv_1 = a_\theta p + aq + s. \tag{3.4.5}$$

Here, m, p, q, s are vectors, and P, Q are matrices. The explicit expressions for the coefficients are

$$C^\alpha = l \cdot A^\alpha r,$$
$$\mathcal{G} = l \cdot \kappa_\alpha D_u^2 f^\alpha(x, u_0, 0) \cdot (r, r),$$
$$\mathcal{V} = l \cdot \kappa_\alpha D_{uv}^2 f^\alpha(x, u_0, 0) \cdot (r, v_0),$$
$$\mathcal{H} = l \cdot (A^\alpha r)_{x^\alpha},$$
$$m = \frac{1}{2} l \cdot \kappa_\alpha D_v f^\alpha(x, u_0, 0),$$
$$p = -\kappa_\alpha D_u g^\alpha(x, u_0, v_0) \cdot r,$$
$$q = D_u h(x, u_0, v_0) \cdot r,$$
$$s = -g^\alpha(x, u_0, v_0)_{x^\alpha},$$
$$P = \kappa_\alpha D_v g^\alpha(x, u_0, v_0),$$
$$Q = -D_v h(x, u_0, v_0).$$

The asymptotic equations therefore consist of an inviscid Burgers equation for a coupled with a linear dynamical system for v_1. Gradient flows for v_1 lead to dissipation, while Hamiltonian flows lead to dispersion. The simplest normalized asymptotic equations are (3.4.13) in the dissipative case and (3.4.15) in the dispersive case.

A non-intuitive feature of these equations is that the dynamical system for v_1 involves derivatives of the space-like variable θ, rather than the time-like ray derivative, $\partial_s = C^\alpha \partial_{x^\alpha}$. This reflects the fact that the time scale of the internal state dynamics is comparable with the wave period, and not with the longer nonlinear timescale.

One example which does not quite fit into the scheme above arises in a model of detonation waves, where the simplest weakly nonlinear asymptotic equations [231, 233] are of the form

$$u_t + \left(\frac{1}{2}u^2 \right)_x = q\Phi(u)v,$$
$$v_x = -\Phi(u)v.$$

Here, $u(x,t)$ is the normalized amplitude of a sound wave, driven by an exothermic chemical reaction, while $v(x,t)$ measures the concentration of the reactant. The constitutive function Φ gives the reaction rate, for example $\Phi(u) = \exp(u/b)$, and q and b are constants. This system consists of a relaxation equation for v coupled to an inviscid Burgers equation for u, but here the coupling is through the source terms rather than the fluxes.

For a general numerical method of solving quasi-linear hyperbolic conservation laws based on a stiff relaxation limit, see [136]. The idea is to replace the quasi-linear conservation law,

$$u_t + \sum_{\alpha=1}^{n} f^\alpha(u)_{x^\alpha} = 0,$$

by the semi-linear system

$$u_t + \sum_{\alpha=1}^{n} v_{x^\alpha}^\alpha = 0,$$
$$v_t^\alpha + A^\alpha u_{x^\alpha} = -\frac{1}{\varepsilon} \left(v^\alpha - f^\alpha(u) \right), \qquad \alpha = 1, \ldots, n.$$

The intermediate long wave equation. Another example of weak dispersion which is neither long or short wave dispersion is provided by internal gravity waves in a two layer fluid. Dispersion is weak if the one of the layers is much thinner than the wavelength of the internal wave. The thickness of the other layer may be comparable with the wavelength, so this is not a pure long wave expansion, and the asymptotic dispersion relation is more complicated than the usual long wave forms. The resulting weakly nonlinear equation is the intermediate long wave equation [1, 3, 96],

$$u_t + uu_x - \frac{\partial^2}{\partial x^2} \left[\text{p.v.} \int_{-\infty}^{\infty} \coth \left(\frac{\pi(x-z)}{2\delta} \right) u(\frac{z}{\delta}, t) \, dz \right] = 0.$$

This equation is integrable and contains both the KdV and the BO equation in suitable limits.

3.4.1.1. Sound waves in a relaxing gas. The energy in a gas molecule is divided among several different modes, such as translational modes, vibrational modes, and rotational modes. The time which it takes a given mode to attain equilibrium is called the relaxation time of the mode. The relaxation time of translational modes is very short, since only a few collisions with other molecules are required to bring translational modes into equilibrium. Relaxation times for rotational degrees of freedom are also short, typically less than ten collision times. On the other hand, thousands of collisions may be required to bring internal vibrational modes into thermal equilibrium. If a sound wave has period much longer than the vibrational relaxation time, then the gas always adjusts to local thermodynamic equilibrium, and the sound wave propagates at the relaxed (or equilibrium) sound speed. If the period is much less than the vibrational relaxation time (but larger than the collision time, otherwise a continuum description of the sound wave breaks down), the vibrational modes are not excited, and the sound wave propagates at the frozen sound speed. For intermediate frequencies, there is a partial approach of the vibrational modes to equilibrium. This process is irreversible, and therefore the sound wave is damped.

The vibrational relaxation of N_2 and CO_2 leads to significant broadening of shock fronts in the atmosphere. For example, it is largely responsible for spreading out the initial "crack" of a lightning strike into the low rumble of thunder.

Weak dissipation limits. There are several limits in which one can develop a perturbative treatment of relaxing gases. One limit is for waves whose period is much greater than the vibrational relaxation time. This long wave limit leads to a Burgers or Navier–Stokes equation. Relaxation is the physical origin of bulk viscosity, which is viscosity associated with volume changes of a fluid (see the Chapman–Enskog expansion below).

A second limit is one in which the wave period is much less than the vibrational relaxation time. This limit leads to a Varley–Rodgers equation with a frequency-independent rate of dissipation, and applies to very high frequency ultrasonic waves.

A third limit is when relaxation effects are small, meaning that the internal energy of the slowly relaxing modes is small compared with the total internal energy of the gas. In this case, dissipative effects are weak even for sound waves whose period is comparable with the vibrational relaxation time, so that a balance with weakly nonlinear effects is possible. The resulting asymptotic equation is the relaxing gas equation [235].

Primitive relaxing gas equations. A simple set of equations for the isentropic flow of a gas with a single relaxation process is [268]

$$\rho_t + \nabla \cdot (\rho u) = 0,$$
$$(\rho u)_t + \nabla \cdot (\rho u \otimes u + p(\rho, \xi)I) = 0, \qquad (3.4.6)$$
$$(\rho \xi)_t + \nabla \cdot (\rho u \xi) = \frac{\rho \left(\xi_*(\rho) - \xi \right)}{\tau},$$

Here $\rho(x, t)$ and $u(x, t)$ are the density and velocity of the gas, and the internal state variable $\xi(x, t)$ measures the departure of the relaxation process from local equilibrium, $\xi = \xi_*(\rho)$. The constant τ is the relaxation time. The relaxation process is coupled to the fluid equations through the dependence of the pressure p on ξ. Nonisentropic flows are easily treated in a similar way [49].

The Chapman–Enskog expansion. First, we consider the Chapman–Enskog expansion of (3.4.6) as $\tau \to 0$. Following the general procedure described in Subsection 3.3.2.1, we write

$$\rho = \rho(x, t; \tau)$$
$$u = u(x, t; \tau),$$
$$\xi = \xi_*(\rho) + \tau\xi'(x, t; \tau).$$

From the first two equations in (3.4.6), we have

$$\rho_t + \nabla \cdot (\rho u) = 0,$$
$$(\rho u)_t + \nabla \cdot \left(\rho u \otimes u + \left[p_*(\rho) + \tau \frac{\partial p}{\partial \xi}\bigg|_\rho \xi'\right] I\right) = 0(\tau^2),$$

where the relaxed pressure function, p_*, is given by

$$p_*(\rho) = p(\rho, \xi_*(\rho)). \qquad (3.4.7)$$

Expanding the third equation in (3.4.6), and using the equation of mass conservation to replace time derivatives by spatial derivatives in the result, gives

$$\begin{aligned}
\xi' &= -(\xi_{*t} + u \cdot \nabla\xi_*) + O(\tau) \\
&= -\frac{d\xi_*}{d\rho}(\rho_t + u \cdot \nabla\rho) + O(\tau) \qquad (3.4.8) \\
&= \rho\frac{d\xi_*}{d\rho}\nabla \cdot u + O(\tau).
\end{aligned}$$

Eliminating ξ' from these equations, and neglecting terms of the order τ^2, gives a reduced system of the form

$$\rho_t + \nabla \cdot (\rho u) = 0,$$
$$(\rho u)_t + \nabla \cdot (\rho u \otimes u + [p_*(\rho) - \mu_b \nabla \cdot u] I) = 0.$$

Relaxation effects thus lead to bulk viscosity but no shear viscosity. The effective bulk viscosity μ_b given by

$$\mu_b(\rho) = -\tau\rho\frac{\partial p}{\partial \xi}\bigg|_\rho (\rho, \xi_*(\rho)) \frac{d\xi_*}{d\rho}(\rho).$$

We can write μ_b in terms of the frozen sound speed, c, and the relaxed sound speed, c_*, which are defined by

$$c^2(\rho) = \frac{\partial p}{\partial \rho}\bigg|_\xi (\rho, \xi_*(\rho)), \qquad (3.4.9)$$

$$c_*^2(\rho) = \frac{dp_*}{d\rho}(\rho).$$

Differentiating equation (3.4.7) with respect to ρ shows that

$$\mu_b(\rho) = \tau\rho\left(c^2 - c_*^2\right).$$

Long waves are damped provided that $c > c_*$. This stability condition implies that wavefronts carrying the first signal of a disturbance propagate faster than the larger scale, equilibrated part of the wave. Whitham [272] gives a related discussion of wave hierarchies, and [49] analyzes conditions for long wave stability in general systems of relaxing conservation laws.

If $c < c_*$, then the long wave equations are ill-posed, and long waves will rapidly generate oscillations with periods of the order τ.

The dispersion relation. The linearized dispersion relation of (3.4.6) at $\rho = \rho_0$, $\xi = \xi_0 = \xi_*(\rho_0)$, $u = 0$ is

$$\left(\frac{\tau\omega + i}{\tau\omega + ic_*^2/c^2} \right) \omega^2 = c^2 |k|^2. \tag{3.4.10}$$

Here, the frozen sound speed c and the equilibrium sound speed c_* are evaluated at $\rho = \rho_0$ and $\xi = \xi_0$.

The long wave expansion of (3.4.10) as $k \to 0$ is

$$\omega = c_*|k| - \frac{1}{2} i\tau \left(c^2 - c_*^2 \right) |k|^2 + O(|k|^3).$$

Long waves propagate at the relaxed sound speed, c_*. The viscosity coefficient agrees with the one obtained by the Chapman–Enskog expansion.

The short wave expansion of (3.4.10) as $k \to \infty$ is

$$\omega = c|k| - i\frac{c^2 - c_*^2}{2\tau c_*^2} + O(|k|^{-1}).$$

Short waves propagate at the frozen sound speed c and are damped at a rate which is independent of their wavenumber.

These two expansions show that relaxation leads to weak damping for long and short waves. The third limit in which damping is small is that of weak relaxation effects, meaning that

$$\delta = \frac{c^2 - c_*^2}{c^2} \ll 1.$$

The corresponding expansion of (3.4.10) is

$$\omega = c|k| - \frac{1}{2} i\delta \frac{c|k|}{\tau c|k| + i} + O(\delta^2).$$

To obtain this limit, we assume a constitutive relation of the form

$$p = p(\rho, \zeta), \qquad \zeta = \varepsilon\xi. \tag{3.4.11}$$

Also, adopting a nondimensionalization in which the nonlinear timescale is of order one and the wave period is of order ε, we write

$$\tau = \varepsilon\sigma, \tag{3.4.12}$$

where σ is an order one parameter. This means that the relaxation time is of the order of the wave period.

The asymptotic equation. We consider the system (3.4.6) with the constitutive relation (3.4.11) and the scaling (3.4.12). Restricting to one space dimension for simplicity, the equations of motion are

$$\rho_t + (\rho u)_x = 0,$$
$$(\rho u)_t + \left(\rho u^2 + p(\rho, \varepsilon\xi)I\right)_x = 0,$$
$$(\rho\xi)_t + (\rho u\xi)_x = \frac{\rho\left(\xi_*(\rho) - \xi\right)}{\sigma\varepsilon}.$$

Expanding as $\varepsilon \to 0$, with $x, t = O(1)$, gives the following asymptotic solution,

$$
\begin{bmatrix} \rho \\ u \\ \xi \end{bmatrix} = \begin{bmatrix} \rho_0 \\ 0 \\ \xi_0 \end{bmatrix} + \varepsilon a \left[\varepsilon^{-1}(kx - \omega t), x, t\right] \begin{bmatrix} \rho_0 \\ c_0 \\ 0 \end{bmatrix}
$$
$$
+ \varepsilon b \left[\varepsilon^{-1}(kx - \omega t), x, t\right] \begin{bmatrix} 0 \\ 0 \\ \rho_0\xi'_*(\rho_0) \end{bmatrix} + O(\varepsilon^2),
$$

where $\omega = c_0 k$, $c_0 = c(\rho_0)$, and $\xi_0 = \xi(\rho_0)$.

The sound wave amplitude, $a(\theta, x, t)$, and the amplitude of the internal state variable, $b(\theta, x, t)$, satisfy

$$a_t + c_0 a_x + \left(\frac{1}{2}\omega\mathcal{G}a^2\right)_\theta - \frac{1}{2}\omega\alpha b_\theta = 0,$$
$$\omega\sigma b_\theta - b = a,$$

where

$$\alpha = -p_\zeta(\rho_0, 0)\frac{d\xi_*}{d\rho}(\rho_0).$$

The constant α measures the difference between the relaxed and frozen sound speeds defined in (3.4.9) and (3.4.11):

$$\frac{c_0^2 - c_*^2}{c_0^2} = \varepsilon\alpha + O(\varepsilon^2).$$

After a change of variables, these equations take the normalized form

$$u_t + \left(\frac{1}{2}u^2\right)_x + v_x = 0, \tag{3.4.13}$$
$$v_x + v - u = 0.$$

Eliminating v gives the relaxing gas equation,

$$(\partial_x + 1)\left[u_t + \left(\frac{1}{2}u^2\right)_x\right] + u_x = 0.$$

In the long wave limit,

$$(\partial_x + 1)^{-1} \sim 1 - \partial_x,$$

and this equation reduces to the Burgers equation. In the high-frequency limit,

$$\partial_x + 1 \sim \partial_x,$$

and we recover the Varley–Rodgers equation (3.3.78) after an integration with respect to x.

If $u(x, t) \to 0$ sufficiently rapidly as $x \to +\infty$, this equation can be reformulated as an integro-differential equation,

$$u_t + (\tfrac{1}{2}u^2)_x + Ku = 0,$$

where

$$Ku(x, t) = \int_x^\infty e^{y-x} u(y, t)\, dy.$$

For small amplitude initial data, relaxation prevents shock formation. However, if the initial data is sufficiently steep, the integral term cannot prevent the convective nonlinearity from forming shocks. Thus, one can have a wide shock front smoothed out by relaxation processes which contains a viscous subshock inside the profile. Including normal viscosity as well as relaxation effects leads to a relaxing Burgers equation,

$$(\partial_x + 1)\left[u_t + \left(\frac{1}{2}u^2\right)_x - \mu u_{xx}\right] + u_x = 0.$$

3.4.1.2. Sound waves in a bubbly fluid. The presence of a small number of gas bubbles in a fluid has a large effect on the propagation of sound waves. The mixture has the compressibility of the gas and the inertia of the fluid, so the sound speed is dramatically reduced.

Bubbly fluids are one of the few media in which sound waves are dispersive. A gas bubble in a fluid has a natural resonant frequency of oscillation, and the speed of a sound wave depends on its frequency with respect to this resonant frequency. From the point of view of the present discussion, we regard bubbly fluids as an effective fluid in which the bubble radius is an internal state variable. In contrast with a relaxing gas, where the internal state variable decays to a stable equilibrium, in a bubbly fluid the internal state variable is oscillatory in time. This leads to dispersion. Again, long and short wave expansions (compared with the period of resonant bubble oscillations) can be carried out. For long waves, one obtains the KdV equation [267] and for short waves one obtains the Ostrovsky equation [64, 257].

For a comparison between experimental observations and theoretical predictions of waves in bubbly fluids, see [204].

Bubbly fluid equations. We will use the simplest effective equations for a bubbly fluid which include nonlinear acoustic effects. These are

$$\begin{aligned}
&\rho_t + \nabla \cdot (\rho u) = \varepsilon\, [\rho V(R)]_t\,, \\
&(\rho u)_t + \nabla \cdot [\rho u \otimes u + p(\rho)I] = 0, \\
&RR_{tt} + \frac{3}{2}R_t^2 = \frac{1}{\rho}\,[Q(R) - p(\rho)]\,,
\end{aligned} \tag{3.4.14}$$

The first two equations are mass and momentum balance for the liquid. Bubble oscillations act as a mass source for the liquid. The third equation is the Rayleigh–Plesset equation for an oscillating bubble [202]. This could be rewritten as a first order system, as in the general expansion, but it is simpler not to do so. The constant parameter ε is the number of bubbles per unit volume. We assume the following constitutive relations for the bubble volume $V(R)$, the liquid pressure $p(\rho)$ and the pressure of gas inside the bubble $Q(R)$,

$$V(R) = \frac{4}{3}\pi R^3,$$
$$p(\rho) = \kappa\rho^\gamma,$$
$$Q(R) = \kappa_g R^{-3\gamma_g}.$$

One can begin with more general effective equations than (3.4.14), but this does not alter the form of the final asymptotic equation [257].

The asymptotic equation. We consider a limit in which the volume fraction of the bubbles is small. The dimensionless perturbation parameter is

$$\frac{4}{3}\pi\varepsilon R_0^3 \ll 1,$$

where R_0 is the unperturbed bubble radius. The asymptotic solution for plane waves in one space dimension is

$$\begin{bmatrix} \rho \\ u \\ R \end{bmatrix} = \begin{bmatrix} \rho_0 \\ 0 \\ R_0 \end{bmatrix} + \varepsilon a\left[\varepsilon^{-1}(kx - \omega t), x, t\right] \begin{bmatrix} \rho_0 \\ c_0 \\ 0 \end{bmatrix}$$
$$+ \varepsilon b\left[\varepsilon^{-1}(kx - \omega t), x, t\right] \begin{bmatrix} 0 \\ 0 \\ R_0 \end{bmatrix} + O(\varepsilon^2),$$

as $\varepsilon \to 0$ with $x, t = O(1)$. Here, $\omega = c_0 k$, and

$$c_0^2 = \frac{dp}{d\rho}(\rho_0).$$

The expansion above leads to the following system of equations for $a(\theta, x, t)$ and $b(\theta, x, t)$:

$$a_t + c_0 a_x + \left(\frac{\gamma+1}{4}\omega a^2\right)_\theta + \frac{\omega}{2}b_\theta = 0,$$
$$b_{\theta\theta} + b + a = 0.$$

The normalized form of this system is

$$u_t + \left(\frac{1}{2}u^2\right)_x + v_x = 0, \qquad\qquad (3.4.15)$$
$$v_{xx} + v + u = 0.$$

Eliminating v gives an equation for the normalized sound wave amplitude, u,

$$(\partial_x^2 + 1)\left[u_t + \left(\frac{1}{2}u^2\right)_x\right] = u_x. \qquad (3.4.16)$$

All of the nonlinearity in this limit derives from nonlinearity in the liquid. The bubble dynamics add linear dispersion. Fornberg and Whitham [79] studied the analog of (3.4.16) with the singular operator (∂_x^2+1) replaced by the nonsingular operator $(-\partial_x^2 + 1)$. (The resulting equation differs from the BBM equation (3.3.4) only because the nonlinear term is "inside" the operator instead of outside — this example illustrates the inherent ambiguity in modifications of systematically derived asymptotic equations.)

Traveling waves. For long waves we can use the approximation

$$(1 + \partial_x^2)^{-1} \sim 1 - \partial_x^2,$$

and (3.4.16) reduces to the KdV equation. For short waves, (3.4.16) reduces to the Ostrovsky equation (3.3.86). Equation (3.4.16) has traveling waves of KdV type — families of periodic waves which approach a solitary wave as their wavelength tends to infinity — and of the Ostrovsky equation type — periodic waves which approach a maximum amplitude wave with a corner at a crest or trough [257]. A similar phenomena occurs in [79] and also for the Camassa–Holm equation [42, 43],

$$u_t + 2\kappa u_x + 3uu_x = (u_t + uu_x)_{xx} - \frac{1}{2}\left(u_x^2\right)_x. \qquad (3.4.17)$$

This equation is a completely integrable model equation for shallow water waves, and it has traveling wave solutions, including solitons or "peakons," with corners in their crests.

Compactons. An interesting explicit traveling wave solution of (3.4.16) is the finite pulse,

$$u = \begin{cases} 8/3\cos^2(x/4), & \text{if } |x| \leq 2\pi \\ 0, & \text{if } |x| > 2\pi. \end{cases}$$

This has zero speed with respect to the variables in (3.4.16). In terms of the original space–time variables, this corresponds to a wave propagating at the linearized sound speed. For bubbly fluids, the wave consists of a pressure pulse which compresses the bubbles and then returns them to equilibrium as it propagates through the fluid. This is somewhat reminiscent of self-induced transparency in nonlinear optics.

   Similar compactly supported traveling waves, or "compactons," have been found in [234] for equations with nonlinear KdV type dispersion of the form

$$u_t + (u^m)_x + (u^m)_{xxx} = 0. \qquad (3.4.18)$$

Equations (3.4.16) and (3.4.18) allow compacton solutions for different reasons. In the case of (3.4.16), the dispersive term is lower order, and the equation is hyperbolic. Therefore, the weak discontinuities at the edge of a compacton can propagate without smoothing. Compactly supported solutions of (3.4.18) are possible because, although the dispersion term is higher order, its coefficient vanishes at $u = 0$. A related phenomenon is that of "wetting" fronts for the porous medium equation [12],

$$u_t = (u^m)_{xx}.$$

Modulational stability. Another interesting property of (3.4.16) is that the modulational stability of weakly nonlinear waves depends on their wavenumber: they are modulationally stable at long and short wavelengths and modulationally unstable in the wavenumber band

$$1.451 \approx \sqrt{\frac{7 + \sqrt{97}}{8}} < k < \sqrt{3} \approx 1.732. \qquad (3.4.19)$$

One way to show this fact is to consider small amplitude periodic solutions of (3.4.16),

$$u = \varepsilon A(\xi, \tau) e^{ikx - i\omega t} + \text{c.c.} + O(\varepsilon^2),$$

where $\xi = \varepsilon(x - Ct)$, with $C(k)$ equal to the group velocity, and $\tau = \varepsilon^2 t$. The wave amplitude A solves a nonlinear Schrödinger equation [257],

$$i A_\tau + \beta A_{\xi\xi} + \gamma |A|^2 A = 0,$$

with coefficients

$$\beta = \frac{k(k^2 + 3)}{(k^2 - 1)^3},$$

$$\gamma = -\frac{(k^2 - 1)(4k^4 - 7k^2 - 3)}{6k(k^2 - 3)}.$$

If $\beta\gamma > 0$, then the equation for A the focusing nonlinear Schrödinger equation, and the wave is modulationally unstable. This occurs in the wavenumber band (3.4.19).

Periodic solutions of the KdV and the Ostrovsky equations are always modulationally stable. Thus, one cannot study the modulational instability of weakly dispersive waves using a long or short wave limit. Some other limit, such as the one considered here, has to be used.

Resonant NLSE. Coherent bubble oscillations in bubbly fluid with small volume fraction are described by a resonant NLSE (nonlinear Schrödinger equation) which has the same relation to the usual NLSE that (3.4.16) has to the KdV equation. We look for an asymptotic solutions of (3.4.14) which describes spatially modulated bubble oscillations,

$$R = R_0 + \varepsilon^{1/2} A(x, \varepsilon t) e^{-i\omega_0 t} + \text{c.c.} + O(\varepsilon),$$
$$p = \varepsilon^{3/2} B(x, \varepsilon t) e^{-i\omega_0 t} + \text{c.c.} + O(\varepsilon^2).$$

This ansatz leads to the following normalized equations for the bubble oscillation amplitude $A(x, \tau)$ and pressure oscillation amplitude $B(x, \tau)$:

$$i A_\tau + |A|^2 A + sB = 0,$$
$$B_{xx} + B = A.$$

For bubbly fluids, the non-removable sign s has the value $s = 1$. Eliminating B gives

$$(\partial_x^2 + 1) \left[i A_\tau + |A|^2 A \right] + sA = 0. \qquad (3.4.20)$$

For long waves ($\partial_x^2 \ll 1$), this equation reduces to the usual NLSE.

One interesting property of (3.4.20) is that periodic waves have an amplitude threshold for modulational instability. For example, consider weakly nonlinear bubble oscillations with no spatial modulations,

$$A = A_0 e^{-i\omega t}, \qquad \omega = 1 - |A_0|^2.$$

A linearized stability analysis shows that this wave is modulational stable if $|A_0|^2 < 1/2$ and modulationally unstable if $|A_0|^2 > 1/2$.

Convected waves. The reason for the somewhat unusual structure of the amplitude equation (3.4.20) is that, in the zero volume fraction limit, the linearized dispersion relation of bubbles is $\omega = \omega_0$, where the constant ω_0 is the resonant frequency of bubble oscillations. Hayes [108] refers to waves with linearized dispersion relations of the form

$$\omega = \omega_0 + U \cdot k, \qquad \omega_0 \neq 0,$$

as convected waves. Such waves are nondispersive, since $\nabla_k^2 \omega(k) = 0$, but they are not hyperbolic, since ω is not a homogeneous function of degree one of k. As a result, weakly nonlinear convected waves are not described either by a NLSE or by an inviscid Burgers equation.

In the small volume fraction limit, the bubble oscillations described by (3.4.20) above are weakly dispersive perturbations of convective waves (just as waves described by the KdV equation are weakly dispersive perturbations of hyperbolic waves).

A simple model equation for purely convective waves is the system

$$u_t + uu_x - v = 0,$$
$$v_t + vv_x + u = 0.$$

This system describes the resonant reflection of sound waves off an entropy wave in gas dynamics [192]. Its linearized dispersion relation is $\omega^2 = 1$. The weakly nonlinear expansion,

$$u = \varepsilon A(x, \varepsilon^2 t) e^{-it} + \text{c.c.},$$

leads to the normalized amplitude equation,

$$iA_\tau + \left(|A|^2 A_x\right)_x = 0.$$

Asymptotic equations for general systems with convected waves can be derived in a similar way.

3.4.2.   *Mean Field Interactions*

Nonlinearity often causes a wave to generate a slowly varying mean field. This does not happen for pure conservation laws because they are in divergence form, but mean fields can be generated if source terms are present. The interesting limit for conservation laws occurs when the source term is a rapidly varying, nonlinear function of the densities.

We therefore consider a quasi-linear hyperbolic system of conservation laws with a source term of the form

$$f^\alpha(x, u)_{x^\alpha} = \varepsilon g\left(x, \frac{u}{\varepsilon}\right). \tag{3.4.21}$$

We suppose that

$$f^\alpha(x, 0)_{x^\alpha} = 0,$$
$$g(x, 0) = 0.$$

It follows that $u = 0$ is a solution of (3.4.21). For example, these conditions are satisfied when the coefficients in (3.4.21) are independent of x and $g = 0$ at $u = 0$.

The weakly nonlinear expansion is

$$u = \varepsilon u_1\left(\varepsilon^{-1}\varphi(x), x\right) + \varepsilon^2 u_2\left(\varepsilon^{-1}\varphi(x), x\right) + O(\varepsilon^3), \tag{3.4.22}$$

as $\varepsilon \to 0$ with $x, t = O(1)$. We use (3.4.22) in (3.4.21), Taylor expand, and equate coefficients of ε^0 and ε. Using the same notation as in subsection 3.2.3.2, we obtain that

$$\kappa_\alpha A^\alpha u_{1\theta} = 0, \tag{3.4.23}$$

$$\kappa_\alpha A^\alpha u_{2\theta} + (A^\alpha u_1)_{x^\alpha} + \frac{1}{2}\kappa_\alpha D_u^2 f^\alpha \cdot (u_1, u_1)_\theta = g(x, u_1). \tag{3.4.24}$$

A solution of (3.4.23) is

$$u_1 = \bar{u}(x) + a(\theta, x)r(x).$$

It is now essential to include a mean field $\bar{u}$ in the leading order approximation as well as the wave amplitude a. We choose $\langle a \rangle = 0$ without loss of generality, where $\langle \cdot \rangle$ denotes the mean with respect to the phase variable θ,

$$\langle u \rangle(x) = \lim_{T \to \infty} \frac{1}{T} \int_{T_0}^{T+T_0} u(\theta, x)\, d\theta.$$

We assume that the dependence of u on θ is ergodic, so that these averages are well-defined and are independent of T_0. For example, periodic or almost periodic dependence on the phase is sufficient.

Averaging (3.4.24) with respect to θ, and assuming that u_2 is a sublinear function of θ, gives a mean field equation

$$(A^\alpha \bar{u})_{x^\alpha} = \langle g(x, \bar{u} + ar) \rangle. \tag{3.4.25}$$

If the source term is a nonlinear function of u, the mean field is driven by averages of the wave amplitude.

Subtracting (3.4.25) from (3.4.24) and taking the scalar product of the result with the left null vector l gives an equation for the wave amplitude

$$C^\alpha a_{x^\alpha} + \left(\frac{1}{2}\mathcal{G}a^2 + \mathcal{V}a\right)_\theta + \mathcal{H}a = g(x, \bar{u} + ar) - \langle g(x, \bar{u} + ar) \rangle, \tag{3.4.26}$$

where C^α, $\mathcal{G}$, and $\mathcal{H}$ are given in (3.2.42), and

$$\mathcal{V} = l \cdot k_\alpha D_u^2 f^\alpha(0) \cdot (\bar{u}, r).$$

The wave amplitude in (3.4.26) is driven by a zero mean source term depending on the mean field. The mean field also leads to an order ε shift in the characteristic velocity which is described by the advection term $\mathcal{V}a_\theta$.

The coupled equations (3.4.25) and (3.4.26) cannot be solved analytically except in very simple special cases. For oscillatory waves there is therefore a complex coupling between wave and mean field. In the case of pulses — where a is a compactly supported function of θ — we have $\langle f(a) \rangle = f(0)$ for any function $f(a)$, so the wave mean field interaction disappears.

An important physical application of these expansions is to combustion problems. For a survey of the applications of weakly nonlinear geometrical optics in combustion, including mean-field interactions, see [52], which gives further references. In particular, [6] studies the combined effects of wave-wave and wave-mean field interactions in initiating detonation.

Next, we consider some examples of wave-mean field interactions.

3.4.2.1. Semi-linear systems. Consider a hyperbolic system of semi-linear PDEs,

$$(A^\alpha(x)v)_{x^\alpha} = g(x, v), \tag{3.4.27}$$

where $g(x, 0) = 0$. The change of variables $u = \varepsilon v$ puts (3.4.27) in the form (3.4.21) with $f^\alpha(x, u) = A^\alpha(x)u$. Thus a special case of the above expansion gives large amplitude, high-frequency solutions of (3.4.27). The asymptotic solution is

$$v = \bar{v}(x) + a(\theta, x)r(x) + O(\varepsilon),$$

where

$$(A^\alpha \bar{v})_{x^\alpha} = \langle g(x, \bar{v} + ar) \rangle,$$
$$C^\alpha a_{x^\alpha} + \mathcal{H}a = g(x, \bar{u} + ar) - \langle g(x, \bar{u} + ar) \rangle.$$

Carlemann equations. A simple example of a semi-linear system is the Carlemann system of equations, which is a model of the Boltzmann equation,

$$u_t + u_x + u^2 - v^2 = 0,$$
$$v_t - v_x - u^2 + v^2 = 0.$$

The generation of a mean field by high frequency oscillations for this equation was analyzed by McLaughlin, Papanicolaou, and Tartar [199].

In this problem, it is simple to include waves propagating in both directions. The asymptotic solution is

$$\begin{bmatrix} u \\ v \end{bmatrix} = \begin{bmatrix} \bar{u}(x, t) \\ \bar{v}(x, t) \end{bmatrix} + a\left(\frac{x-t}{\varepsilon}, x, t\right) \begin{bmatrix} 1 \\ 0 \end{bmatrix} + b\left(\frac{x+t}{\varepsilon}, x, t\right) \begin{bmatrix} 0 \\ 1 \end{bmatrix} + O(\varepsilon^2).$$

The equation for the mean field is

$$\bar{u}_t + \bar{u}_x + \bar{u}^2 - \bar{v}^2 + \langle a^2 \rangle - \langle b^2 \rangle = 0,$$
$$\bar{v}_t - \bar{v}_x - \bar{u}^2 + \bar{v}^2 - \langle a^2 \rangle + \langle b^2 \rangle = 0. \tag{3.4.28}$$

The equations for the wave amplitudes are

$$a_t + a_x + 2\bar{u}a + a^2 - \langle a^2 \rangle = 0,$$

$$b_t - b_x + 2\bar{v}b + b^2 - \langle b^2 \rangle = 0.$$

We can obtain a "homogenized" [23] system of equations for the mean field $\bar{u}$, $\bar{v}$, and the moments of the wave amplitudes $\langle a^n \rangle$, $\langle b^n \rangle$ ($n = 2, 3, ...$). Multiplying the equations for the wave amplitudes by the wave amplitudes and averaging shows that

$$\langle a^n \rangle_t + \langle a^n \rangle_x + 2n\bar{u}\langle a^n \rangle + n\langle a^{n+1} \rangle - n\langle a^2 \rangle\langle a^{n-1} \rangle = 0,$$

$$\langle b^n \rangle_t - \langle b^n \rangle_x + 2n\bar{v}\langle b^n \rangle + n\langle b^{n+1} \rangle - n\langle b^2 \rangle\langle b^{n-1} \rangle = 0. \qquad (3.4.29)$$

Equations (3.4.28) and (3.4.29) determine the evolution of the mean field.

For three by three, or higher order systems resonant wave interactions occur between waves in different characteristic families, and this greatly complicates the dynamics (see Subsection 3.6.1). In particular, the effect of the waves on the mean field typically depends on phase correlations between the interacting high frequency waves. When this happens, the "microscopic" wave dynamics cannot be averaged out to obtain "macroscopic" mean field equations.

An extension of this kind of expansion applies to linearly degenerate wave-fields for quasilinear hyperbolic equations [72, 242, 244].

3.4.2.2. Gravitational waves. An interesting example of wave-mean field interactions of hyperbolic waves occurs in general relativity [53, 130]. The average energy density of a rapidly oscillating gravitational wave causes a mean background curvature in space time. Gravitational waves are linearly degenerate, since in a suitably defined frame of reference they propagate at the speed of light independently of their amplitude. Thus, in contrast with gas dynamics, where the nonlinear self-interaction of a sound wave is the dominant nonlinear process and mean field interactions are negligible, in general relativity the self-interaction of a gravitational wave is negligible and the mean field interaction is dominant.

A detailed description of asymptotic equations for gravitational waves is given in [8]. For an introduction to rigorous work on general relativity, and other classical field theories, see [155].

In a vacuum, the Einstein field equations are

$$R_{\alpha\beta} = 0, \qquad (3.4.30)$$

where $R_{\alpha\beta}$ is the Ricci tensor which is defined in terms of the (symmetric) metric tensor $g_{\alpha\beta}$ and the connection coefficients $\Gamma^{\alpha}_{\beta\gamma}$ by

$$R_{\alpha\beta} = \frac{\partial \Gamma^{\lambda}_{\alpha\beta}}{\partial x^{\lambda}} - \frac{\partial \Gamma^{\lambda}_{\beta\lambda}}{\partial x^{\alpha}} + \Gamma^{\lambda}_{\alpha\beta}\Gamma^{\mu}_{\lambda\mu} - \Gamma^{\mu}_{\alpha\lambda}\Gamma^{\lambda}_{\beta\mu},$$

$$\Gamma^{\alpha}_{\beta\gamma} = \frac{1}{2}g^{\alpha\lambda}\left[\frac{\partial g_{\beta\lambda}}{\partial x^{\gamma}} + \frac{\partial g_{\gamma\lambda}}{\partial x^{\beta}} - \frac{\partial g_{\beta\gamma}}{\partial x^{\lambda}}\right].$$

The field equations (3.4.30) are invariant under gauge transformations,

$$\tilde{x}^\alpha = \tilde{x}^\alpha(x),$$

$$\tilde{g}_{\alpha\beta} = \frac{\partial x^\lambda}{\partial \tilde{x}^\alpha} \frac{\partial x^\mu}{\partial \tilde{x}^\beta} g_{\lambda\mu}.$$

The solutions $g_{\alpha\beta}$ and $\tilde{g}_{\alpha\beta}$ are physically equivalent. This gauge invariance introduces completely different features from gas dynamics. If a solution of (3.4.30) develops a singularity, it may be possible to continue the solution past the singularity by making an appropriate change of coordinates; in that case the singularity does not correspond to an intrinsic singularity in space–time, but simply to a coordinate singularity.

When the gauge is fixed in an appropriate way, the Einstein field equations (3.4.30) are a second order hyperbolic system for the metric tensor $g_{\alpha\beta}(x)$.

We look for an asymptotic solution of (3.4.30) of the form

$$g_{\alpha\beta} = \bar{g}_{\alpha\beta}(x) + \varepsilon h_{\alpha\beta} \left(\varepsilon^{-1}\varphi(x), x\right) + O(\varepsilon^2).$$

Here, $\bar{g}(x)$ is a slowly varying mean gravitational field and $h(\theta, x)$ is the perturbation due to a gravitational wave.

The eikonal equation is

$$\kappa^\alpha \kappa_\alpha = 0,$$

where $\kappa_\alpha = \varphi_{x^\alpha}$. We raise and lower indices using the mean metric tensor $\bar{g}^{\alpha\beta}$, so that

$$\kappa^\alpha = \bar{g}^{\alpha\beta} \kappa_\beta.$$

The wave amplitude h satisfies the algebraic constraint

$$\frac{1}{2} h^\lambda_\lambda \kappa_\alpha - h_{\alpha\lambda} \kappa^\lambda = 0.$$

There are several unknown components in h since the Einstein field equations are non-strictly hyperbolic; physically this corresponds to the fact that gravitational waves are polarized.

It is convenient to use adapted coordinates, in which $(\kappa_\alpha) = (1, 0, 0, 0)$. We sum latin indices $i, j, k, \ldots$ over $1, 2, 3$. The first order perturbation in the metric tensor then satisfies

$$h^{00} = 0,$$

$$\kappa^j h_{ij} = 0.$$

In addition, after a suitable gauge transformation, we can set

$$h_{0\alpha} = 0.$$

In adapted coordinates, the transport equation for the wave amplitude h_{ij} is

$$\kappa^k \left(\partial_{x^k} h_{ij} - \bar{\Gamma}^l_{ik} h_{lj} - \bar{\Gamma}^l_{jk} h_{li}\right) + \frac{1}{2} h_{ij} \left(\partial_{x^\lambda} \kappa^\lambda + \bar{\Gamma}^\mu_{\mu\lambda} \kappa^\lambda\right) = 0.$$

where the dot denotes a derivative with respect to the phase variable θ. We use $\bar{R}_{\alpha\beta}$ and $\bar{\Gamma}^{\alpha}_{\beta\gamma}$ to denote the Ricci tensor and connection coefficients evaluated at the mean field, $\bar{g}$.

The equation for the mean field is

$$\bar{R}_{\alpha\beta} = \tau \kappa_{\alpha}\kappa_{\beta}. \tag{3.4.31}$$

Here, the scalar function $\tau(x)$ is an average of the gravitational wave amplitude,

$$\tau = -\frac{1}{4}\langle \dot{h}^{\alpha\beta}\dot{h}_{\alpha\beta}\rangle$$

$$= -\lim_{T\to\infty}\frac{1}{4T}\int_{0}^{T}\dot{h}^{\alpha\beta}\dot{h}_{\alpha\beta}\,d\theta.$$

Equation (3.4.31) shows that the mean energy density of the gravitational wave generates a slowly varying gravitational field. The wave is then refracted by this mean field.

3.4.3. Hyperbolic Variational Principles

An interesting class of nonlinear hyperbolic PDEs is given by variational principles of the form

$$\delta \int A^{\alpha\beta}_{pq}(x,u)\frac{\partial u^p}{\partial x^\alpha}\frac{\partial u^q}{\partial x^\beta}\,dx = 0, \tag{3.4.32}$$

where we use the summation convention. Here, $x \in R^{n+1}$ are the space–time variables, and $u : R^{n+1} \to R^m$ are the dependent variables. Without loss of generality, we assume that

$$A^{\alpha\beta}_{pq} = A^{\beta\alpha}_{pq} = A^{\alpha\beta}_{qp}.$$

One physical system which gives rise to this type of equation is wave propagation in a massive nematic liquid crystal director field [122].

The Euler–Lagrange equations of (3.4.32) are

$$\frac{\partial}{\partial x^\alpha}\left(A^{\alpha\beta}_{kp}\frac{\partial u^p}{\partial x^\beta}\right) = \frac{1}{2}\frac{\partial A^{\alpha\beta}_{pq}}{\partial u^k}\frac{\partial u^p}{\partial x^\alpha}\frac{\partial u^q}{\partial x^\beta}. \tag{3.4.33}$$

We assume that this equation is strictly hyperbolic. Equation (3.4.33) is scale invariant since the Lagrangian density in (3.4.32) is a homogeneous function of ∇u. The waves are therefore nondispersive.

The nonlinearity in (3.4.33) is somewhat unusual. The equation is not semi-linear because the coefficients of the second order derivatives depend on u. However, it does not have the usual quasi-linear form (of nonlinear elasticity, for example), since the coefficients are independent of ∇u.

We use an asymptotic expansion of exactly the same form as the one used in Subsection 3.2.3.2 for a first order system of quasi-linear hyperbolic conservation laws [122]. This is a different expansion, because here we are applying it to a second order system. We will obtain a different asymptotic equation for the wave amplitude as well as wave-mean field interactions.

We therefore consider asymptotic solutions of the form

$$u(x;\varepsilon) = u_0(x) + \varepsilon u_1\left(x, \varepsilon^{-1}\varphi(x)\right) + \varepsilon^2 u_2\left(x, \varepsilon^{-1}\varphi(x)\right) + O(\varepsilon^3). \quad (3.4.34)$$

In (3.4.34), $\varphi(x)$ is a phase variable, and we introduce the wavenumber

$$\kappa_\alpha = \varphi_{x^\alpha},$$

as usual. We will see that, because of mean field interactions, u_0 is *not* a solution of (3.4.33).

Using (3.4.34) in (3.4.33), expanding the coefficients in Taylor series, and equating coefficients of ε^{-1} and ε^0 gives the equations

$$\kappa_\alpha \kappa_\beta A_{kp}^{\alpha\beta} \frac{\partial^2 u_1^p}{\partial\theta^2} = 0, \quad (3.4.35)$$

$$\kappa_\alpha \kappa_\beta A_{kp}^{\alpha\beta} \frac{\partial^2 u_2^p}{\partial\theta^2} + 2\kappa_\alpha A_{kp}^{\alpha\beta} \frac{\partial^2 u_1^p}{\partial\theta\partial x^\beta} + \frac{\partial}{\partial x^\beta}\left(\kappa_\alpha A_{kp}^{\alpha\beta}\right) \frac{\partial u_1^p}{\partial\theta}$$

$$+\kappa_\alpha\kappa_\beta \frac{\partial A_{kp}^{\alpha\beta}}{\partial u^q} \frac{\partial}{\partial\theta}\left(u_1^q \frac{\partial u_1^p}{\partial\theta}\right) + \frac{\partial}{\partial x^\alpha}\left(A_{kp}^{\alpha\beta} \frac{\partial u_0^p}{\partial x^\beta}\right) \quad (3.4.36)$$

$$= \frac{1}{2}\kappa_\alpha\kappa_\beta \frac{\partial A_{pq}^{\alpha\beta}}{\partial u^k} \frac{\partial u_1^p}{\partial\theta} \frac{\partial u_1^q}{\partial\theta} + \kappa_\alpha \frac{\partial A_{pq}^{\alpha\beta}}{\partial u^k} \frac{\partial u_1^p}{\partial\theta} \frac{\partial u_0^q}{\partial x^\beta} + \frac{1}{2} \frac{\partial A_{pq}^{\alpha\beta}}{\partial u^k} \frac{\partial u_0^p}{\partial x^\alpha} \frac{\partial u_0^q}{\partial x^\beta}.$$

Here, and below, all coefficients and their derivatives are evaluated at $u = u_0$.

Equation (3.4.35) implies that the wavenumber vector κ satisfies the local dispersion relation,

$$\det\left(\kappa_\alpha \kappa_\beta A_{pq}^{\alpha\beta}\right) = 0, \quad (3.4.37)$$

and the phase $\varphi(x)$ satisfies the corresponding eikonal equation.

By strict hyperbolicity, there is a one-dimensional nullspace associated with κ. A solution of (3.4.35) for u_1 is

$$u_1(\theta, x) = \bar{u}_1(x) + a(\theta, x)r(x), \quad (3.4.38)$$

where $\bar{u}_1$ is a perturbation to the mean field, which does not enter into the final equations, a is a scalar wave amplitude, and

$$r(x) = (r^1, \dots, r^m) \in R^m$$

is a null vector such that

$$\kappa_\alpha\kappa_\beta A_{kp}^{\alpha\beta} r^p = 0.$$

The equation for u_0 follows on averaging (3.4.36) with respect to θ, with the assumption that u_2 grows sublinearly in θ. The result is

$$\frac{\partial}{\partial x^\alpha}\left(A_{kp}^{\alpha\beta} \frac{\partial u_0^p}{\partial x^\beta}\right) = \frac{1}{2} \frac{\partial A_{pq}^{\alpha\beta}}{\partial u^k} \frac{\partial u_0^p}{\partial x^\alpha} \frac{\partial u_0^q}{\partial x^\beta} + \frac{1}{2}\kappa_\alpha\kappa_\beta \frac{\partial A_{pq}^{\alpha\beta}}{\partial u^k} r^p r^q \left\langle\left(\frac{\partial a}{\partial\theta}\right)^2\right\rangle.$$

Here, the angular brackets denote the average with respect to θ. This equation is the same as the original equation with an additional forcing term proportional to $\langle a_\theta^2 \rangle$.

The second solvability condition for (3.4.36) is obtained by taking the inner product with r. After subtracting out the mean, this gives

$$\frac{\partial}{\partial \theta}\left(\frac{\partial a}{\partial s} + \Gamma a \frac{\partial a}{\partial \theta} + \Lambda a \right) = \frac{1}{2}\Gamma\left\{ \left(\frac{\partial a}{\partial \theta}\right)^2 - \left\langle \left(\frac{\partial a}{\partial \theta}\right)^2 \right\rangle \right\}. \qquad (3.4.39)$$

In (3.4.39),

$$\frac{\partial}{\partial s} = C^\alpha(x)\frac{\partial}{\partial x^\alpha}$$

is a derivative along the rays associated with φ, the coefficient $\Gamma(x) \in R$ measures the strength of quadratically nonlinear effects, and $\Lambda(x) \in R$ describes the effects of focusing and nonuniformities on the wave. The coefficients are given explicitly by

$$
\begin{aligned}
C^\alpha &= 2\kappa_\beta A_{pq}^{\alpha\beta} r^p r^q \\
\Gamma &= \kappa_\alpha \kappa_\beta \frac{\partial A_{pq}^{\alpha\beta}}{\partial u^k} r^k r^p r^q, \\
\Lambda &= \frac{\partial}{\partial x^\alpha}\left(\kappa_\beta A_{pq}^{\alpha\beta} r^p r^q \right) - \kappa_\alpha \frac{\partial A_{pq}^{\alpha\beta}}{\partial u^k}\frac{\partial u_0^q}{\partial x^\beta} r^k r^p.
\end{aligned}
\qquad (3.4.40)
$$

To normalize (3.4.39), we define

$$E(s) = \exp\left(\int \Lambda(s)\,ds \right).$$

Then the change of variables

$$u(x,t) = E(s)a(\theta, s), \qquad t(s) = \int \frac{\Gamma(s)}{E(s)}\,ds, \qquad x = \theta,$$

transforms (3.4.39) into

$$(u_t + uu_x)_x = \frac{1}{2}\left(u_x^2 - \langle u_x^2 \rangle \right). \qquad (3.4.41)$$

In this equation, x is a normalized phase variable (not the original space–time variables), and angular brackets denote the average with respect to x. Equation (3.4.41) is the high-frequency limit of the Camassa–Holm equation (3.4.17).

For solutions in which u_x is compactly supported the mean field interaction drops out. The asymptotic equation for the wave amplitude is then

$$(u_t + uu_x)_x = \frac{1}{2}u_x^2.$$

This is a very interesting equation in itself. Calogero [40] observed that this type of equation can be solved exactly by a linearizing transformation. It can also be solved directly by the method of characteristics. Moreover, the equation is a completely integrable bi-Hamiltonian system [126]. Smooth solutions break down in finite time, but the equation has global, continuous weak solutions [125].

3.4.3.1. A wave equation with nonlinear wave speed. The simplest example of the hyperbolic variational equations considered above is the following variational principle for a scalar function $u(x, t)$ in $(1 + 1)$-space–time dimensions:

$$\delta \int \left\{ \frac{1}{2} u_t^2 - \frac{1}{2} c^2(u) u_x^2 \right\} \, dx \, dt = 0.$$

The associated Euler–Lagrange equation is a wave equation with a nonlinear wave speed,

$$u_{tt} = c(u) \left(c(u) u_x \right)_x.$$

This equation arises as a special case for waves in a massive director field [122]. Asymptotic equations for wave-mean field interactions for this wave equation were derived in [153].

We expand solutions as

$$u(x, t; \varepsilon) = u_0(x, t) + \varepsilon u_1(x, t, \varepsilon^{-1} \varphi(x, t)) + O(\varepsilon^2).$$

At order ε^{-1} we find that φ solves the eikonal equation,

$$\varphi_t^2 = c_0^2 \varphi_x^2,$$

where $c_0 = c(u_0)$. At order one, we find the equation

$$
\begin{aligned}
u_{0tt} - c_0 \left(c_0 u_{0x} \right)_x - 2\omega u_{1\theta t} - \omega_t u_{1\theta} \\
= 2c_0^2 k u_{1\theta x} + c_0^2 k_x u_{1\theta} + 2c_0 c_0' k u_{0x} u_{1\theta} \\
+ 2k^2 c_0 c_0' \left[(u_1 u_{1\theta})_\theta - \frac{1}{2} u_{1\theta}^2 \right],
\end{aligned}
$$

where $k = \varphi_x$, $\omega = -\varphi_t$, and

$$c_0' = \frac{dc}{du}(u_0).$$

Averaging this equation with respect to θ implies that

$$u_{0tt} = c_0 \left(c_0 u_{0x} \right)_x - k^2 c_0 c_0' \langle u_{1\theta}^2 \rangle.$$

Here, angular brackets denote averages with respect to θ. Assuming that $\omega = c_0 k$, the remaining equation for u_1 is

$$
\left(u_{1t} + c_0 u_{1x} + k c_0' u_1 u_{1\theta} + \left\{ \frac{\omega_t + c_0^2 k_x}{2 c_0 k} + c_0' u_{0x} \right\} u_1 \right)_\theta \\
= \frac{1}{2} k c_0' \left\{ u_{1\theta}^2 - \langle u_{1\theta}^2 \rangle \right\}.
$$

The simplest case is when there are no spatial modulations, so that $u_0 = u_0(t)$ and $\varphi = x - \psi(t)$, with $k = 1$ and $\omega = c_0$. Then these equations reduce to

$$u_{0tt} + c_0 c_0' \langle u_{1\theta}^2 \rangle = 0, \tag{3.4.42}$$

$$\left(u_{1t} + c_0' u_1 u_{1\theta} + \frac{c_{0t}}{2 c_0} u_1 \right)_\theta = \frac{1}{2} c_0' \left\{ u_{1\theta}^2 - \langle u_{1\theta}^2 \rangle \right\}. \tag{3.4.43}$$

For smooth solutions, it follows from (3.4.43) that

$$\frac{1}{2}c_0 \langle u_{1\theta}^2 \rangle = E = \text{constant}.$$

Using this in (3.4.42) shows that the mean field $u_0(t)$ satisfies

$$u_{0tt} + V'(u_0) = 0,$$

where

$$V(u_0) = Ec(u_0).$$

The mean field therefore evolves like a particle moving in a potential proportional to c. The strength of the potential is proportional to the energy in the wave field.

Neu has derived similar mean field equations for nonlinear Klein–Gordon equations [206].

3.5.   DIFFRACTION

All the asymptotic equations derived in previous sections involve only one space dimension. Although geometrical optics applies to multidimensional problems, it is based on the assumption that locally the wave can be approximated by a plane wave. In many circumstances, such as focusing, this quasi-one-dimensional approximation breaks down and diffraction effects become of crucial importance.

The two-dimensional inviscid Burgers equation. One would like to have a nonlinear theory analogous to the Geometrical Theory of Diffraction for linear geometrical optics [150], but nonlinear hyperbolic wave diffraction seems to be a very hard problem. The simplest canonical equation is the following two-dimensional generalization of the inviscid Burgers equation,

$$\left[u_t + \left(\frac{1}{2}u^2 \right)_x \right]_x + u_{yy} = 0. \tag{3.5.1}$$

Equation (3.5.1) can also be written as a system,

$$u_t + \left(\frac{1}{2}u^2 \right)_x + v_y = 0,$$
$$u_y - v_x = 0. \tag{3.5.2}$$

This equation is almost intractable from an analytical point of view. Very few exact solutions containing shocks are known, and global existence of weak solutions is a completely open question. Short time existence for smooth solutions is shown in [171].

Equation (3.5.1) captures some of the essential difficulties of hyperbolic systems of conservation laws in several space dimensions — a subject which has resisted fundamental analysis for the last thirty years.

Since (3.5.1) is a universal asymptotic equation, it has been derived in many different contexts under many different names: in transonic flow, it is called the unsteady transonic small disturbance equation [57, 58, 265]; in the diffraction of nonlinear acoustic beams it is called the Zaboltskaya–Khokhlov equation [276]. It arises in the focusing of nearly plane shock waves [45, 62, 254, 255]; in the analysis of singular rays [106, 110, 277]; in describing the transition from regular to irregular reflection for weak shocks [37, 114, 254, 255]; and in weakly nonlinear hyperbolic waves at a cusped caustic [111]. For a systematic treatment of weakly nonlinear hyperbolic waves at a smooth convex caustic, see [119].

Dissipation and dispersion. There are also dissipative and dispersive versions of (3.5.1). The simplest examples are the two-dimensional Burgers equation [169],

$$(u_t + uu_x - \mu u_{xx})_x + u_{yy} = 0, \ .$$

and the two-dimensional KdV, or Kadomtsev–Petviashvilli (KP) equation [146],

$$(u_t + uu_x + \nu u_{xxx})_x + u_{yy} = 0.$$

The KP-equation is one of the few known completely integrable equations in two space variables — see [2] for the theory of completely integrable $(2 + 1)$-equations and for references to the enormous literature on this topic.

The original motivation for deriving the KP-equation was to study the stability of KdV solitons to long transverse perturbations. They are stable for $\nu = +1$ (the KPII equation) and unstable for $\nu = -1$ (the KPI equation).

Experimental comparisons between observations of shallow water waves solutions of the KP equation show excellent agreement [104]. Numerical solutions of the KP equation appear in [149, 270]. For global existence theory, see [31, 238] and the references given there.

Paradoxically, (3.5.1) is in some ways harder to study than the KP-equation which has an additional dispersive term. In this context it should be noted that, once shocks form, equation (3.5.1) is not the zero-dispersion limit of the KP-equation: the formation of oscillations in the KP-solution prevents strong convergence as $\nu \to 0$. For an analysis of the integrability properties of the dispersionless KP equation, see [158, 159].

It is presumably true that (3.5.1) is the zero viscosity limit, $\mu \to 0+$, of the two-dimensional Burgers equation. Unfortunately, there is no two-dimensional analog of the Cole–Hopf transformation, and the two-dimensional Burgers equation does not seem to be integrable [55].

More generally, introducing several transverse variables leads, in the plane wave case, to an equation of the form

$$\left[u_t + \left(\frac{1}{2}u^2\right)_x\right]_x + \mathcal{D}^{\alpha\beta}u_{y^\alpha y^\beta} = 0.$$

For isotropic wave motions,

$$\mathcal{D}^{\alpha\beta}u_{y^\alpha y^\beta} = \Delta_y u,$$

is the transverse Laplacian. This form is used to describe the diffraction of nonlinear sound beams, often assuming cylindrical symmetry in the transverse directions [16, 235].

3.5.1. The Parabolic Approximation

We begin with a heuristic discussion of the parabolic approximation for the linear wave equation to explain the structure of the terms which appear in (3.5.1). The parabolic approximation is originally due to Leontovich and Fock [176]. For additional information about the linear parabolic approximation, see [15, 258].

We consider solutions of the wave equation,

$$\frac{\partial^2 u}{\partial t^2} = c_0^2 \left(\frac{\partial^2 u}{\partial x^2} + \frac{\partial^2 u}{\partial y^2} \right),$$

which are nearly unidirectional and propagate in the positive x-direction. We rewrite the wave equation as

$$\left(\frac{\partial}{\partial t} - c_0 \frac{\partial}{\partial x} \right) \left(\frac{\partial}{\partial t} + c_0 \frac{\partial}{\partial x} \right) u = c_0^2 \frac{\partial^2 u}{\partial y^2}.$$

By the unidirectional assumption, $u_t \approx -c_0 u_x$, so we can approximate this equation by

$$-2c_0 \frac{\partial}{\partial x} \left(\frac{\partial}{\partial t} + c_0 \frac{\partial}{\partial x} \right) u = c_0^2 \frac{\partial^2 u}{\partial y^2}.$$

After a Galilean transformation, $x \to x - c_0 t$, and a rescaling, $y \to (2/c_0)^{1/2} y$, this equation becomes

$$u_{xt} + u_{yy} = 0. \tag{3.5.3}$$

Equation (3.5.3) is the linearization of (3.5.1) at $u = 0$. The nonlinear equation is derived by including weakly nonlinear effects in this parabolic approximation and choosing the scaling so that the effects of nonlinearity and diffraction balance.

Another way to deduce the form of the terms in (3.5.3) heuristically is to expand the linearized dispersion relation of the wave motion. This approach is illustrated in some of the examples below.

Looking for exponential solutions, $u = A(y, t) \exp(ikx)$, of (3.5.3) leads to a Schrödinger equation for A,

$$ikA_t + A_{yy} = 0.$$

This is the usual form of the parabolic equation for linear waves, but the reduction to exponential solutions is not valid for weakly nonlinear hyperbolic waves.[5]

Equation (3.5.1) is a hyperbolic equation in which (x, t) are characteristic coordinates. Thus, $t = 0$ is a characteristic surface, and one has to be careful in formulating supplementary conditions. A reasonable initial-boundary value problem for (3.5.1) is

$$u(x, y, 0) = u_0(x, y),$$
$$v(x, y, t) \to 0 \quad \text{as} \quad x \to +\infty. \tag{3.5.4}$$

---

[5] For weakly nonlinear dispersive waves, solutions are locally sinusoidal, and the weakly nonlinear version of the parabolic equation is the nonlinear Schrödinger equation.

In addition, we need an entropy condition to pick out admissible shocks. For example, we can require the following entropy inequality to hold in the sense of distributions,

$$\left(u^2\right)_t + \left(\frac{2}{3}u^3 - v^2\right)_x + (2uv)_y \leq 0.$$

When solving (3.5.1) forward in time, it is essential to impose the boundary condition as $x \to +\infty$, since the linearized equation (3.5.3) has Fourier solutions which decay exponentially as $x \to -\infty$ but which grow arbitrarily fast in t,

$$u = e^{k^2 t + x + iky}.$$

The reverse time final value problem requires boundary conditions as $x \to +\infty$.

If we assume that $v \to 0$ as $x \to +\infty$, then equation (3.5.1) can be written as a nonlocal evolution equation for $u(x, y, t)$,

$$u_t + \left(\frac{1}{2}u^2\right)_x = \int_x^{+\infty} u_{yy}(x', y, t)\, dx'.$$

The nonlocality in this equation has a simple physical explanation. We use dimensionless variables in which the wavelength and wave period are of the order one. Equation (3.5.1) describes the long time behavior of a solution in a reference frame moving with a wave propagating in the x-direction. The velocity of oblique waves differs from the velocity of the reference frame by an order one amount. Therefore, these waves can propagate information arbitrarily large distances as the perturbation parameter tends to zero and the time scale of validity of the expansion tends to infinity. This leads to the nonlocal effects in the asymptotic equation.

We will concentrate on deriving equation (3.5.1) for hyperbolic systems. Dispersive and dissipative versions follow by adding transverse variables to the one dimensional expansions in a very straightforward way, with diffraction adding linearly onto any other effects. Thus, if the one dimensional asymptotic equation is

$$u_t + \left(\frac{1}{2}\mathcal{G}u^2\right)_x + \mathcal{H}u + \mathcal{L}[u] = 0,$$

where $\mathcal{L}$ is an operator in x, the weakly diffractive version is typically

$$\left[u_t + \left(\frac{1}{2}\mathcal{G}u^2\right)_x + \mathcal{H}u + \mathcal{L}[u]\right]_x + \mathcal{J}u_{yy} = 0.$$

For an exceptional case, see the Zakharov–Kuznetsov equation discussed in Subsection 3.5.2.2.

3.5.2. *The Two-Dimensional Burgers Equation*

In this subsection we give a formal derivation of (3.5.1), following [110]. We consider a hyperbolic system of conservation laws,

$$f^\alpha(x, u)_{x^\alpha} + g(x, u) = 0. \tag{3.5.5}$$

We look for an asymptotic solution depending on three different scales, a "fast" phase, θ, "slow" space–time variables x, and an "intermediate" transverse variable, η:

$$
\begin{aligned}
u(x;\varepsilon) &= u_0(x) + \varepsilon u_1(\theta,\eta,x) + \varepsilon^{3/2} u_2(\theta,\eta,x) \\
&\quad + \varepsilon^2 u_3(\theta,\eta,x) + O(\varepsilon^{5/2}), \\
\theta &= \varepsilon^{-1}\varphi(x), \qquad \eta = \varepsilon^{-1/2}\psi(x),
\end{aligned}
\tag{3.5.6}
$$

where $\varepsilon \to 0$ with $x = O(1)$. Here u_0 is a smooth solution of (3.5.5). Using (3.5.6) in (3.5.5) and expanding gives

$$
\kappa_\alpha A^\alpha u_{1\theta} = 0,
\tag{3.5.7}
$$

$$
\kappa_\alpha A^\alpha u_{2\theta} + \lambda_\alpha A^\alpha u_{1\eta} = 0,
\tag{3.5.8}
$$

$$
\begin{aligned}
&\kappa_\alpha A^\alpha u_{3\theta} + \lambda_\alpha A^\alpha u_{2\eta} \\
&\quad + (A^\alpha u_1)_{x^\alpha} + \frac{1}{2}\kappa_\alpha D^2 f^\alpha \cdot (u_1,u_1)_\theta + G u_1 = 0.
\end{aligned}
\tag{3.5.9}
$$

Here, we write

$$
\psi_{x^\alpha} = \lambda_\alpha,
$$

and the other notation is the same as in Subsection 3.2.3.2. Equation (3.5.7) implies that κ satisfies the local dispersion relation (3.2.39). A solution for u_1 is

$$
u_1 = a(\theta,\eta,x)r(x).
$$

The solvability condition for (3.5.8) is obtained by taking the scalar product with respect to the left null vector l. If $u_{1\eta} \neq 0$, this implies that

$$
l \cdot \lambda_\alpha A^\alpha r = 0.
\tag{3.5.10}
$$

Using (3.2.42), this equation can be written in the equivalent form

$$
C^\alpha \psi_{x^\alpha} = 0.
$$

In other words,

$$
\psi \text{ is constant along the rays associated with } \varphi.
$$

A solution of (3.5.8) for u_2 is then

$$
u_2 = b(\theta,\eta,x)s(x),
$$

where the scalar amplitude b satisfies

$$
a_\eta - b_\theta = 0,
\tag{3.5.11}
$$

and the vector s is a solution of

$$
\kappa_\alpha A^\alpha s + \lambda_\alpha A^\alpha r = 0.
$$

This equation has a solution even though $\kappa_\alpha A^\alpha$ is singular, because λ satisfies (3.5.10). We can add an arbitrary homogeneous solution, $c(\theta,\eta,x)r + \bar{u}_2(\eta,x)$, to u_2. Since this does not enter into the final result, we omit it.

The solvability condition for (3.5.9) is

$$C^\alpha a_{x^\alpha} + \left(\frac{1}{2}\mathcal{G}a^2\right)_\theta + \mathcal{H}a + \mathcal{J}b_\eta = 0, \tag{3.5.12}$$

where

$$\mathcal{J} = l \cdot \lambda_\alpha A^\alpha s. \tag{3.5.13}$$

Equations (3.5.11) and (3.5.12) form a system of equations for a and b. The coefficients may depend on s but are independent of θ and η.

To write these equations in normalized form, we assume that $\mathcal{G} \neq 0$, and define new variables,

$$\bar{u}(\bar{x}, \bar{y}, \bar{t}) = \mathcal{E}(s)a(\theta, \eta, s),$$
$$\bar{v}(\bar{x}, \bar{y}, \bar{t}) = \mathcal{E}(s)b(\theta, \eta, s),$$
$$\bar{x} = \theta, \qquad \bar{y} = \eta, \qquad \bar{t} = \mathcal{T}(s),$$
$$\alpha(\bar{t}) = \frac{\mathcal{E}^2(s)}{\mathcal{G}(s)}.$$

Here $\mathcal{E}$ and $\mathcal{T}$ are given in (3.2.51). The transformed equations are then

$$\bar{u}_{\bar{t}} + \left(\frac{1}{2}\bar{u}^2\right)_{\bar{x}} + \alpha(\bar{t})\bar{v}_{\bar{y}} = 0,$$
$$\bar{u}_{\bar{y}} - \bar{v}_{\bar{x}} = 0. \tag{3.5.14}$$

Several transverse variables. A slight generalization of this expansion is to use several transverse variables and introduce a mean field. We write

$$\eta = (\eta^1, \cdots, \eta^m),$$
$$\eta^k = \frac{\psi^k(x)}{\varepsilon^{1/2}},$$
$$\lambda_\alpha^k = \psi_{x^\alpha}^k,$$

where each ψ^k satisfies (3.5.10). A solution of (3.5.7) for u_1 is then

$$u_1 = a(\theta, \eta, x)r(x) + \bar{u}(\eta, x).$$

Equation (3.5.8) has a solution for u_2 provided that $\bar{u}$ solves

$$\lambda_\alpha^k A^\alpha \bar{u}_{\eta^k} = 0.$$

The solution for u_2 is then

$$u_2 = b_k(\theta, \eta, x)s^k(x) + \bar{u}_2(\eta, x),$$

where

$$\kappa_\alpha A^\alpha s^k + \lambda_\alpha^k A^\alpha r = 0,$$
$$b_{k\theta} = a_{\eta^k}.$$

The solvability condition for u_3 implies that

$$C^\alpha a_{x^\alpha} + \left(\frac{1}{2}\mathcal{G}a^2 + \mathcal{V}a\right)_\theta + \mathcal{H}a + \mathcal{J}^{pq}b_{q\eta^p} = 0,$$

where

$$\mathcal{V}(\eta, x) = l \cdot D_u^2 \kappa_\alpha f^\alpha \cdot (r, \bar{u}), \tag{3.5.15}$$
$$\mathcal{J}^{pq}(x) = l \cdot \lambda_\alpha^p A^\alpha s^q.$$

Eliminating b_k gives the equation

$$\left[C^\alpha a_{x^\alpha} + \left(\frac{1}{2}\mathcal{G}a^2\right)_\theta + \mathcal{V}a_\theta + \mathcal{H}a\right]_\theta + \mathcal{J}^{pq}a_{\eta^p\eta^q} = 0. \tag{3.5.16}$$

The equation for the mean field is independent of the wave amplitude, while the mean field leads to a θ-independent change in the wave speed in the equation for a.

To explain the significance of the diffractive terms, we write $x^0 = t$ again. We normalize $l \cdot A^0 r = 1$, which can always be done if the group velocity is finite. This puts (3.5.16) in the form

$$\left[a_t + C^\alpha a_{x^\alpha} + \left(\frac{1}{2}\mathcal{G}a^2\right)_\theta + \mathcal{V}a_\theta + \mathcal{H}a\right]_\theta + \frac{1}{2}\mathcal{D}^{pq}a_{\eta^p\eta^q} = 0,$$

where C^α, with $\alpha = 1, \ldots, n$, is the group velocity vector and

$$\mathcal{D}^{pq} = \mathcal{J}^{pq} + \mathcal{J}^{qp}.$$

The linearized dispersion relation $\omega = W(k)$ satisfies

$$\left[k_\alpha A^\alpha - W(k)A^0\right] r(k) = 0,$$

for an appropriate right null vector r. Differentiating this equation twice with respect to $k = (k_1, \ldots, k_n)$ and contracting the result with λ^p and λ^q shows that

$$\mathcal{D}^{pq} = \lambda_\alpha^p \lambda_\beta^q \frac{\partial^2 \omega}{\partial k_\alpha \partial k_\beta}. \tag{3.5.17}$$

Thus, $\mathcal{D}$ consists of the transverse components of the dispersion tensor $\nabla_k^2 W$.

As this calculation illustrates, diffraction corresponds to dispersion of a wave packet composed of Fourier modes whose wavenumber vectors are in different directions.

We can write the expression for $\mathcal{V}$ in terms of the nonlinear dispersion relation, $\omega = W(u, k)$, as

$$\mathcal{V} = D_u W(u, k) \cdot \bar{u}\big|_{u=u_0}.$$

Thus, if $|k| = 1$, it follows that $\mathcal{V}$ is the change in the wave speed with respect to u in the direction of the mean field.

For plane waves in an isotropic medium, the normalized form of (3.5.16) is

$$\left[u_t + \left(\frac{1}{2}u^2\right)_x + V(y, t)u_x\right]_x + \Delta_y u = 0.$$

For harmonic solutions, $u = A(y,t)e^{ix}$, the linearized equation reduces to the Schrödinger equation for A, with potential V determined from the mean field $\bar{u}$ by (3.5.15),

$$iA_t + \Delta A + V(y,t)A = 0.$$

This version of the parabolic approximation has been used very successfully in random wave propagation [236, 250].

3.5.2.1. Nonlinear acoustics (4).

Plane waves. For plane sound waves propagating in the x direction through a medium at rest, the weakly nonlinear, weak diffraction solution is

$$
\begin{bmatrix} \rho \\ u \\ v \\ w \\ s \end{bmatrix} = \begin{bmatrix} \rho_0 \\ 0 \\ 0 \\ 0 \\ s_0 \end{bmatrix} + \varepsilon a(\theta,\eta,\zeta,x,y,z,t) \begin{bmatrix} \rho_0 \\ c_0 \\ 0 \\ 0 \\ 0 \end{bmatrix} + \varepsilon \begin{bmatrix} \bar{\rho} \\ \bar{u} \\ \bar{v} \\ \bar{w} \\ \bar{s} \end{bmatrix} (\eta,\zeta,x,y,z,t)
$$
$$
+ \varepsilon^{3/2} b(\theta,\eta,\zeta,x,y,z,t) \begin{bmatrix} 0 \\ 0 \\ c_0 \\ 0 \\ 0 \end{bmatrix} + \varepsilon^{3/2} c(\theta,\eta,\zeta,x,y,z,t) \begin{bmatrix} 0 \\ 0 \\ 0 \\ c_0 \\ 0 \end{bmatrix} + O(\varepsilon^2).
$$

Here, the phase variable and the transverse variables are evaluated at

$$\theta = \frac{kx - \omega t}{\varepsilon}, \qquad \eta = \frac{y}{\varepsilon^{1/2}}, \qquad \zeta = \frac{z}{\varepsilon^{1/2}},$$

and $\omega = c_0 k$. Acoustic waves in a non-moving medium are isotropic, so y and z are constant on the rays associated with $\varphi = kx - \omega t$. Thus, the transverse variables satisfy (3.5.10).

The mean field satisfies the linear equations,

$$\bar{v}_\eta + \bar{w}_\zeta = 0,$$
$$c_0^2 \bar{\rho} + d_0 \bar{s} = 0,$$

where d_0 is defined in (3.2.16). The components $\bar{u}$ and $\bar{\rho}$ are arbitrary functions. The mean field therefore consists of a longitudinal shear flow, an incompressible transverse velocity field, and density perturbations at constant pressure.

Including weak viscous effects, the asymptotic equations for the velocity perturbations are

$$a_t + c_0 a_x + \left(\frac{\gamma+1}{4} \omega a^2 + \omega V a \right)_\theta + \frac{c_0}{2} (b_\eta + c_\zeta) = \frac{1}{2} \hat{\delta} |k|^2 a_{\theta\theta},$$
$$k b_\theta - a_\eta = 0,$$
$$k c_\theta - a_\zeta = 0,$$

where $\hat{\delta}$ is the scaled diffusivity of sound, defined in (3.3.11), and V is given by

$$V = \frac{\bar{u}}{c_0} + \left(\frac{\gamma-1}{2} \right) \frac{\bar{\rho}}{\rho_0}.$$

The derivative, $\mathcal{V}$, of the wave speed in the direction of the mean field is independent of the mean transverse velocity field, $(\bar{v}, \bar{w})$. Eliminating the transverse velocity components, b and c, and taking $\mathcal{V} = 0$, gives a two-dimensional Burgers equation for a

$$\left[a_t + c_0 a_x + \left(\frac{\gamma + 1}{4} \omega a^2 \right)_\theta - \frac{1}{2} \hat{\delta} |k|^2 a_{\theta\theta} \right]_\theta$$
$$+ \frac{c_0^2}{2\omega} \left(a_{\eta\eta} + a_{\zeta\zeta} \right) = 0. \tag{3.5.18}$$

The normalized form of this equation is the three-dimensional Burgers equation,

$$\left(u_t + u u_x - \mu u_{xx} \right)_x + u_{yy} + u_{zz} = 0.$$

It is not necessary to carry out all the algebra to write down equation (3.5.18). The coefficients of the new diffractive terms can be read off directly from the linearized dispersion relation. We denote the x, y, and z components of the wavenumber vector by k, l, and m, respectively. The expansion of the dispersion relation $\omega = W(k, l, m)$ for a wave propagating close to the x direction is:

$$W = c_0 \left(k^2 + l^2 + m^2 \right)^{1/2}$$
$$\sim c_0 k + c_0 \frac{l^2 + m^2}{2k}$$
$$\sim \omega + \frac{c_0^2}{2\omega} \left(l^2 + m^2 \right),$$

where $\omega = c_0 k$. The coefficient of $l^2 + m^2$ gives the coefficient of $a_{\eta\eta} + a_{\zeta\zeta}$ in equation (3.5.18).

Magneto-acoustic waves. Three-dimensional Burgers and KdV–Burgers equations for plane magneto-acoustic waves in MHD are derived in [271]. Magneto-acoustic waves are anisotropic, and the signs of the diffractive terms depend on the angle of the wavenumber vector to the background magnetic field. In some ranges of the angle, the diffraction terms have opposite signs, and the normalized KdV–Burgers equation is

$$\left(u_t + u u_x - \mu u_{xx} + \nu u_{xxx} \right)_x + u_{yy} - u_{zz} = 0.$$

When the diffraction terms are of opposite sign, the linearized, nondissipative, nondispersive equation,

$$u_{xt} + u_{yy} - u_{zz} = 0,$$

is of ultrahyperbolic type [59]. It is therefore not clear what boundary conditions should be imposed to obtain a well-posed problem.

Nonplanar waves. For nonplanar waves, and using the same notation as in Subsection 3.2.3.4, a transverse variable $\psi(x, t)$ is constant along the rays associated with $\varphi(x, t)$ provided that

$$\psi_t + C \cdot \nabla \psi = 0.$$

where C is the group velocity,

$$C = u_0 + \frac{c_0^2}{\Omega} k.$$

For a single transverse variable, the asymptotic solution is given by

$$
\begin{bmatrix} \rho \\ u \\ s \end{bmatrix} = \begin{bmatrix} \rho_0 \\ u_0 \\ s_0 \end{bmatrix} + \varepsilon a(\varepsilon^{-1}\varphi, \varepsilon^{-1/2}\psi, x, t) \begin{bmatrix} \rho_0 \\ c_0^2 \Omega^{-1} k \\ 0 \end{bmatrix} + O(\varepsilon^{3/2}).
$$

The equation for the wave amplitude $a(\theta, \eta, x, t)$ is a generalized two-dimensional Burgers equation,

$$
\partial_\theta \left\{ a_t + C \cdot \nabla a + \left(\frac{\gamma+1}{4} \Omega a^2 \right)_\theta \right.
$$
$$
+ \frac{\Omega}{2\rho_0 c_0^2} \left[\left(\frac{\rho_0 c_0^2}{\Omega} \right)_t + \nabla \cdot \left(\frac{\rho_0 c_0^2}{\Omega} C \right) \right] a - \frac{1}{2} \hat{\delta} |k|^2 a_{\theta\theta} \left. \right\}
$$
$$
+ \frac{c_0^2}{2\Omega} |\nabla \psi|^2 a_{\eta\eta} = 0.
$$

As an example, we consider cylindrical waves in two space dimensions. We use nondimensional variables in which $c_0 = 1$. The appropriate phase and transverse variable are

$$
\varphi = r - t,
$$
$$
\psi = \tan^{-1}\left(\frac{y}{x} \right),
$$

where $r = (x^2 + y^2)^{1/2}$ is the radial polar variable. The resulting two-dimensional, cylindrical Burgers equation is

$$
\left(a_t + a_r + \frac{\gamma+1}{2} a a_\theta + \frac{1}{2r} a - \frac{1}{2} \hat{\delta} a_{\theta\theta} \right)_\theta + \frac{1}{2r^2} a_{\eta\eta} = 0. \qquad (3.5.19)
$$

The change of variables (3.2.51) puts this equation in the normalized form

$$
(u_t + u u_x - t u_{xx})_x + t^{-3} u_{yy} = 0.
$$

Many other specific equations can be obtained from the general equation (spherical waves, waves in a stratified atmosphere, etc.) but we will not write them out here.

3.5.2.2. *Ion acoustic waves.* The derivation of the KP equation for ion acoustic waves (see Subsection 3.3.2.4) is completely analogous to the above, so we will be very brief.

For nondimensionalized plane waves, the phase is $\varphi = x - t$ and the transverse variable is $\psi = y$. In two space dimensions, the KP-expansion for (3.3.69) is

$$
\begin{bmatrix} n \\ u \\ v \\ E \\ n_e \end{bmatrix} = \begin{bmatrix} 1 \\ 0 \\ 0 \\ 0 \\ 1 \end{bmatrix} + \varepsilon^{2/3} a(\theta, \eta, x, y, t) \begin{bmatrix} \omega \\ k \\ 0 \\ 0 \\ \omega \end{bmatrix} + O(\varepsilon).
$$

where

$$
\theta = \frac{x-t}{\varepsilon^{2/3}}, \qquad \eta = \frac{y}{\varepsilon^{1/3}}.
$$

The wave amplitude solves the KPII equation,

$$\left(a_t + a_x + aa_\theta + \frac{1}{2}a_{\theta\theta\theta}\right)_\theta + \frac{1}{2}a_{\eta\eta} = 0.$$

In the case of cylindrical waves, instead of the dissipative equation (3.5.19), one obtains the cylindrical KPII equation,

$$\left(a_t + a_r + aa_\theta + \frac{1}{2}a_{\theta\theta\theta} + \frac{1}{2r}a\right)_\theta + \frac{1}{2r^2}a_{\eta\eta} = 0,$$

whose normalized form is

$$(u_t + uu_x + tu_{xxx})_x + t^{-3}u_{yy} = 0.$$

Magnetized plasmas. The diffractive equation for weakly nonlinear ion acoustic waves in a magnetized plasma, described by equations (3.3.71), is not the KP equation, because the coefficient of the KP-type diffractive term turns out to vanish. The linearized dispersion relation of (3.3.71) is [128]

$$|k|^2 + 1 - \frac{|k_\perp|^2}{\omega^2 - \Omega^2} - \frac{(e \cdot k)^2}{\omega^2} = 0.$$

Here, $k_\perp$ is the component of the wavenumber vector orthogonal to e, and we set $\varepsilon = 1$ in (3.3.71). The linearized long wave expansion (for Ω of order one) is therefore

$$\omega \sim e \cdot k \left[1 + |k|^2 + |k_\perp|^2 \Omega^{-2}\right]^{-1/2}$$
$$\sim e \cdot k - \frac{1}{2}e \cdot k \left[(e \cdot k)^2 + |k_\perp|^2 \left(1 + \Omega^{-2}\right)\right]. \tag{3.5.20}$$

The long wave limit of the dispersion relation is

$$\omega \sim e \cdot k.$$

Long waves are therefore fully nondispersive, since $\nabla_k^2 \omega = 0$. It follows from (3.5.17) that the coefficient of the diffraction term is zero.

The appropriate scaling to retain diffraction effects in this example has the transverse wave number of the same order of magnitude as the longitudinal wavenumber. This scaling leads to the Zakharov–Kuznetsov equation [128, 278], which has the form

$$u_t + uu_x + \frac{1}{2}u_{xxx} + \frac{1}{2}\left(1 + \Omega^{-2}\right)u_{xyy} = 0. \tag{3.5.21}$$

We can "interpolate" between the KP ($\Omega = 0$) and Zakharov–Kuznetsov ($\Omega = O(1)$) equations by taking a limit in which both Ω and k tend to zero. This limit is chosen so that the KP and Zakharov–Kuznetsov types of dispersion are both weak and of the same order of magnitude. The resulting equation describes long waves in a weakly magnetized plasma [170].

To obtain the usual KP balance between diffraction and KdV-dispersion, we need the orders of magnitude of the transverse and longitudinal wavenumbers to be related by $k_\perp = O((e \cdot k)^2)$. From (3.5.20), the dispersion due to the magnetic field will balance these dispersive effects if $|k_\perp|/\Omega = O(e \cdot k)$, which implies that $\Omega = O(e \cdot k)$. The long wave expansion of the linearized dispersion relation is then

$$\omega \sim e \cdot k - \frac{1}{2} e \cdot k \left[(e \cdot k)^2 + \frac{|k_\perp|^2}{\Omega^2 - (e \cdot k)^2} \right].$$

The asymptotic equation for the wave amplitude has the form

$$\left(\partial_x^2 + \Omega^2 \right) \left(u_t + u u_x + \frac{1}{2} u_{xxx} \right) + \frac{1}{2} u_{xyy} = 0.$$

For large Ω this reduces to (3.5.21), while for small Ω it reduces to the KP equation after an integration with respect to x.

The books [9, 128] give further information on asymptotic equations in plasma physics, and compare the results of the asymptotic theory with experimental observations of solitons in plasmas.

3.6. OTHER TOPICS

In this Section we give a brief guide to the literature on subjects not treated in the main part of this survey.

3.6.1. Resonant Wave Interactions

A major topic which we have not described is resonantly interacting hyperbolic waves. The corresponding asymptotic equations involve multiple phases, and require a non-trivial generalization of the single phase expansions discussed up to this point.

The fundamental example of resonant interactions arises when three plane periodic waves, with wavenumbers k_j and frequencies $\omega_j = W_j(k_j)$, satisfy the tri-resonance condition,

$$\omega_1 + \omega_2 + \omega_3 = 0,$$
$$k_1 + k_2 + k_3 = 0.$$

Assuming advective nonlinearity, the normalized equations for the wave amplitudes consist of a coupled system of integro-differential equations,

$$u_t + \left(\lambda u^2 \right)_x + \alpha \frac{1}{2\pi} \left(\int_0^{2\pi} v(-x-y,t) w(y,t) \, dy \right)_x = 0,$$

$$v_t + \left(\mu v^2 \right)_x + \beta \frac{1}{2\pi} \left(\int_0^{2\pi} w(-x-y,t) u(y,t) \, dy \right)_x = 0, \qquad (3.6.1)$$

$$w_t + \left(\nu w^2 \right)_x + \gamma \frac{1}{2\pi} \left(\int_0^{2\pi} u(-x-y,t) v(y,t) \, dy \right)_x = 0.$$

Here, $u(x,t), v(x,t), w(x,t)$ are 2π-periodic functions of x and $\alpha, \dots, \nu$ are constants.

An introductory account of resonant hyperbolic wave interaction is given in Section 3.11 of [9] and in [115, 187]. The general equations for resonant hyperbolic waves were derived in [121, 190] (see also [195, 196, 197]). Dissipative [190], dispersive, and diffractive [114] effects can be included in the resonant interaction expansion.

The simplest physical application is to the compressible Euler equations. In that case, equation (3.6.1) describes the resonant reflection of sound waves (amplitudes u and v) off an entropy wave (amplitude w). The coefficients ν and γ are zero, so w is constant, and (3.6.1) reduces to a simpler form [192]:

$$u_t + \left(\frac{1}{2}u^2\right)_x - Kv = 0, \qquad (3.6.2)$$

$$v_t + \left(\frac{1}{2}v^2\right)_x + K^\dagger u = 0.$$

Here K is an integral operator, with adjoint $K^\dagger$, whose kernel is proportional to the derivative of w. These equations consist of a pair of inviscid Burgers equations coupled by a lower order, skew-symmetric, integral operator.

For the derivation and study of resonant interaction equations in some other physical systems see: Almgren, Majda, and Rosales [6] (combustion, including mean field interactions); Artola and Majda [14] (waves on a supersonic vortex sheet); and Hunter [117] (elasticity).

Conservation of wave action. Smooth solutions of (3.6.1) satisfy the averaged conservation laws [111],

$$\int_0^{2\pi} \left\{\beta u^2(x,t) - \alpha v^2(x,t)\right\} \, dx = \text{constant}, \qquad (3.6.3)$$

and cyclic permutations. As usual, shocks introduce additional dissipative terms. These conservation laws are analogous to the Manley–Rowe relations for the three wave resonant interaction of dispersive waves [61].

When the interaction coefficients α, β, and γ do not all have the same sign, equation (3.6.3) shows that solutions remain bounded in L^2. It is then possible to have smooth traveling wave solutions (see [192, 220] in the case of gas dynamics), in which wave interactions oppose nonlinear distortion and prevent the formation of shocks, at least over the time scale for which the weakly nonlinear asymptotics is valid. There also seem to be time periodic solutions as well [192]. It would be interesting to carry out "long-long" time asymptotics to determine the ultimate fate of these solutions. The long time behavior of solutions of periodic solutions of conservation laws when resonant interactions occur is not clear. For two by two systems in one space dimension, where resonant interactions do not occur, Glimm and Lax [89] showed that the total variation of periodic solutions decays like t^{-1}. Nothing is known for general higher order systems.

When the interaction coefficients have the same sign, the resonant interaction equations can have solutions which blow up in finite time [113, 131, 143]. Physically this situation corresponds to negative wave energies, which occur in explosive instabilities of resonantly interacting dispersive waves [61, 217]. There do not seem to be any known physical examples of hyperbolic conservation laws where this phenomenon occurs. It cannot happen when the conservation law has an entropy function which is convex at the unperturbed state, because then the asymptotic equations would inherit a conserved convex quadratic functional.

Randall and Rosales [224, 225] carry out a numerical and analytical study of the general equations (3.6.1). They classify regimes with different long time asymptotics, and discuss sawtooth wave solutions and "humpback" waves containing square-root singularities.

Equations for oblique plane wave interactions in several space dimensions are derived in [121] (this paper actually treats the slightly more general case of "coherent" waves). One curiosity is that successive resonant interactions starting from three oblique waves can generate new waves propagating in a dense set of directions [115, 121, 144, 142].

Interaction of nonplanar waves. The interaction of nonplanar ("incoherent") oscillatory waves leads to some difficult problems involving passage through resonance — see [9] and [141] for a detailed discussion. In particular, [141] gives examples of "hidden focusing," in which a resonantly generated wave whose amplitude is formally of lower order focuses, leading to the L^∞-blow up of the wave amplitude.

Soliton interactions. Pulses propagating on different families of rays only overlap for short distances, so the cumulative nonlinear effect of the wave interaction is negligible. Nonlinear dispersive wave equations, such as the Boussinesq equation,

$$u_{tt} = \left(u + \frac{1}{2}u^2 \right)_{xx} + u_{xxxx},$$

are sometimes used to model the collision of solitons propagating in opposite directions. From the point of view of systematic weakly nonlinear asymptotics, this is not correct. Rather, to leading order, the oppositely propagating solitons satisfy a pair of decoupled KdV equations and the effects of the interaction can be neglected. It is only when the soliton speeds are almost equal that weakly nonlinear interaction effects between solitons become important.

3.6.2. Rigorous Results

Most of the asymptotic equations described here have been derived using formal singular perturbation methods. There has been significant progress in providing rigorous proofs of these formal expansions. A detailed description of this work is not possible here, but we will briefly list some of the relevant papers.

DiPerna and Majda [69] proved the validity of the one wave, single phase, weakly nonlinear geometrical optics solution in one space dimension. Their proof applies to weak solutions containing shocks. A similar result is also implicit in the work of Liu [181]. Joly, Metevier, and Rauch [139] proved the validity of the resonant, multiphase expansion for smooth solutions in one space dimension, using the method of characteristics and stationary phase estimates. Schochet [239] proved the validity of the resonant expansion for periodic weak solutions in one space dimension. Schochet's proof is based on Glimm's scheme, and applies to solutions which contain shocks but requires exact spatial periodicity.

For rigorous justifications of the expansion for smooth oscillatory multidimensional waves, see [68, 98, 140, 145, 240], and the references given there. Single-phase and multi-phase cases have been treated for both quasi-linear and semi-linear systems. Since there is no existence theory for weak solutions of multidimensional systems of hyperbolic conservation laws, rigorous proof of the expansions for weak solutions containing shocks in several space dimensions looks rather hopeless.

Rigorous results for the propagation of singularities in linear equations can be obtained using microlocal analysis, and these techniques been extended to treat the interaction of singularities (but not shocks) in some nonlinear equations [19, 263].

Despite the similarities between the resonant interaction of high-frequency oscillatory waves (weakly nonlinear geometrical optics) and the interaction of shocks (solutions of Riemann problems), the analogies are not exact. For linear problems, the propagation of singularities and oscillations are related via the Fourier transform. For nonlinear problems, there is no precise relationship between the two.

There do not seem to be any rigorous justifications of the weak diffraction equations, such as the two-dimensional Burgers equation (3.5.1), even for smooth solutions.

Young [275] has given a fundamental improvement of Glimm's scheme for proving global existence of weak solutions of arbitrary conservation laws in one space dimension with small enough initial data. Although this approach differs from the asymptotic methods described here, it is worth noting that Young uses a careful analysis of quadratically nonlinear interactions between waves in different families. This bears a close resemblance to the asymptotic analysis of resonant interactions.

Compensated compactness. There are a number of connections between singular perturbation methods like the ones described here and the use of compensated compactness [262] and other weak convergence ideas in nonlinear partial differential equations. For example, in Subsection 3.4.2.1 we mentioned the rigorous derivations of homogenized equations for the Carlemann system.

Rapid oscillations in weakly converging sequences can be characterized by using Young measures. However, the Young measure contain no information about the waveform of the oscillations. Young measures are therefore not appropriate for problems in which phase correlations are important. For surveys of the application of weak convergence methods to nonlinear partial differential equations, see [76, 188].

3.6.3. Hyperbolic Surface Waves

We have emphasized the fundamental consequences of scale invariance for hyperbolic waves. Hyperbolic equations posed on a half-space retain this scale invariance, since the geometry does not define any length scales. This leads to some interesting problems.

Nonlinear Rayleigh waves. One example of a hyperbolic surface wave is a Rayleigh wave, or surface acoustic wave (SAW), on the surface of an elastic solid. Although originally of interest in seismology, there has been recent interest in ultrasonic SAW devices because of their use in signal processing applications.

The derivation of weakly nonlinear equations for surface waves uses essentially the same ideas as for hyperbolic waves in free space. The main difference is that the algebraic eigenvalue problem (3.2.6) for the wave speeds is replaced by an eigenvalue problem for an ordinary differential equation in the space variable normal to the half space.

The main new effect that arises is nonlocal nonlinearity. The canonical asymptotic equation for this type of surface wave is of the form [112]

$$u_t(x,t) + \left(\int K(x - y, x - z)u(y,t)u(z,t)\,dydz \right)_x = 0, \qquad (3.6.4)$$

where K is the kernel of a quadratically nonlinear singular integral operator. Scale invariance requires that

$$K(\alpha y, \alpha z) = \alpha^{-2} K(y, z),$$

for all $\alpha > 0$. For detailed expressions of (the Fourier transform of) K in the case of nonlinear elasticity, see [219]. One interesting property of the asymptotic equation for nonlinear elasticity is that it has smooth traveling waves. Thus wave breaking does not always occur in these nonlocal equations. How to extend solutions of (3.6.4) after breaking occurs is not entirely clear.

Other effects, like weak dissipation, dispersion, diffraction, and resonant interactions, can be combined with the nonlocal nonlinearity in the standard way. For example, in unpublished work with David Parker, we have derived a Benjamin–Ono type equation for a nonlinear SAW on an elastic half-space where dispersion is caused by the presence of a thin layer of substrate on the surface of the half-space. The normalized asymptotic equation has the form

$$u_t + \left(\int K(x - y, x - z)u(y,t)u(z,t)\,dydz \right)_x + \mathcal{H}[u_{xx}] = 0,$$

where $\mathcal{H}$ is the Hilbert transform, and K is the usual kernel for SAWs.

More generally, any boundary value problem for hyperbolic waves in which the geometry is self-similar will not break their scale invariance. An example from elasticity is the propagation of nonlinear elastic waves along the tip of a wedge, which leads to a cubically nonlinear version of (3.6.4) [168].

Nonlinear waves on supersonic vortex sheets. Another interesting hyperbolic surface wave problem concerns waves ("kink modes") on a supersonic vortex sheet. Kink modes propagating along the sheet radiate acoustic waves into the interior of the fluid; thus the disturbance does not decay exponentially away from the surface as happens for SAWs, nor is the interior field determined by solving an elliptic equation. One consequence of this fact is that a single, weakly nonlinear surface wave satisfies the usual local inviscid Burgers equation [13]. Much more surprisingly, the asymptotic equations for resonantly interacting waves on the vortex sheet adopt almost exactly the form as those for resonant gas dynamics, (3.6.2); the only difference is that the lower order integral terms are symmetric instead of skew symmetric [14]. This leads to growth in small amplitude kink modes (although the instability saturates, so the amplitudes do not blow up).

Nonlinear stability of shock waves. A third kind of problem involving hyperbolic surface waves, concerns the stability of shocks. Here the unperturbed state is a plane shock wave, and one is interested in the interaction of the shock with weakly nonlinear waves, and especially with the stability of the shock. As the vortex sheet problem illustrates, even if the shock is linearly neutrally stable, it may be unstable when nonlinear effects are included. Particularly interesting and important examples occur in combustion, where detonation waves are observed to develop complex transverse instabilities. For an analysis of such situations, see [191, 230].

An introduction to weakly nonlinear geometrical optics for hyperbolic half-space problems is given in [189].

Acknowledgments. This work was partially supported by the National Science Foundation. I wish to thank J. B. Keller, A. Majda, R. R. Rosales, M. Brio, J. Rauch, G. Papanicolaou, and an anonymous reviewer for their comments on earlier versions of this paper.

REFERENCES

[1] L. Abdelouhab, J. L. Bona, M. Felland, and J. C. Saut, "Nonlocal models for nonlinear dispersive waves," *Physica D*, **40**, 360–392 (1989).

[2] M. J. Ablowitz and P. A. Clarkson, *Solitons, Nonlinear Evolution Equations and Inverse Scattering*, London Mathematical Society Lecture Note Series **149**, Cambridge University Press, Cambridge, 1991.

[3] M. J. Ablowitz and H. Segur, *Solitons and the Inverse Scattering Transform*, SIAM, Philadelphia, 1981.

[4] M. S. Alber, R. Camassa, D. D. Holm, and J. E. Marsden, "The geometry of peaked solitons and billiard solutions of a class of integrable PDEs," to appear in: *Lett. Math. Phys.*

[5] R. Almgren, "Modulated high-frequency waves," *Stud. Appl. Math.*, **83**, 159–181 (1990).

[6] R. Almgren, A. Majda, and R. R. Rosales, "Rapid initiation in condensed phases through resonant nonlinear acoustics," *Phys. Fluids A*, **2**, 1014–1029 (1990).

[7] C. J. Amick and J. F. Toland, "Solitary waves with surface tension I. Trajectories homoclinic to periodic orbits in four dimensions," *Arch. Rat. Mech. Anal.*, **118**, 37–69 (1992).

[8] A. M. Anile, *Relativistic Fluids and Magneto-fluids*, Cambridge University Press, Cambridge, 1989.

[9] A. M. Anile, J. K. Hunter, P. Pantano, and G. Russo, *Ray Methods for Nonlinear Waves in Fluids and Plasmas*, Longman, London, 1993.

[10] A. M. Anile and G. Russo, "Generalized wavefront expansion I. Higher order corrections for the propagation of weak shock waves," *Wave Motion* **8**, 243 (1986).

[11] V. I. Arnold, *Mathematical Methods of Classical Mechanics*, 2nd Edition, Springer-Verlag, New York, 1989.

[12] D. G. Aronson, "The porous medium equation," in: *Nonlinear Diffusion Problems*, ed. A. Fasano, and M. Primicerio, Lecture Notes in Math. **1224**, Springer-Verlag, New York, 1–44 (1985).

[13] M. Artola and A. J. Majda, "Nonlinear development of instabilities in supersonic vortex sheets I: the basic kink modes," *Physica D*, **28**, 253–281 (1987).

[14] M. Artola and A. J. Majda, "Nonlinear development of instabilities in supersonic vortex sheets II: resonant interactions among kink modes," *SIAM J. Appl. Math.*, **49**, 1311–1349 (1989).

[15] V. M. Babič and V. S. Buldyrev, *Short Wavelength Diffraction Theory*, Springer Series on Wave Phenomena, Vol. **4**, Springer-Verlag, New York, 1991.

[16] N. S. Bakhvalov, Ya. M. Zhileikin, and E. A. Zabolotskaya, *Nonlinear Theory of Sound Beams*, American Institute of Physics, Translation Series, New York, 1987.

[17] J. D. Bdzill and D. S. Stewart, "Modeling two-dimensional detonations with detonation shock dynamics," *Phys. Fluids A*, **1**, 1261–1267 (1989).

[18] C. M. Bender and S. A. Orszag, *Advanced Mathematical Methods for Scientists and Engineers*, McGraw-Hill, New York, 1978.

[19] M. Beals, *Propagation and Interaction of Singularities in Nonlinear Hyperbolic Problems*, Birkhäuser, Boston, 1989.

[20] E. S. Benilov and E. M. Pelinovskii, "Dispersionless wave propagation in nonlinear fluctuating media," *Sov. Phys. JETP*, **67**, 1040–1052 (1986).

[21] T. B. Benjamin, Internal waves of finite amplitude and permanent form, *J. Fluid. Mech.*, **25**, 559–592 (1966).

[22] T. B. Benjamin, J. G. Bona, and J. J. Mahoney, "Model equations for long waves in nonlinear dispersive systems," *Phil. Trans. Roy. Soc. London A*, **272**, 47–78 (1972).

[23] A. Bensoussan, J. L. Lions, and G. C. Papanicolaou, *Asymptotic Analysis for Periodic Structures*, North-Holland, Amsterdam, 1978.

[24] R. T. Beyer and S. V. Letcher, *Physical Ultrasonics*, Academic Press, New York, 1969.

[25] R. T. Beyer, *Nonlinear Acoustics*, Naval Ship Systems Command, Washington, 1974.

[26] J. J. Binney, N. J. Dowrick, A. J. Fisher, and M. E. J. Newman, *The Theory of Critical Phenomena*, Oxford University Press, Oxford, 1992.

[27] N. Bleistein, *Mathematical Methods for Wave Phenomena*, Academic Press, Orlando, 1984.

[28] J. L. Bona, V. A. Dougalis, O. A. Karakashian, and W. R. McKinney, "Computations of blow-up and decay for periodic solutions of the generalized Korteweg–de Vries–Burgers equation," *Applied Numerical Mathematics*, **10**, 335–355 (1992).

[29] J. L. Bona and J. C. Saut, "Dispersive blowup of solutions of generalized Korteweg–de Vries equations," *J. Diff. Eq.*, **103**, 3–57 (1993).

[30] J. L. Bona and M. Schonbek, "Travelling wave solutions to Korteweg–de Vries–Burgers equations," *Proc. Roy. Soc. Edinburgh A*, **101**, 207–226 (1985).

[31] J. Bourgain, "On the Cauchy problem for the Kadomtsev-Petviashvili equation," *Geometric and Functional Analysis*, **3**, 315–341 (1993).

[32] J. P. Boyd, "New directions in solitons and nonlinear periodic waves," in: *Advances in Applied Mechanics*, Vol. **27**, ed. J. W. Hutchinson, and T. Y. Wu, Academic Press, New York, 1–82 (1989).

[33] M. A. Breazeale and J. Philip, in: *Physical Acoustics*, Vol. XVII, ed. W. P. Mason, and R. N. Thurston, Academic Press, New York, 1–60 (1984).

[34] F. P. Bretherton, "The general linearized theory of wave propagation," in: *Mathematical Problems in the Geophysical Sciences*, ed. W. H. Reid, Amer. Math. Soc., Providence, 61–102 (1971).

[35] M. Brio and J. K. Hunter, "Rotationally invariant hyperbolic waves," *Comm. Pure Appl. Math.*, **43**, 1037–1053 (1990).

[36] M. Brio and J. K. Hunter, "Asymptotic equations for conservation laws of mixed type," *Wave Motion*, **16**, 57–64 (1992).

[37] M. Brio and J. K. Hunter, "Mach reflection for the two-dimensional Burgers equation," *Physica D*, **60**, 194–207 (1992).

[38] J. M. Burgers, *The Nonlinear Diffusion Equation*, Dordrecht-Holland, Amsterdam, 1974.

[39] F. Calogero and A. Degasperis, *Spectral Transform and Solitons. I.*, North-Holland, Amsterdam, 1982.

[40] F. Calogero, "A solvable nonlinear wave equation," *Stud. Appl. Math.*, **70**, 189–199 (1984).

[41] F. Calogero, "Why are certain nonlinear PDEs both widely applicable and integrable?," in: *What is Integrability?*, ed. V. E. Zakharov, Springer-Verlag, Berlin, 1991, 1–62.

[42] R. Camassa and D. Holm, "An integrable shallow water equation with peaked solitons," *Phys. Rev. Lett.*, **71**, 1661 (1993).

[43] R. Camassa, D. D. Holm, and J. M. Hyman, "A new integrable shallow water equation," to appear in: *Adv. in Appl. Mech.*

[44] J. Canosa and J. Gazdag, "The Korteweg–de Vries–Burgers equation," *J. Comp. Phys.*, **23**, 393–403 (1977).

[45] A. T. Cates and D. G. Crighton, "Nonlinear diffraction and caustic formation," *Proc. Roy. Soc. Lond. A*, **430**, 69–88 (1990).

[46] P. Cehelsky and R. R. Rosales, "Resonantly interacting weakly nonlinear waves in the presence of shocks: a single space variable in a homogeneous time independent medium," *Stud. Appl. Math.* **74**, 117–138 (1986).

[47] C. Cercignani, *The Boltzmann Equation and Its Applications*, Applied Mathematical Sciences Vol. **67**, Springer-Verlag, New York, 1988.

[48] F. F. Chen, *Introduction to Plasma Physics. Vol. 1. Plasma Physics*, 2nd ed., Plenum, New York, 1984.

[49] G.-Q. Chen, D. Levermore, and T. P. Liu, "Hyperbolic conservation laws with stiff relaxation and entropy," *Comm. Pure Appl. Math.*, **47**, 787–830 (1994).

[50] P. Chen, *Selected Topics in Wave Propagation*, Nordhoff, Leyden, 1976.

[51] R. C. Y. Chin, J. C. Garrison, C. D. Levermore, and J. Wong, "Weakly nonlinear acoustic instabilities in inviscid fluids," *Wave Motion*, **8**, 537–559 (1986).

[52] Y. S. Choi and A. Majda, "Amplification of small-amplitude high-frequency waves in a reactive mixture," *SIAM Rev.*, **31**, 401–427 (1989).

[53] Y. Choquet-Bruhat, "Construction de solutions radiative approchées des equations d'Einstein," *Commun. Math. Phys.* **12**, 16–35 (1969).

[54] Y. Choquet-Bruhat, "Ondes asymptotique et approchées pour systèmes nonlineaires d'équations aux dérivées partielles nonlinéaires," *J. Math. Pure et Appl.*, **48**, 117–158 (1969).

[55] P. A. Clarkson and S. Hood, "Nonclassical symmetry reductions and exact solutions of the Zabolotskaya–Khokhlov equation," *Euro. J. Appl. Math.*, **3**, 381–415 (1992).

[56] J. D. Cole, "On a quasilinear parabolic equation occurring in aerodynamics," *Quart. Appl. Math.*, 9 , 225–236 (1951).

[57] J. D. Cole and L. P. Cook, *Transonic Aerodynamics*, Elsevier, Amsterdam, 1986.

[58] L. P. Cook, ed., *Transonic Aerodynamics: Problems in Asymptotic Theory*, Frontiers in Applied Mathematics, SIAM, Philadelphia, 1993.

[59] R. Courant and D. Hilbert, *Methods of Applied Mathematics*, Vol. 2, Interscience, New York, 1962.

[60] W. Craig, "Water waves theory," *Comm. Part. Diff. Eq.*, **10**, 787–1003 (1985).

[61] A. D. D. Craik, *Wave Interactions and Fluid Flows*, Cambridge University Press, Cambridge, 1985.

[62] M. S. Cramer and A. R. Seebass, "Focusing of a weak shock wave at an arête," *J. Fluid. Mech.* **88**, 209–222 (1978).

[63] D. G. Crighton, "Model equations for nonlinear acoustics," *Ann. Rev. Fluid. Mech.*, **11**, 11–33 (1979).

[64] D. G. Crighton, "Basic theoretical nonlinear acoustics," *Frontiers in Physical Acoustics*, Proc. Int. School of Physics "Enrico Fermi, " Course **93**, Elsevier, Amsterdam, 1986.

[65] D. G. Crighton, "Nonlinear acoustics of bubbly fluids," *Nonlinear Waves in Real Fluids*, ed. A. Kluwick, CISM Courses and Lectures no. **315**, Springer-Verlag, New York, 45–68 (1991).

[66] D. G. Crighton and I. P. Lee-Bapty, "Spherical nonlinear wave propagation in a dissipative stratified atmosphere," *Wave Motion*, **15**, 315–331 (1992).

[67] C. M. Dafermos, "Hyperbolic systems of conservation laws," in: *Systems of Nonlinear Partial Differential Equations*, ed. J. M. Ball, Reidel, Boston, 25–70 (1983).

[68] J. M. Delort, "Oscillations semi-lineaires multiphasées compatibles en dimension 2 et 3 d'espace," *J. Diff. Equations*, **16**, 845–872 (1991).

[69] R. DiPerna and A. J. Majda, "The validity of nonlinear geometrical optics for weak solutions of conservation laws," *Commun. Math. Phys.* **98**, 313–347 (1985).

[70] R. K. Dodd, J. C. Eilbeck, J. D. Gibbon, and H. C. Morris, *Solitons and Nonlinear Wave Equations*, Academic Press, London, 1982.

[71] P. Drazin and R. Johnson, *Solitons: an Introduction*, Cambridge University Press, Cambridge, 1989.

[72] W. E, "Propagation of oscillations in the solutions of 1-D compressible fluid equations," *Comm. Partial Differential Equations*, **17**, 347–370 (1992).

[73] K. S. Eckhoff, "On the stability for symmetric hyperbolic systems," I, *Journal of Differential Equations*, **40**, 94–115 (1981).

[74] P. Embid, J. K. Hunter, and A. Majda, "Simplified asymptotic equations for the transition to detonation in reactive granular materials," *SIAM J. Appl. Math.*, **52**, 1199–1237 (1992).

[75] P. Embid and A. Majda, "An asymptotic theory for hot spot formation and transition to detonation for reactive granular material," *Comb. and Flame*, **89**, 17–36 (1992).

[76] L. C. Evans, *Weak Convergence Methods for Nonlinear Partial Differential Equations*, Regional Conf. Series in Mathematics, Amer. Math. Soc., Providence, 1990.

[77] L. C. Evans, *Partial Differential Equations*, Berkeley Mathematics Lecture Notes, Vol. 3A–3B, Berkeley, 1993.

[78] H. Flaschka, M. G. Forest, and D. W. McLaughlin, "Multiphase averaging and the inverse spectral solution of the Korteweg–de Vries equation," *Comm. Pure Appl. Math.*, **33**, 739–784 (1980).

[79] B. Fornberg and G. B. Whitham, "A numerical and theoretical study of certain nonlinear wave phenomena," *Phil. Trans. R. Soc. Lond.* **289**, 373–404 (1978).

[80] H. Freistühler, Dynamical stability and vanishing viscosity: a case study of a non-strictly hyperbolic system, *Comm. Pure Appl. Math.*, **XLV**, 561–582 (1992).

[81] H. Freistühler, "Some remarks on the structure of intermediate magnetohydrodynamic shocks," *J. Geophys. Res.*, **96**, 3825–3827 (1991).

[82] H. Fujita, "On the blowing up of solutions of the Cauchy problem for $u_t = \Delta u + u^{1+\alpha}$," *J. Fac. Sci. Tokyo, Sect. IA, Math.*, **13**, 109–124 (1966).

[83] S. A. Gabov, "On Whitham's equation," *Sov. Math. Dokl.*, **19**, 1225–1229 (1978).

[84] C. S. Gardner, J. M. Greene, M. D. Kruskal, and R. M. Miura, "Method for solving the Korteweg–de Vries equation," Phys. Rev. Lett., **19**, 1095–1097 (1967).

[85] C. S. Gardner and G. M. Morikawa, "Similarity in the asymptotic behavior of collision free hydromagnetic wave and water waves," Report NYO-**9082**, Courant Institute of Mathematical Sciences, 1962.

[86] P. Germain, "Progressive waves," *Jahrbuch DGLR*, 11–30 (1971).

[87] S. Giambò, A. Greco, and P. Pantano, "Sur la méthode perturbative et réductive à n-dimensions: le cas général," *Comptes Rendus Acad. Sci. Paris*, Ser. A, **289**, 553–556 (1979).

[88] M.-J. Gionnoni, A. Voros, and J. Zinn-Justin, Eds, *Chaos and Quantum Physics*, Les Houches École d'Été de Physique Théorique, Session LII 1989, North-Holland, Amsterdam, 1991.

[89] J. Glimm and P. Lax, "Decay of solutions of systems of nonlinear hyperbolic conservation laws," *Mem. Amer. Math. Soc.*, **101**, (1970).

[90] N. Goldenfeld, *Lectures in Phase Transitions and the Renormalization Group*, Frontiers in Physics, Vol. **85**, Addison-Wesley, Reading Mass., 1992.

[91] J. Goodman, "Stability of viscous scalar shock fronts in several space dimensions," *Trans. Amer. Math. Soc.*, **311**, 683–695 (1989).

[92] J. Goodman, Stability of the Kuramoto–Sivashinsky and related systems, *Commun. Pure Appl. Phys.*, *Comm. Pure Appl. Math.*, **47**, 293–306 (1994).

[93] K. A. Gorshkov, L. A. Ostrovsky, and A. S. Pikovsky, "On the existence of stationary multisolitons," *Phys. Lett. A*, **74**, 177–179 (1979).

[94] R. Grimshaw, "Wave action and wave-mean flow interaction, with application to stratified shear flows," in: *Ann. Rev. Fluid Mech.*, **16**, 11–44 (1984).

[95] R. Grimshaw, "Evolution equations for weakly nonlinear long internal waves in a rotating fluid," *Stud. Appl. Math.*, **73**, 1–33 (1985).

[96] R. Grimshaw, "Theory of solitary waves in shallow fluids," in: *Encyclopedia of Fluid Mechanics*, Vol. **2**, ed. N. P. Cheremisinoff, Gulf Publication Co., Houston, 3–25 (1986).

[97] R. Grimshaw and S. Tang, "The rotation modified Kadomtsev–Petviashvilli equation: an analytical and numerical study," *Stud. Appl. Math.*, **83**, 223–248 (1990).

[98] O. Gues, "Développements asymptotiques de solutions exactes de systèmes hyperboliques quasi-linéaires," *Asymptotic Anal.*, **6**, 241–269 (1993).

[99] J.-P. Guiraud, "Transonic degeneracy in systems of conservation laws," *Symposium Transonicum* III. eds. J. Sierep and H. Oertel, Springer-Verlag, Berlin, 171–178 (1989).

[100] S. Gurbatov, A. Malakhov, and A. Saichev, *Nonlinear Random Waves and Turbulence in Nondispersive Media: Waves, Rays, and Particles*, Manchester University Press, Manchester, 1991.

[101] B. Gurevich, A. Jeffrey, and E. N. Pelinovsky, "A method for obtaining evolution equations for nonlinear waves in a random medium," *Wave Motion*, **17**, 287–296 (1993).

[102] M. E. Gurtin, *Introduction to Continuum Mechanics*, Academic Press, New York, 1981.

[103] M. F. Hamilton, Fundamentals and applications of nonlinear acoustics, *Nonlinear Wave Propagation in Mechanics*, ed. T. W. Wright, AMD-**77**, 1–28 (1986).

[104] J. L. Hammack, N. Scheffner, and H. Segur, "Periodic waves in shallow water," in: *Nonlinear Topics in Ocean Physics*, Proceedings of the International School of Physics "Enrico Fermi," Course CIX, North-Holland, Amsterdam, 891–914 (1991).

[105] J. L. Hammack and H. Segur, "The Korteweg–de Vries equation and water waves part 2. Comparison with experiments," *J. Fluid Mech.*, **84**, 289–314 (1974).

[106] E. Harabetian, "Diffraction of a weak shock by a wedge," *Comm. Pure Appl. Math.*, **40**, 849–863 (1987).

[107] H. Hasimoto, *Kagato*, **40**, 401 (1970).

[108] W. D. Hayes, Introduction to wave propagation, in: *Nonlinear Waves*, ed. S. Leibovich and A. R. Seebass, Cornell Univ. Press, Ithica, 1–43 (1974).

[109] E. J. Hinch, *Perturbation Methods*, Cambridge Texts in Applied Mathematics, Cambridge Univ. Press, Cambridge, 1991.

[110] J. K. Hunter, Transverse diffraction of nonlinear waves and singular rays, *SIAM J. Appl. Math*, **48**, 1–37 (1988).

[111] J. K. Hunter, "Hyperbolic waves and nonlinear geometrical acoustics," ARO Report 89-**1**, in: *Transactions of the Sixth Army Conference on Applied Mathematics and Computing* Boulder, 527–570 (1989).

[112] J. K. Hunter, Nonlinear surface waves, in: *Contemporary Mathematics* Vol. **100**, ed. B. Lindquist, American Mathematical Society, 185–202 (1989).

[113] J. K. Hunter, "Strongly nonlinear hyperbolic waves," in Proceedings of the Second Intenational Conference on Nonlinear Hyperbolic Problems, Aachen, *Notes on Numerical Fluid Mechanics*, Vol. **24**, ed. J. Ballmann and R. Jeltsch, Braunschweig, 257–268 (1989).

[114] J. K. Hunter, "Nonlinear geometrical optics," in: *Multidimensional Hyperbolic Problems and Computations*, Vol. **29**, IMA Volumes in Mathematics and its Applications, ed. A. J. Majda and J. Glimm, Springer-Verlag, New York, 179–197 (1991).

[115] J. K. Hunter, "Interacting weakly nonlinear hyperbolic and dispersive waves," in: *Microlocal Analysis and Nonlinear Waves*, Vol. **30**, IMA Volumes in Mathematics and its Applications, ed. M. Beals, R. Melrose, and J. Rauch, Springer-Verlag, New York, 83–112 (1991).

[116] J. K. Hunter, "Numerical solutions of some nonlinear dispersive wave equations," in: *Lectures in Applied Mathematics* **26**, 301–316 (1990).

[117] J. K. Hunter, "Interaction of elastic waves," *Stud. Appl. Math.*, **86**, 281–314 (1992).

[118] J. K. Hunter and J. B. Keller, "Weakly nonlinear high frequency waves," *Comm. Pure Appl. Math.*, **36**, 547–569 (1983).

[119] J. K. Hunter and J. B Keller, "Caustics of nonlinear waves," *Wave Motion* **9**, 429–443 (1987).

[120] J. K. Hunter and J. B. Keller, "Strongly nonlinear hyperbolic waves," *Proc. Roy. Soc. Lond. A*, **417**, 299–308 (1988).

[121] J. K. Hunter, A. J. Majda, and R. R. Rosales, "Resonantly interacting, weakly nonlinear, hyperbolic waves. II Several space variables," *Stud. Appl. Math.*, **75**, 187–226 (1986).

[122] J. K. Hunter and R. Saxton, "Dynamics of director fields," *SIAM J. Appl. Math.*, **51**, 1498–1521 (1991).

[123] J. K. Hunter and J. Scheurle, "Perturbed solitary wave solutions of a model equation for water waves," *Physica D*, **32**, 253–268 (1988).

[124] J. K. Hunter and K. Tan, "Weakly dispersive short waves," in: *Proceedings of the IVth international Congress on Waves and Stability in Continuous Media*, Sicily, 181–209 (1987).

[125] J. K. Hunter and Y. Zheng, "A nonlinear hyperbolic variational equation. I. Global existence of weak solutions," *Arch. Rat. Mech. Anal.*, **129**, 305–353 (1995).

[126] J. K. Hunter and Y. Zheng, "A completely integrable nonlinear hyperbolic variational equation," *Physica D*, **79**, 361–386 (1994).

[127] J. M. Hyman and B. Nicolaenko, "The Kuramoto–Sivashinsky equation: a bridge between pdes and dynamical systems," *Physica D*, **18**, 113–126 (1986).

[128] E. Infeld and G. Rowlands, *Nonlinear Waves, Solitons and Chaos*, "Cambridge University Press," Cambridge, 1990.

[129] E. Isaacson, D. Marchesin, B. Plohr, and B. Temple, "The Riemann problem near a hyperbolic singularity: the classification of solutions of quadratic Riemann problems I," *SIAM J. Appl. Math.*, **48**, 1009–1032 (1988).

[130] R. A. Isaacson, "Gravitational radiation in the limit of high frequency," II: nonlinear terms and the effective stress tensor. *Phys. Rev.*, **166**, 1272–1280 (1968).

[131] A. Jeffrey, "Breakdown of the solution to a completely exceptional system of hyperbolic equations," *J. Math. Anal. and Applics.*, **45**, 375–381 (1974).

[132] A. Jeffrey, *Quasilinear Hyperbolic Systems and Waves*, Research Notes in Mathematics **5**, Pitman, London, 1976.

[133] A. Jeffrey and T. Kakutani, "Weak nonlinear dispersive waves: a discussion centered around the Korteweg–de Vries equation," *SIAM Rev.*, **14**, 582–643 (1972).

[134] A. Jeffrey and T. Kawahara, *Asymptotic Methods in Nonlinear Wave Theory*, Pitman, London, 1982.

[135] A. Jeffrey and Sq. Xu, "Exact solutions to the Korteweg–de Vries–Burgers equation," *Wave Motion*, **11**, 559–564 (1989).

[136] S. Jin and Z. Xin, "Relaxing schemes for systems of conservation laws in arbitrary space dimensions," *Comm. Pure Appl. Math.*, **48**, 235–276 (1995).

[137] F. John, *Plane Waves and Spherical Means Applied to Partial Differential Equations*, 2nd ed., Springer-Verlag, New York, 1981.

[138] F. John, *Nonlinear Wave Equations, Formation of Singularities*, University Lecture Series, Vol. **2**, Amer. Math. Soc., Providence, 1990.

[139] J.-L. Joly, G. Metivier, and J. R. Rauch, "Resonant one dimensional nonlinear geometrical optics," *J. Funct. Analy.*, **114**, 106–231 (1993).

[140] J.-L. Joly, G. Metivier, and J. R. Rauch, "Generic rigorous asymptotic expansions for weakly nonlinear multidimensional oscillatory waves," *Duke Math. J.*, **70**, 373–404 (1993).

[141] J.-L. Joly, G. Metivier, and J. R. Rauch, "Coherent and focusing multidimensional nonlinear geometrical optics," *Centre de Recherche en Mathématiques de Bordeaux*, Preprint no. 9205 (1992).

[142] J.-L. Joly, G. Metivier, and J. R. Rauch, "Dense oscillations for the compressible 2d-Euler equations," in: *Nonlinear Partial Differential Equations and their Applications*, College de France Seminar, 1993–1994, ed. H. Brezis and J. L. Lions, Longman.

[143] J.-L. Joly, G. Metivier, and J. R. Rauch, "Nonlinear instability for 3×3 systems of conservation laws," *Comm. Math. Phys.*, **162**, 147–149 (1994).

[144] J.-L. Joly and J. Rauch, "Nonlinear resonance can create dense oscillations," *Microlocal Analysis and Nonlinear Waves*, Vol. **30**, IMA Volumes in Mathematics and its Applications, ed. M. Beals, R. Melrose, and J. Rauch, Springer-Verlag, New York, 113–124 (1991).

[145] J.-L. Joly and J. Rauch, "Justification of multidimensional single phase semilinear geometric optics," *Trans. Amer. Math. Soc.*, **330**, 599–623 (1992).

[146] B. B. Kadomtsev and V. I. Petviashvilli, "On the stability of solitary waves in a weakly dispersive media," *Sov. Phys. Dokl.* **15**, 539–541 (1970).

[147] T. Kakutani and H. Ono, "Weak nonlinear hydromagnetic waves in a cold collision free plasma," *J. Phys. Soc. Japan* **26**, 1305–131 (1969).

[148] T. Kato, "On the Cauchy problem for the (generalized) Korteweg-de Vries equation," *Adv. in Math. Suppl. Studies, Studies in Appl. Math.*, **8**, 93–128 (1983).

[149] C. Katsis and T. R. Akylas, "On the excitation of long nonlinear water waves by a moving pressure distribution," Part 2. Three-dimensional effects, *J. Fluid Mech.*, **177**, 49–65 (1987).

[150] J. B. Keller, "Rays waves and asymptotics," *Bulletin of the AMS*, **84**, 727–750 (1978).

[151] J. B. Keller, "Semiclassical mechanics," *SIAM Review*, **27**, 485–504 (1985).

[152] C. F. Kennel, R. P. Blandford, and C. C. Wu, "Structure and evolution of small amplitude intermediate shock waves," *Phys. Fluids B*, **2**, 253 (1990).

[153] R. Kersell, *Interaction between large and small scales in nonlinear dynamics*, Masters Thesis, University of California at Berkeley, 1988.

[154] J. Kevorkian and J. D. Cole, *Perturbation Methods in Applied Mathematics*, Springer-Verlag, New York, 1980.

[155] S. Klainerman, "Mathematical theory of classical fields and general relativity," in: *Mathematical Physics X*, ed. K. Schmüdgen, Springer-Verlag, New York, 213–236 (1992).

[156] A. Kluwick, "The analytical method of characteristics," *Prog. Aerospace Sc.* **19**, 197–313 (1981).

[157] A. Kluwick ed., *Nonlinear Waves in Real Fluids*, CISM Courses and Lectures no. **315**, Springer-Verlag, New York, 1991.

[158] Y. Kodama, A solution method for the dispersionless KP equation and its exact solutions, *Phys. Lett. A*, **135**, 167–170 (1989).

[159] Y. Kodama and J. Gibbons, "A solution method for the dispersionless KP equation and its exact solutions II," *Prog. Theor. Phys. Supp.*, **94**, 223–226 (1988).

[160] R. A. Kogelman and J. B. Keller, "Asymptotic theory of nonlinear wave propagation," *SIAM J. Appl. Math.*, **24**, 352–361 (1973).

[161] V. E. Korepin, N. M. Bogoliubov, and A. G. Izegin, *Quantum Inverse Scattering Method and Correlation Functions*, Cambridge University Press, Cambridge, 1993.

[162] D. J. Korteweg and G. de Vries, "On the change of form of long waves advancing in a rectangular canal and on a new type of long stationary waves," *Philos. Mag.* **39**, 422–443 (1895).

[163] Yu. A. Kravtsov and Yu. I. Orlov, *Geometrical Optics of Inhomogeneous Media*, Springer Series on Wave Phenomena, Vol. **6**, Springer-Verlag, New York, 1990.

[164] H.-O. Kreiss and J. Lorentz, *Initial-Boundary Value Problems and the Navier–Stokes Equations*, Academic Press, San Diego, 1989.

[165] M. D. Kruskal, "Asymptotology," in: *Mathematical Models in Physical Sciences*, ed. S. Drobot, Prentice-Hall, New Jersey, 1963.

[166] M. D. Kruskal, "Nonlinear wave equations," in: *Dynamical Systems, Theory and Applications*, Lecture Notes in Physics, Vol. **38**, ed. J. Moser, Springer-Verlag, Berlin, 1975.

[167] M. D. Kruskal and H. Segur, "Asymptotics beyond all orders in a model of crystal growth," *Stud. Appl. Math.*, **85**, 129–181 (1991).

[168] V. V. Krylov and D. F. Parker, "Harmonic generation and parametric mixing in wedge acoustic waves," *Wave Motion*, **15**, 185–200 (1992).

[169] V. P. Kuznetsov, "Equations of nonlinear acoustics," *Sov. Phys. Acoustics*, **16**, 467–470 (1971).

[170] E. W. Laedke and K. H. Spatschek, "Nonlinear ion acoustic waves in weak magnetic fields," *Phys. Fluids*, **25**, 985–989 (1982).

[171] N. A. Larkin, *Smooth Solutions for Equations of Transonic Gas Dynamics*, Nauka, Novosibirsk, 1991 (in Russian).

[172] P. D. Lax, *Hyperbolic Systems of Conservation Laws and the Mathematical Theory of Shock Waves*, CBMS-NSF Regional Conference Series in Applied Mathematics, Vol. 11, SIAM, Philadelphia, 1973.

[173] P. D. Lax, "Shock waves, increase of entropy, and loss of information," in: *Wave Motion: Theory, Modelling, and Computation*, ed. A. J. Chorin and A. J. Majda, Math. Sci. Res. Inst. Publications 7, Springer-Verlag, New York, 129–171 (1987).

[174] P. D. Lax and C. D. Levermore, "The small dispersion limit of the Korteweg–de Vries equation," *Comm. Pure Appl. Math.*, 36: I, 253–290; II, 571–593; III, 809–829 (1983).

[175] S. B. Leble, *Nonlinear Waves in Waveguides*, Research Reports in Physics, Springer-Verlag, New York, 1991.

[176] M. A. Leontovich and V. A. Fock, "Solution of the problem of propagation of electromagnetic waves along the earth's surface by the method of parabolic equations," Chapter 11, in: *Electromagnetic Diffraction and Propagation Problems*, ed. V. A. Fock, Pergamon Press, Oxford, 1965.

[177] R. J. LeVeque, *Numerical Methods for Conservation Laws*, Lectures in Mathematics, ETH Zürich, Birkhäuser, Basel, 1992.

[178] M. J. Lighthill, "A technique for rendering approximate solutions to physical problems uniformly valid," *Phil. Mag.*, 44, 1179–1201 (1949).

[179] M. J. Lighthill, Viscosity effects in sound waves of finite amplitude, *Surveys in Mechanics*, eds. G. K. Batchelor and R. M. Davis, 255–351, Cambridge University Press, 1956.

[180] M. J. Lighthill, *Waves in Fluids*, Cambridge University Press, Cambridge, 1978.

[181] T. P. Liu, Linear and nonlinear large-time behavior of solutions of general systems of conservation laws, *Comm. Pure Appl. Math.*, 30, 767–796 (1977).

[182] T. P. Liu, "Nonlinear stability of shock waves for viscous conservation laws," *Mem. Amer. Math. Soc.*, 328, Amer. Math. Soc., Providence, 1985.

[183] T. P. Liu, "Hyperbolic conservation laws with relaxation," *Commun. Math. Phys.*, 108, 153–175 (1987).

[184] T. P. Liu, "On the viscosity criterion for hyperbolic conservation laws," in: *Viscous Profiles and Numerical Methods for Shock Waves*, ed. M. Shearer, SIAM, Philadelphia, 105–114 (1991).

[185] D. Ludwig, "Uniform asymptotic expansions at a caustic," *Comm. Pure Appl. Math.*, 19, 215–250 (1966).

[186] A. J. Majda, *Compressible Fluid Flow and Conservation Laws in Several Space Dimensions*, Appl. Math. Sci. Vol. 53, Springer-Verlag, New York, 1984.

[187] A. J. Majda, "Nonlinear geometrical optics for hyperbolic systems of conservation laws," in: *Oscillation Theory, Computation, and methods of Compensated Compactness*, IMA Volume 2, Springer-Verlag, New York, 115–16 (1986)5

[188] A. J. Majda, "The interaction of nonlinear analysis and modern applied mathematics," in: *Proceedings of the International Congress of Mathematicians, Kyoto, Japan*, Springer-Verlag, Tokyo, 175–192 (1991).

[189] A. J. Majda and M. Artola, "Nonlinear geometrical optics for hyperbolic mixed problems," in: *Analyse Mathématique et Applications*, (Volume in honor of J.-L. Lions 60th birthday), ed. C. Agmon, A. V. Ballakrishnan, J. M. Ball, and L. Caffarelli, Gauthier-Villars, Paris, 229–262 (1988).

[190] A. J. Majda and R. R. Rosales, "Resonantly interacting weakly nonlinear hyperbolic waves. I. A single space variable," *Stud. Appl. Math.*, 71, 149–179 (1984).

[191] A. J. Majda and R. R. Rosales, "A theory for spontaneous Mach stem formation in reacting shock fronts," *SIAM J. Appl. Math.*, 43, 1310–1334 (1983).

[192] A. J. Majda, R. R. Rosales, and M. Schonbek, "A canonical system of integro-differential equations arising in resonant nonlinear acoustics," *Stud. Appl. Math.*, 79, 205–262 (1988).

[193] P. Manneville, *Dissipative Structures and Weak Turbulence*, Perspectives in Physics, Academic Press, San Diego, 1990.

[194] V. P. Maslov, "Propagation of shock waves in an isentropic nonviscous gas," *J. Soviet Math.*, **13**, 119–163 (1980).

[195] V. P. Maslov, *Mathematical Aspects of Integral Optics*, Moscow Institute of Electronic Machine-building, Moscow, 1983.

[196] V. P. Maslov, *Resonance Processes in the Wave Theory and Self-focalization*, Moscow Institute of Electronic Machine-building, Moscow, 1983.

[197] V. P. Maslov, Coherent structures, resonances, and asymptotic non-uniqueness for Navier–Stokes equations with large Reynolds numbers, *Russ. Math. Surveys*, **41**, 23–42 (1986).

[198] W. D. McComb, *The Physics of Fluid Turbulence*, Oxford University Press, Oxford, 1990.

[199] D. W. McLaughlin, G. Papanicolaou, and L. Tartar, "Weak limits of semilinear hyperbolic systems with oscillating initial data," *Lecture Notes in Physics* **230**, 277–289 (1985).

[200] R. Menikoff and B. J. Plohr, "The Riemann problem of fluid flow of real materials," *Rev. Mod. Phy.* **61**, 75–130 (1989).

[201] R. E. Meyer and D. V. Ho, "Notes on nonuniform shock propagation," *J. Acous. Soc. Amer.*, **35**, 1126–1132 (1963).

[202] M. Miksis and L. Ting, "Effective equations for multiphase flow — waves in a bubbly liquid," *Adv. Appl. Mech.* **28**, 141–260 (1992).

[203] M. Mulase, "Algebraic theory of the KP equation," to appear in: *Interface Between Mathematics and Physics*, ed. S. T. Yau, International Press Co., Harvard.

[204] V. E. Nakoryakov, B. G. Pokusaev, and I. R. Sheiber, *Wave Propagation in Gas–Liquid Media*, CRC Press, Boca Raton, 1993.

[205] K. A. Naugolnykh and L. A. Ostrovsky, *Nonlinear Wave Processes in Acoustics*, Cambridge University Press, Cambridge, to appear.

[206] J. C. Neu, "Kinks and minimal surfaces in Minkowski space," *Physica D*, **43**, 421–434 (1990).

[207] A. C. Newell, *Solitons in Mathematics and Physics*, SIAM, Philadelphia, 1985.

[208] A. C. Newell and J. V. Moloney, *Nonlinear Optics*, Addison-Wesley, Redwood City, 1992.

[209] A. C. Newell, T. Passot, and J. Lega, "Order parameter equations for patterns," *Ann. Rev. Fluid. Mech.* **25**, 399–453 (1993).

[210] B. Nicolaenko, B. Scheurer, and R. Temam, "Some global dynamical properties of the Kuramoto–Sivashinsky equations: Nonlinear stability and attractors," *Physica D*, **16**, 155–183 (1985).

[211] J. J. C. Nimmo and D. G. Crighton, "Bäcklund transformations for nonlinear parabolic equations: the general results," *Proc. Roy. Soc. Lond.* A 384 381 (1982).

[212] J. J. C. Nimmo and D. G. Crighton, "Geometrical and diffusive effects in nonlinear acoustic propagation over long ranges," *Phil. Trans. R. Soc. Lond. A*, **320**, 1–35 (1986).

[213] B. K. Novikov, O. V. Rudenko, and V. I. Timoshenko, *Nonlinear Underwater Acoustics*, American Institute of Physics, Translation Series, New York, 1987.

[214] S. P. Novikov, S. V. Manakov, L. P. Pitaevskii, and V. E. Zakharov, *Theory of Solitons. The Inverse Scattering Method*, Plenum, New York, 1984.

[215] L. A. Ostrovsky, Short wave asymptotics for weak shock waves and solitons in mechanics, *Int. J. Non-Linear Mechanics*, **129**, 401–416 (1976).

[216] L. A. Ostrovsky, "Nonlinear internal waves in a rotating ocean," *Oceanology*, **18**, 181–191 (1978).

[217] L. A. Ostrovsky, S. A. Rybak, and L. Sh. Tsimring, "Negative energy waves in hydrodynamics," *Sov. Phys. Usp.*, **29**, 1040–1052 (1986).

[218] L. A. Ostrovsky and Yu. A. Stepanyants, "Nonlinear surface and internal waves in rotating fluids," in: *Nonlinear Waves 3*, ed. A. V. Gaponov-Grekhov, M. I. Rabinovich, and J. Engelbrecht, Research Reports in Physics, Springer-Verlag, New York, 106–128 (1990).

[219] D. Parker, "Waveform evolution for nonlinear surface waves," *Int. J. Engng. Sci.* **26**, 59–75 (1988).

[220] R. L. Pego, "Some explicit resonating waves in weakly nonlinear gas dynamics," *Stud. Appl. Math.* **79**, 263–269 (1988).

[221] Y. Pomeau, A. Ramani, and B. Grammaticos, "Structural stability of the Korteweg–de Vries solitons under a singular perturbation," *Physica D*, **31**, 127–134 (1988).

[222] P. Prasad, *Propagation of a Curved Shock and Nonlinear Ray Theory*, Pitman Research Notes No. **292**, Longman, 1993.

[223] P. Prasad, "A nonlinear ray theory," *Wave Motion*, **20**, 21–31 (1994).

[224] E. Randall and R. R. Rosales, "A numerical and analytical study of the Resonant interaction equations for hyperbolic systems I: Classification of solutions," to appear in: *Stud. Appl. Math.*

[225] E. Randall and R. R. Rosales, "A numerical and analytical study of the Resonant interaction equations for hyperbolic systems II: A class of large time solutions," to appear in: *Stud. Appl. Math.*

[226] J. Rauch, *Partial Differential Equations*, Springer-Verlag, New York, 1991.

[227] M. Renardy and R. C. Rogers, *An Introduction to Partial Differential Equations*, Texts in Applied Mathematics, Vol. **13**, Springer-Verlag, New York, 1993.

[228] R. Ravindran and P. Prasad, "Kinematics of a shock front and resolution of a hyperbolic caustic," in: *Advances in Nonlinear Waves*, ed. L. Debnath, Pitman Research Notes in Mathematics **111**, Pitman, Boston, 77–99 (1985).

[229] F. Rezakhanlou, "Hydrodynamic limit for attractive particle systems on $\mathbf{Z}^d$," *Commun. Math. Phys.*, 140 (1991), 417–448 (1991).

[230] R. R. Rosales, "Stability theory for shocks in reacting media: Mach stems in detonation waves," in: *AMS Lectures in Applied Mathematics*, **24**, ed. G. Ludford, 431–465 (1986).

[231] R. R. Rosales, "An introduction to weakly nonlinear geometrical optics," in: *Multidimensional Hyperbolic Problems and Computations*, Vol. **29**, IMA Volumes in Mathematics and its Applications, ed. A. J. Majda and J. Glimm, Springer-Verlag, New York, 281–310 (1991).

[232] R. R. Rosales, "Canonical equations of long wave weakly nonlinear asymptotics," in: *Continuum Mechanics and its Applications*, ed. G. A. C. Graham and S. K. Malik, Hemisphere, New York, 365–397 (1989).

[233] R. R. Rosales and A. J. Majda, "Weakly nonlinear detonation waves," *SIAM J. Appl. Math.*, **43**, 1086–1118 (1983).

[234] P. Roseneau and J. M. Hyman, "Compactons: solitons with finite support," *Phys. Rev. Lett.*, **70**, 564–567 (1993).

[235] O. V. Rudenko and S. I. Soluyan, *Theoretical Foundations of Nonlinear Acoustics*, Consultants Bureau, Plenum, New York, 1977.

[236] S. M. Rytov, Yu. A. Kravtsov, and V. I. Tartarskii, *Principles of Statistical Radiophysics*, Vol. **4**, Springer-Verlag, New York, 1987.

[237] P. L. Sachdev, *Nonlinear Diffusive Waves*, Cambridge University Press, Cambridge, 1987.

[238] J.-C. Saut, "Generalized Kadomtsev-Petviashvili equations," *Indiana Univ. Math. J.*, **42**, 1011–1026 (1993).

[239] S. Schochet, "Resonant nonlinear geometrical optics for weak solutions of conservation laws," *J. Diff. Eq.*, **113**, 473–504 (1994).

[240] S. Schochet, "Fast singular limits of hyperbolic PDEs," *J. Diff. Eq.*, **114**, 476–512 (1994).

[241] D. G. Schaeffer and M. Shearer, "The classification of 2×2 systems of non-strictly hyperbolic conservation laws with applications to oil recovery," *Comm. Pure Appl. Math.*, **15**, 141–178 (1987).

[242] D. Serre, "Oscillations non-linéaires de haute fréquence," in: *Nonlinear Partial Differential Equations and their Applications*, Collège de France Seminar, Vol. **11**, ed. H. Brezis and J.-L. Lions, Pitman Research Notes in Mathematics, Longman, New York, to appear.

[243] D. Serre, "Oscillations non-linéaires de haute fréquence; $\dim = 1$," Report de Recherche no. **18**, Ecole Normale Supérieure de Lyon (1989).

[244] D. Serre, "Oscillations non-linéaires hyperbolic de grande amplitude; $\dim \geq 2$," in: *Nonlinear Variational Problems and Partial Differential Equations*, ed. A. Mariano and M. K. Venkatesha Murthy, Pitman Research Notes in Mathematics 320, Longman, New York, 1995.

[245] D. W. Schwendeman, "A new numerical method for shock wave propagation based on geometrical shock dynamics," *Proc. R. Soc. Lond. A*, **441**, 331–341 (1993).

[246] Z.-S. She, E. Aurell, and U. Frisch, "The inviscid Burgers equation with initial data of Brownian type," *Commun. Math. Phys.*, **148**, 623–641 (1992).

[247] V. A. Shutilov, *Fundamental Physics of Ultrasound*, Gordon and Breach, Glasgow, 1988.

[248] Ya. G. Sinai, "Statistics of shocks in solutions of inviscid Burgers equation," *Commun. Math. Phys.*, **148**, 601–621 (1992).

[249] J. Smoller, *Shock Waves and Reaction-Diffusion Equations*, Springer-Verlag, New York, 1983.

[250] K. Sobczyk, *Stochastic Wave Propagation*, Elsevier, Warszawa, 1985.

[251] W. Strauss, *Nonlinear Wave Equations*, Conference Board of the Mathematical Sciences, AMS, Philadelphia, 1989.

[252] J. C. Strikwerda, *Finite Difference Schemes and Partial Differential Equations*, Wadsworth, Belmont, 1989.

[253] A. Szepessy and Z. Xin, "Nonlinear stability of viscous shock waves," *Arch. Rat. Mech. Anal.*, **122**, 53–103 (1993).

[254] E. G. Tabak, "Focusing of weak shock waves and the von Neumann paradox of oblique shock reflection," Ph. D. thesis, Massachusetts Institute of Technology, 1992.

[255] E. Tabak and R. R. Rosales, "Weak shock focusing and the von Neumann paradox of oblique shock reflection," *Phys. Fluids*, **6**, 1874–1892 (1994).

[256] T. R. Taha and M. J. Ablowitz, "Analytical and numerical aspects of certain nonlinear evolution equations. III Numerical, Korteweg–de Vries equation," *J. Comp. Phys.*, **55**, 231–253 (1984).

[257] K. Tan, *Weakly Dispersive Wave Equations*, Ph. D. thesis, Colorado State University, 1991.

[258] F. D. Tappert, "The parabolic approximation method," in: *Wave Propagation and Underwater Acoustics*, ed. J. B. Keller, and J. S. Papadakis, Chapter V, Springer-Verlag, 1977.

[259] T. Taniuti and C.-C. Wei, "Reductive perturbation method in nonlinear wave propagation," *J. Phys. Soc. Japan*, **24**, 941–946 (1968).

[260] T. Taniuti ed., "Reductive perturbation method for nonlinear wave propagation," *Prog. Theor. Phy. Suppl.*, **55**, 1–306 (1974).

[261] T. Taniuti and K. Nishihara, *Nonlinear Waves*, Pitman, Boston, 1983.

[262] L. Tartar, Compensated compactness and applications to PDEs, *Nonlinear Analysis and Mechanics*, Herriot-Watt symposium, Pitman (1979).

[263] M. E. Taylor, *Pseudodifferential Operators and Nonlinear Partial Differential Equations*, Progress in Mathematics, Vol. **100**, Birkhäuser, Boston, 1991.

[264] P. A. Thompson and K. C. Lambrakis, "Negative shock waves," *J. Fluid Mech.*, **60**, 187–208 (1973).

[265] R. Timman, "Unsteady motion in transonic flow," in: *Symposium Transonicum*, ed. K. Oswatitsch, Springer-Verlag, Aachen, 394–401 (1962).

[266] S. Venakides, "The Korteweg–de Vries equation with small dispersion — higher order Lax-Levermore theory," *Comm. Pure Appl. Math.*, **43**, 335–361 (1990).

[267] L. van Wijngaarden, "One dimensional flow of liquids containing small bubbles," *Ann. Rev. Fluid Mech.*, **4**, 369–396 (1972).

[268] W. G. Vicenti and C. H. Kruger, *Introduction to Physical Gas Dynamics*, R. E. Kreiger Publication Co., 1982.

[269] J. Wagner and J. Keizer, "Effects of rapid buffers on Ca^{2+} diffusion and Ca^{2+} oscillations," *Biophysical Journal*, **67**, 447–456 (1994).

[270] X. P. Wang, M. J. Ablowitz, and H. Segur, "Wave collapse and instability of solitary waves of a generalized Kadomtsev-Petviashvili equation," preprint (1994).

[271] G. M. Webb and G. P. Zank, "Wave diffraction in weak cosmic-ray-modified shocks," *The Astrophysical Journal*, **396**, 549–574 (1992).

[272] G. B. Whitham, *Linear and Nonlinear Waves*, Wiley, New York, 1974.

[273] Z. Xin, "The fluid dynamic limit of the Broadwell model of the nonlinear Boltzmann equation in the presence of shocks," *Comm. Pure Appl. Math.*, **44**, 679–713 (1991).

[274] H. Q. Yang and A. J. Przekwas, "A comparative study of advance shock capturing schemes applied to Burgers equation," *J. Comp. Phys.*, **102**, 139–159 (1992).

[275] R. Young, "Sup-norm stability for Glimm's scheme," *Comm. Pure Appl. Math.*, **46**, 903–948 (1993).

[276] E. A. Zabolotskaya and R. V. Khokhlov, "Quasi-plane waves in the nonlinear acoustics of confined beams," *Sov. Phys. Acoustics*, **15**, 35–40 (1969).

[277] G. I. Zahalak and M. K. Myers, "Conical flow near singular rays," *J. Fluid Mech.*, **63**, 537–561 (1974).

[278] V. E. Zakharov and E. A. Kuznetsov, "On three dimensional solitons," *Zh. Eksp. Teor. Fiz.*, **8**, 285–286 (1974).

[279] V. E. Zakharov, V. S. L'vov, and G. Falkovich, *Kolmogorov Spectra of Turbulence I*, Springer-Verlag, Berlin, 1992.

[280] V. E. Zakharov and A. B. Shabat, "An exact theory of two-dimensional self-focussing and one-dimensional self-modulation of waves in nonlinear media," *Sov. Phys. JETP*, **34**, 62–69 (1972).

Index